FATFOOT

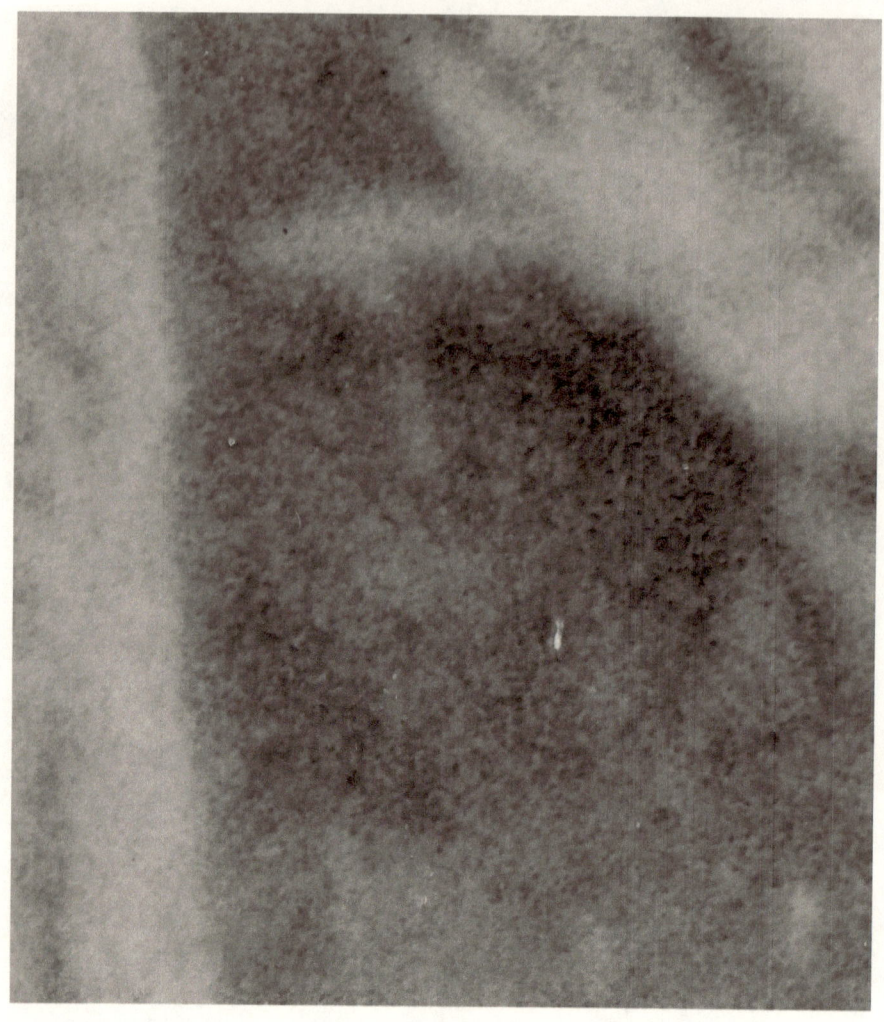

A film photograph of Fatfoot's face captured using an Olympus OM2n camera with an Olympus Zuiko Auto-S 50mm f/1.4 lens set to the maximum aperture and focus at ∞. The photograph was taken using an electro-mechanical, homemade game camera (Fig. 8.3) in the early morning of Monday, 28 June 1993. Assuming that the optical alignment of the photographic elements might have potentially shown the 'red eye' reflection (Fig. 8.2a), there was an approximately one half second mechanical lag between the photo flood lights and flash activating and the cable release taking the photograph, which would have allowed sufficient reflex time for the eyelids to close.

FATFOOT

Encounters with a Dooligahl

An autobiographical journey
of 50+ years investigating
Australia's three marsupial enigmas

Neil Frost

B.A. Dip. Ed. (MQ), Adv. Cert. Tech. (USYD)

COACHWHIP PUBLICATIONS
Greenville, Ohio

This book is dedicated to the ones that I love.
Sandy, Avril and Drew.

In the spirit of reconciliation, the researchers acknowledge the Traditional Custodians of Country throughout Australia and their connections to land, sea and community. We pay our respect to their Elders past and present and extend that respect to all Aboriginal and Torres Strait Islander peoples today.

Also by the author:

© Baker, Mark, and Neil Frost. 1994. *New Senior Computing Studies: The Preliminary Course*. Sydney: McGraw-Hill. ISBN 0 07 470032 4.

© Baker, Mark, and Neil Frost. 1995. *New Senior Computing Studies: The HSC Course*. Sydney: McGraw-Hill. ISBN 0 07 470033 2.

Visit StrangeArk.com/biofortean-notes for more cryptozoology research. Volume 9 includes a survey of historical Yowie reports.

Fatfoot: Encounters with a Dooligahl, by Neil Frost
Edition 2.0 (Black and White)
© 2024 Neil Frost
Cover image: © Daniel Falconer

All rights reserved.
CoachwhipBooks.com

ISBN 1-61646-5-593-X
ISBN-13 978-1-61646-593-3

CONTENTS

In Memoriam	7
Preface: Territorial Maps	9
Prologue	15
1: Bluegum	19
2: Providence	43
3: Home	67
4: Log	85
5: Gondwana	133
6: Convergence	159
7: Octopus	173
8: Dooligahl	229
9: Hominoids	355
10: Quinkan	395
11: Junjudee	451
12: Alibi	527
13: Eyes	599
14: Legs	637
Epilogue	669
About the Author	705
Index	709

Ian Price and the author interviewing a controversial claimant with Tony Healy and North American Bigfoot researcher Daniel Perez (BIGFOOT TIMES) at the Cotter River, Brindabella Range, Canberra, ACT, on 24 September 2000. Daniel boarded with Ian and Cheryl during the Sydney Olympics. (Photo: Tony Healy)

IN MEMORIAM

John Thomas Frost
Flight Sergeant RAAF
(1913-1989)

Alexander Dandie (aka Guraki)
Lance-Sergeant 2/30th AIF
(1912-2008)

Ian Bruce Price
(1958-2016)

Dr. Helmut Hermann Ernst Loofs-Wissowa
Captain, French Foreign Legion
Commandeur des Palmes Académiques
(1927-2018)

John Samuel Appleton
(1954-2023)

"He had been heard of and seen and described so often and by so many reliable liars, that most people agreed that there must be something. The most popular and enduring theory was that he was a gorilla or an ourang-outang which had escaped from a menagerie long ago. He was also said to be a new kind of kangaroo, or the last of a species of Australian animals which hadn't been discovered yet."

Lawson, Henry. 'The Hairy Man.' *Triangles of Life and Other Stories* (1907)

PREFACE: TERRITORIAL MAPS

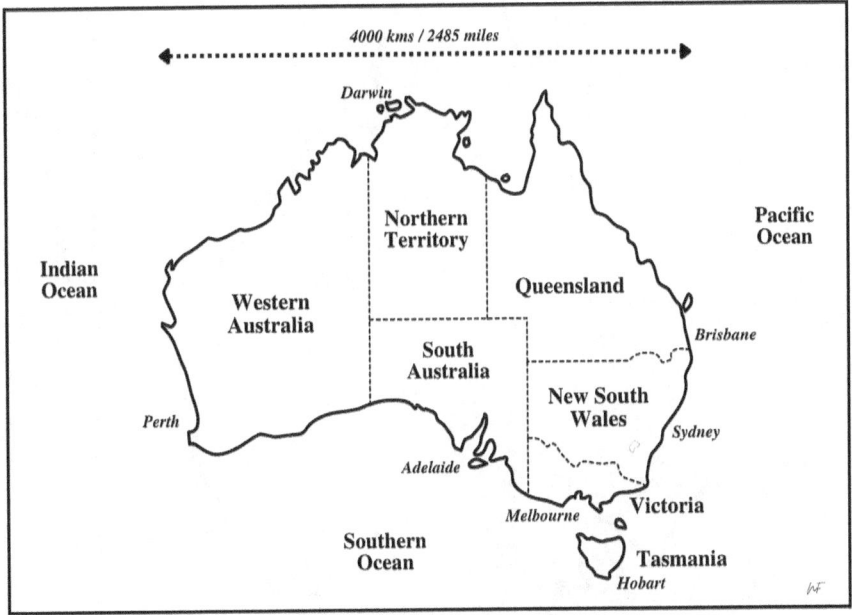

Map 1: The Australian Continent
States and Northern Territory / Capital Cities / Oceans
(Map: Neil Frost)

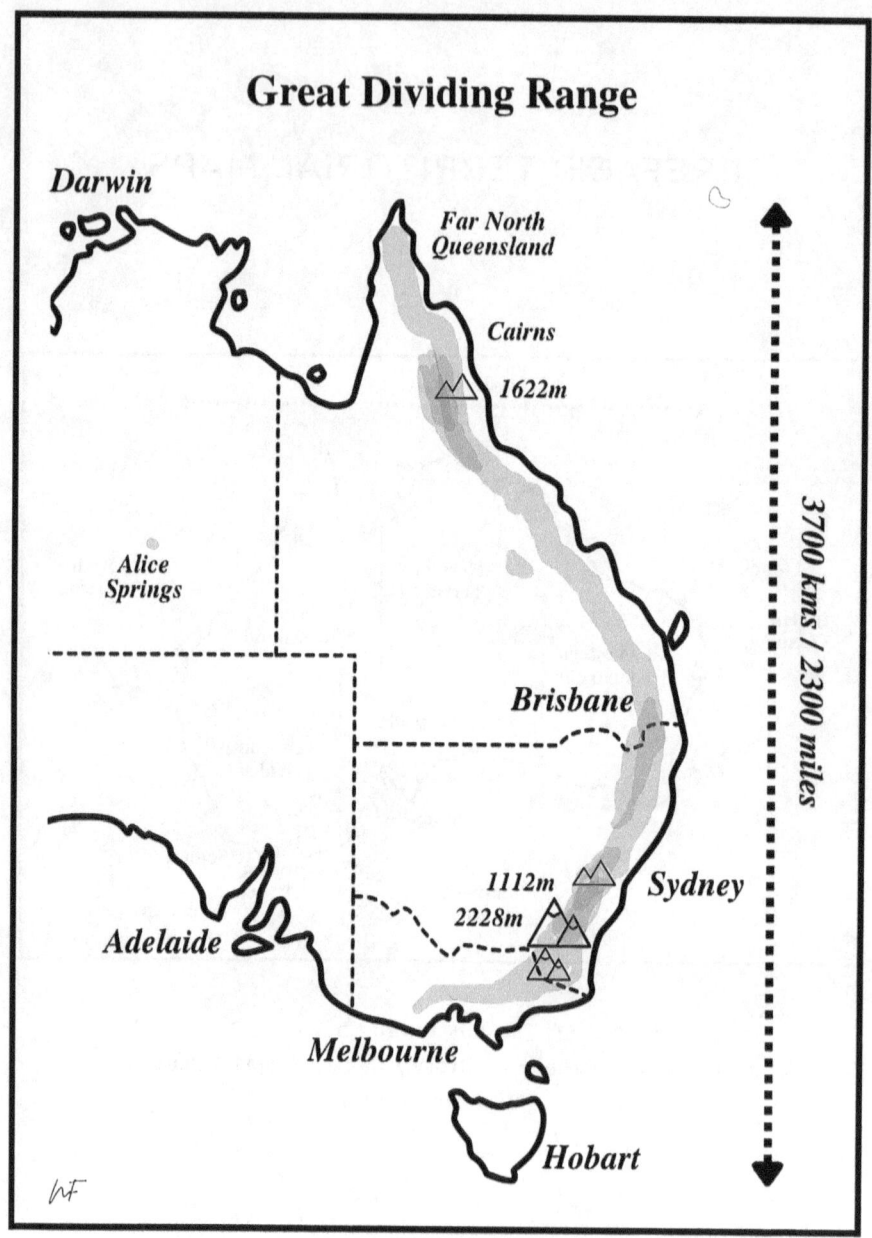

Map 2: Great Dividing Range
(Map: Neil Frost)

Preface: Territorial Maps

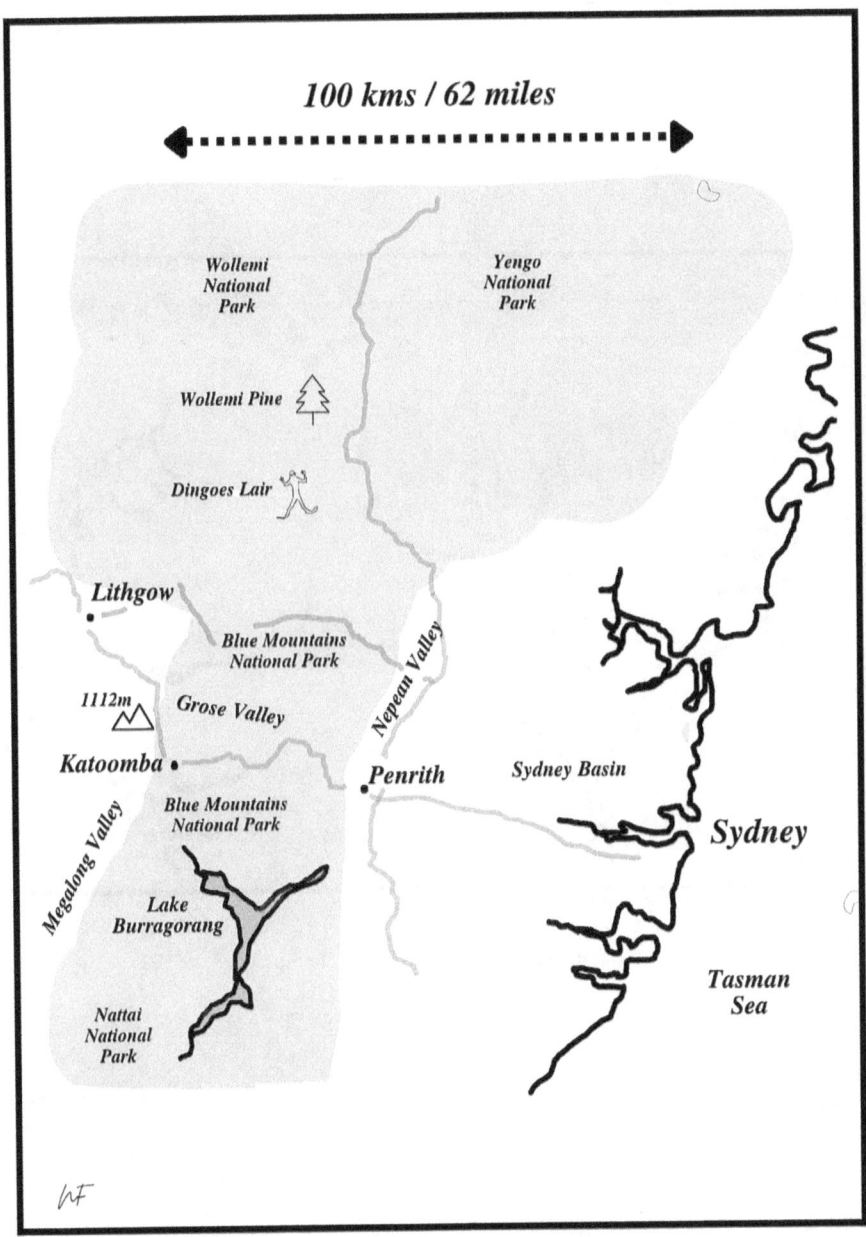

Map 3: Blue Mountains / Sydney Basin
(Map: Neil Frost)

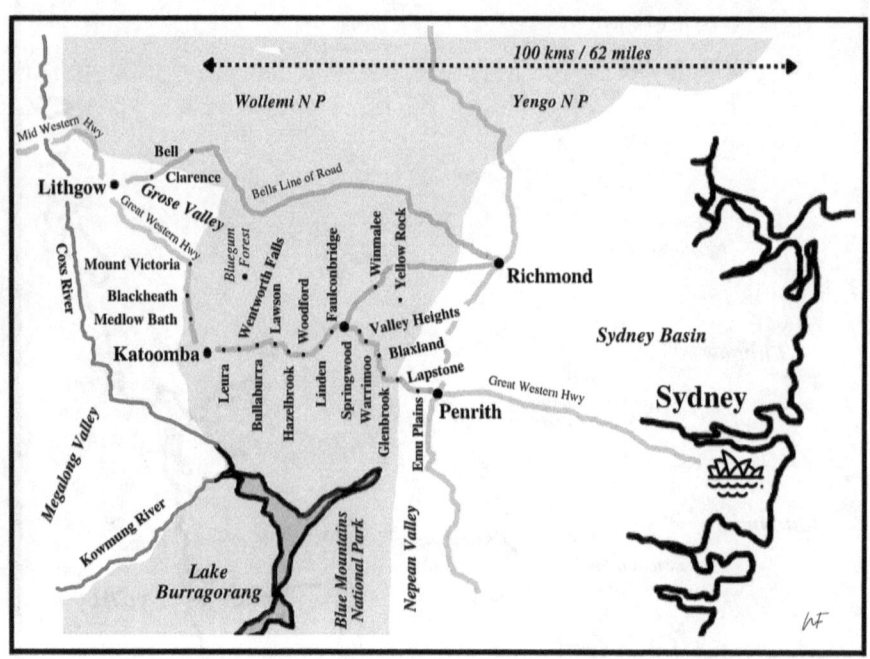

Map 4: Blue Mountains Settlements
(Map: Neil Frost)

Preface: Territorial Maps   13

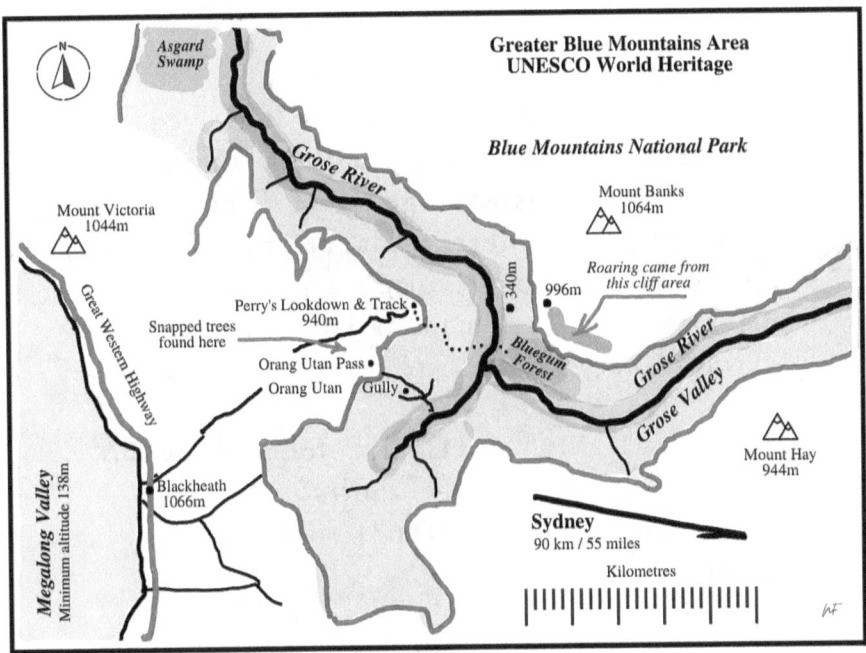

Map 5: Map of the Upper Grose Valley and
Surrounding Bluegum Forest Geography
(Map: Neil Frost)

"It was not, however, until I arrived in the country, and found myself surrounded by objects as strange as if I had been transported to another planet, that I conceived the idea of devoting a portion of my attention to the mammalian class of its extraordinary fauna."

> Gould, John. *Mammals of Australia, Volume 1* (1863)

Prologue

AUSTRALIA IS AN ancient continent that has spent most of its recent geological history surrounded by oceans, which resulted in its biogeographical isolation from the remainder of the planet. Even during periods of glacial maximum, massive drops in the sea level had no ability to reconnect Australia with other continents, which maintained its strict quarantine. The highly eroded landforms of the continent, where nothing rises above 2,228 meters or 7,310 feet above sea level, testify to its abraded antiquity. Being located in the centre of the Indo-Australian tectonic plate, there are no orogenic processes or volcanic activity at work to rebuild and resculpture the vast and open landmass. Since separating tectonically from Antarctica and the other Gondwanan continents many tens of millions of years ago, Australia has been moving northward from the South Pole carrying its contemporaneous mammalian population of monotreme and marsupial animals. Shaped by the environmental conditions, these animals radiated across the continent filling similar zoological niches found on other parts of the planet. Correspondingly, though coming from a similar heritage, each of these early mammals took their own biological path, independent of placental mammals found elsewhere, but still operating within common evolutionary rules and constraints.

The rainforests of the African Rift Valley played a crucial role in the development of arboreal hominins, with climate change probably initiating their transition to the savanna grasslands where bipedalism, the freeing of hands, communication and many other skills led to their increased intelligence. In Australia, the remnant Gondwanan Rainforests permitted a similarly convergent evolutionary path for arboreal marsupials. Climate change resulting in increased aridification was influenced by the glaciation of Antarctica through the establishment of the Antarctic Circumpolar Current, plus other environmental factors, which encouraged the movement of tree-dwelling marsupials to the forest floor and grasslands, where bipedalism, tool usage, communication and other traits have also led to their increased intelligence.

In harmony with the rest of the planet, Australia has its own marsupial megafauna, for example, *Diprotodon optatum*—a giant wombat, *Procoptodon goliah*—a giant short-faced kangaroo and *Thylacoleo carnifex*—a large marsupial lion. Most have become extinct because of climate change and human predation and are not widely known, unlike extant placental megafauna, such as elephants, lions and rhinoceros. More intelligent marsupial megafauna have attempted to adapt to the new circumstances, like the much smaller *Thylacinus cynocephalus*—Tasmanian Tiger, which has become extremely elusive, or 'functionally extinct' as a consequence.

In the placental world, every empty niche has been occupied by at least one animal that has adapted to take advantage of the available opportunities. In the marsupial antipodes, a similar scenario could also have been expected to occur. If this was the case, then where are the marsupial analogues of gorillas, chimpanzees and orangutans? More importantly, where is the Southern Man?

Prologue

The dispersal of *Homo sapiens* throughout Asia by sea voyages significantly expanded the territory of modern humans with the colonisation of Sahul about 65,000 years ago, a journey that had previously been unobtainable for earlier hominins like *Homo erectus,* who were technologically incapable of migrating past Flores and the very formidable biogeographic barriers. After arriving in Sahul, Aboriginal stories, dance and paintings, began to describe beings that were regarded as both spiritual and corporeal. They were "here before us and are not the same as us." There were several different types of 'marsupial hominoids' (a phrase which will be used here colloquially rather than phylogenetically) and every tribe's language had various names for them. The chosen names for these ethno-known animals used in this book come from the South Coast of New South Wales and Far North Queensland. In increasing order of the animal's size, these indigenous names are Junjudee, Dooligahl and Quinkan. Depending upon the species and sex, their temperaments range from mischievous to murderous, requiring vigilance and daily precautions to minimise conflict or personal risk.

This book has recorded many related encounters and draws conclusions based on the evidence. The study mainly centres on the activities of our local Dooligahl which we named "Fatfoot" because of an apparent soft tissue injury to her right foot. Fatfoot is a seven-foot female 'marsupial hominoid' and based on her foot morphology, is most probably a macropod ('big foot') or kangaroo.

Fig. 1.1 The start of the 2 kilometre (1¼ mile) hard grade, six hour return, 600 metre mostly vertical Perry's Track. (Photo: Neil Frost)

Fig. 1.2 View from Perry's Lookdown to Docker's Lookout. The Blue Gum Forest is 600 metres below. (Photo: Neil Frost)

1: Bluegum

THIS CRYPTOZOOLOGICAL JOURNEY began in an isolated forest, within a remote valley of the Blue Mountains, in 1966.[1] It was the second night of a seven-day scouting camp and something on the cliff overlooking the valley, six hundred metres above us, was aggressively vocalising its presence! The following night, whatever it was, was now on the valley floor, walking around our group campsite, before concentrating its attention on our isolated tent for the remainder of the night. None of us slept. No one said anything until dawn. As Ian Price, our future neighbour said in response to a similar situation almost thirty years later, "It was a testicle-tightening experience!"

OUR SCOUT MASTER was Alex Dandie. During World War II, he served as a Lance-Sergeant in the 2/30th Battalion of the Australian Imperial Force (AIF). He was transported to Singapore aboard the Johan Van Oldenbarnevelt, where

---

[1] A *Tourist Map of the Blue Mountains and Burragorang Valley* produced by the Department of Lands, Sydney (A.A. Cooke, 1955), warned that: "A large part of the country represented on this map is wild and rugged. Persons unacquainted with it should not wander from roads or main tracks."

he was captured by the Japanese at the fall of this 'Gibraltar of the East', on February 16, 1942. He was then held in the infamous Changi POW Camp. It is unclear what happened from here, though his regiment was sent to work on the Burma end of the Burma-Thailand railway. His record shows that he was in Nagoya, Japan, on September 6, 1945, undoubtedly having been transported onboard a 'hell ship' to Japan and used as slave labour by the Japanese Army. Like many of his fellow Australian POWs, he never spoke to anyone about his ordeal, although it was clear to everyone that he had been tortured. He had many pink scars on his face, arms, legs, and other exposed skin. Of the thousands of round wounds, most appeared to be cigarette burns. After the war, he was sent back to Japan as part of 'J Force' where efforts were made to assist Japan in recovering from the war and help guide the country towards democracy. As a consequence of his wartime experience, Mr Dandie had a strong dislike of most things Japanese[2] and would often say to us, "Don't buy Jap crap!" My father also told me that Mr Dandie was afraid that Japan might rise again as an ultranationalist military power and that young Australians needed to have very good bush survival skills, in case they were needed in a future Australia-based conflict.

As a Scout Master, Mr Dandie was a very dedicated and experienced man who made extraordinary efforts to educate every one of his scouts. He had a very detailed knowledge of a broad range of bush skills and the ability to apply them. If he found a gap in his knowledge,

---

[2] Alex Dandie's attitude contrasted strongly with maternal relative, Peter Rushforth, who was also at Changi POW camp. After spending five months in Japan in 1963, he embraced many aspects of Japanese culture, particularly ancient Japanese ceramic traditions, of which he became a renowned master potter.

1: Bluegum

he would make the effort to close it, and then teach us.[3] He often brought in experts on various subjects and frequently drew upon Aboriginal experience and knowledge.[4] For this reason, Mr Dandie preferred to be addressed as 'Guraki', which is an Aboriginal word meaning, ". . . to be wise; skilful; teacher."[5]

SOMETIME BEFORE THE school holidays, probably in late spring or early summer, at the age of eleven, I made the transition from Cubs to Scouts, in what was called the 'Going Up' Ceremony. This was symbolically achieved by leaving Akela and the cub pack behind with the initiate walking along a rope monkey bridge suspended over water, before arriving at the scout troop which was hidden from view, on the other side. As a tenderfoot scout, I made my preparations for the upcoming seven-day camp, to be held at the Blue Gum Forest.

More significantly, preparations for the camp had been underway for a much longer period. In particular, the Rovers[6] had played a critical role, having built the extravagant infrastructure[7] that had been put in place at the

---

[3] One skill that I never fully acquired was fire lighting using a fire drill and consequently, I have great admiration for the competitors of the TV reality series, *Survivor*.

[4] As a Troop we made 'bullroarers.' These are a Northern Territory Corroboree instrument made from a flat piece of timber attached to a length of leather and whirled around the head. We used them to determine how far the sound could be heard.

[5] Alex Dandie. 2003. *Scouting in Denistone / West Ryde*. North Ryde: Kwik Kopy Printing.

[6] Adult male scouting members. Denis was a metallurgist and several others were civil engineers.

[7] There is no way that this infrastructure would have been allowed today. Even the use of open fire is forbidden, with gas stoves being required. Considering what we did and how we went about it, current OH&S requirements would have prohibited nearly all aspects of the camp.

campsite during the days when permission to conduct such activities was not required or even remotely considered. There was a flying fox, a pulley-and-cable system, that was used to transport supplies between stations,[8] mainly from the cliff top to the valley floor, six hundred metres below. Along its total length, stations or platforms were constructed to break the run into practical and manageable sections. At the campsite, the Rovers had prepared a large clearing, with a round, rock-edged fire pit at its centre. Within the immediate area, there was a garbage pit where burnt and bashed material could be buried.[9] Similarly, pit toilets were in place. However, the most extravagant build was a dam across one of the tributary creeks that flowed into the Grose River, with a concave rock wall with attached galvanised pipes, that provided a pressurised water supply to the camp. In many ways, the construction work that the Rovers completed could have been the inspiration for the David Lean movie, *Bridge on the River Kwai*, although it wasn't, with Alec Guinness attempting to impersonate the more harrowing wartime role of Alex Dandie, aka Guraki.

---

[8] During construction, cement and a petrol-powered concrete mixer were transported to the valley floor. Later on, backpacks, tents, poles, food and other general supplies were mostly carried by the flying fox, with some items backpacked along the narrow trail. Apart from the quantity of materials, the weight of some items was outrageous. The many eight-man, thick-walled, ex-military canvas tents and flys, each probably weighed about 100 kg. As an eleven-year-old boy, I can only remember that unfolding the tent was at the limit of my ability. As preserving the environment was a strong aspect of our scouting philosophy, we carried in our own 100- to 200-mm-diameter, hardwood tent poles, each a minimum of six metres in length! In most regards, this camp was similar to a military operation.

[9] Our environmental policy in those days for all waste, including metal cans, was 'burn, bash and bury.'

1: Bluegum

WITH THE ARRIVAL of the school holidays, we were excitedly ready to begin our seven-day camp. We left before dawn in the allocated four-wheel drive[10] vehicles and drove for several hours to Perry's Lookdown, eight kilometres from Blackheath in the upper Blue Mountains. Looking nauseatingly down from this vantage point, at an altitude of over a thousand metres, we had an excellent view of the Grose Valley. We could see our approximate destination more than six hundred metres below, amidst the thickly forested canopy and near the meandering Grose River.

After we unpacked the vehicles, the backpacks, fresh food and various other supplies were moved to the beginning of the track, where the flying fox had been set up. Here the Rovers filled the nets with the items and began lowering them down the cliff face to the first waypoint or station platform and so on until the equipment reached the valley floor later in the day.

With nothing to bear except our fear, we made our way down the unimproved track. Along some sections of the path, the Rovers had driven pitons into the rock face and attached ropes so that we had continuous handholds. On other, more dangerous descents, ropes were attached to slings with carabiners, like what we used when rock climbing and abseiling. We would click into and lock the carabiners, while a Rover would belay our drop. After arriving at the bottom of the descent, we collected our gear and remaining nerves. While recovering, we moved to the group campsite, trying very hard not to imagine what the return journey was going to be like.[11] The tents, poles and

---

[10] The 4WD vehicles that were owned by the Rovers, weren't necessary for our journey. The Rovers used them for their extraordinary adventures throughout Australia.

[11] Worthwhile looking at the comments on: https://www.alltrails.com/trail/australia/new-south-wales/perry-s-lookdown

Fig. 1.3 The formidable Grose Valley. (Photo: © Attila Kaldy)

some supplies were waiting for us, having been previously transported by the Rovers using the flying fox sometime before.[12]

After inspecting the group campsite, my patrol leader decided that he wanted a more private location. Graham chose a position that was a long way from the group campsite, about a hundred metres distant, far from the communal fire pit. We first needed to clean the new site, because it wasn't a prepared and cleared area, like the group campsite. The thick and heavy khaki canvas tent was rolled out and orientated to take advantage of nearby trees that could be used as very secure anchors for the guide ropes, instead of using tent pegs. As the tent needed to shelter eight of us, we needed long poles, which were then joined with square lashings, erected with the tent and then tied to the trees. Typical of this type of heavy tent, there was no floor. To stop water from entering during a storm,

---

[12] The logistics involved in transporting these very heavy supplies were insane. It would have equalled a military effort.

1: Bluegum

ditches were dug around the tent wall with entrenching tools and the spoil was mounded high, between the drain and wall. To increase the shelter's water tightness, a canvas fly was stretched over the top of the tent. The fly was larger than the area of the tent roof so that the water run-off would fall outside of the drainage ditch surrounding the tent. A gap between the fly and the tent roof was necessarily maintained to prevent rubbing between the two. Apart from the tent's weight, another major disadvantage was maintaining the waterproofing of the canvas. For this reason, we needed to avoid touching the canvas and minimise rubbing, particularly when condensation collected on the inside during the night.

With the tent erected, we were allocated positions within. With only one door flap, everyone entered through it and so the most desirable sleeping position was at the far end of the tent because it experienced less through traffic and consequently, less disturbance and much less walked-in debris. Rank cultivates flagrant exploitation, so my

Fig. 1.4 A downward cliff face view from a drone. (Photo: © Attila Kaldy)

Fig. 1.5 A drone's view of the Grose River. (Photo: © Attila Kaldy)

Fig. 1.6 Blue valley haze as seen from Govett's Leap. This is produced by atmospheric eucalyptus oil scattering short wavelength light, giving this mountain range its name. (Photo: Neil Frost)

patrol leader set himself up at the end of the tent. Despite being the youngest and smallest, I was given a sleeping position in the warm centre of the tent. This placement may have been influenced by the presence of my older brother George, who was the troop leader, a Queen's Scout[13] and the quartermaster for the camp. His rank was undoubtedly the most important and influential position at the camp because he controlled the allocation of resources like toilet paper[14] and in particular, food.

The first day must have seemed to end quickly because we were camped on the northern side of the river and the valley walls towered six hundred metres above us. This would have resulted in the light fading very early. Despite being early summer, a much more memorable effect was the early onset of the cold, which I have not forgotten! Having eaten before sunset, we joined the other patrols at the main campsite for the nightly entertainment which, fortunately, I cannot remember, though it would have likely included singing, music, storytelling and other activities. Something memorable was the large prepared campfire that the Rovers constructed previously. When lit, I remember that it illuminated the campsite and the bush beyond for a considerable distance, like Territory Day fireworks and the associated bushfire celebrations. It must have been visible from the clifftops for many kilometres and to a similar degree, so would have the smell. Afterwards, we returned to our isolated campsite, by the

---

[13] George was awarded the title of Queen's Scout in 1966. As a part of this award, George designed his own one-person tent made from plastic. It was tapered in two dimensions, from the head end to a broad point at the feet, to dramatically reduce the amount of material used. Compared to alternatives at that time, it was extremely light weight, very compact and highly suitable for hiking.

[14] After Covid, everyone knows the importance of good supply.

fading light of the raging fire. I think that because of our tiredness, we all slept reasonably well, despite the discomfort and rapidly falling temperature. Fortunately, having a large scout on either side of me probably kept things much warmer than they normally would have been.

The second day at the camp passed in much the same way as the first. However, I remember how hard the ground was where I had slept, so the first thing on my agenda was to dig up the soil to soften it, line it with fern leaves and re-lay my ground sheet. After eating dinner, we returned to the main campsite for some more entertainment. It seemed that the Rovers had rebuilt the bonfire, but not to the same excessive extent as the night before. We returned to our tent and went to bed.

Sometime during the early part of the evening, we were woken by a very loud commotion coming from the clifftop, six hundred and fifty-six metres above! The noise was similar to something that was either very angry or in distress. Lying in our sleeping bags, I remember that we listened to the various sounds and discussed what we were hearing for some time. At this early stage, the talk amongst the older scouts seemed to concentrate on generating as much terror as possible for everyone else. I can't remember precisely what was said, but it was very effective for me. Over a period of several hours, the turmoil continued and I remember that it varied in intensity and sometimes it sounded like roaring. As sensitivities normalised, including those of the older scouts, the discussion revisited reality. The speculation amongst the boys was that it was a dingo until someone pointed out that the sound was too loud for a lone dog, so the consensus shifted to "a pack of dingoes." I am sure that I agreed with this opinion because I wouldn't have known any better. After several unnerving hours, the noise ceased and I fell asleep.

1: Bluegum

On the morning of the third day, the topic of discussion around the camp was the loud disturbance that we had all heard for hours during the night. As with the discussion that took place in our tent, the broad consensus was that the animals responsible were a pack of dingoes. With the mystery solved, we all slipped back into our comfortable roles.

The third day at the camp probably passed in much the same way as the first and second, except that our patrol leader decided that it was necessary to rearrange the sleeping order. I was promoted to the most desirable position at the end of the tent, whilst Graham would take my position in the middle. After rearranging the tent and then eating dinner, later on, we returned to the main campsite for some more entertainment. Once again, it seemed that the Rovers had attempted to rebuild the bonfire to its previous glory, despite dwindling combustible resources. We returned to our tent and went to bed.

Sometime during the early part of the evening, well before midnight, we were woken by the loud sound of the bush being severely interfered with. The loud noises of plants being thrashed appeared to be coming from an area behind the main campsite, a hundred metres away. After some time, a series of ill-defined noises began to make their way towards our tent. As these sounds became more recognisable, it was apparent that we were listening to very slow-moving, heavy footsteps. From this point in time onwards, until dawn, nobody spoke in our tent.

Over the following hours, many unusual sounds were heard, though some were recognisable as footsteps, grunts and growls. I remember hearing very large branches breaking, of a diameter that might normally require a chainsaw to dismember. Then, something pulled a tent rope or tripped over it, which caused the very heavy tent assembly to precariously wobble about. Overall, if there was any intent being communicated to us, it was the desire to intimidate![15] This animal was powerful, big, heavy, irritated, provocative and intelligent! It was playing with us!

As the night dragged on, I tried to read the time from my wristwatch. It was a manually wound Oris watch with a green phosphorescent-painted dial and hands. Unfortunately, as the night wore on, its radioactive ability to reemit captured photons ceased and the face became dark. Even by placing the watch face close to my eye, failed to provide any indication of the time. From my last reading, I added an estimate of the elapsed time and guessed that it was about four in the morning. By now the temperature had fallen further and was probably about zero degrees Celsius, or possibly lower, despite being early summer. I was becoming extremely cold, but I wasn't prepared to get any additional clothing out of my backpack because of the attention that it would generate. Instead, I pulled myself into a ball and retrieved the excess half of my sleeping bag, placing it back on top of myself.

Soon after I fell asleep I was woken by the very loud sound of something sniffing and then growling directly into my ear, separated only by a layer of canvas. I don't remember how far I recoiled, but I landed firmly across several other scouts down the line. Still, nobody but me made a sound, no one complained and there was no snoring either!

---

[15] Breaking thick branches is a common intimidatory behaviour. After breaking a large branch, it is typically held and not allowed to fall to the ground. This conveys to the listener that something highly unusual has happened, because it doesn't sound right. Typically, the sound of the branch breaking communicates the use of great strength. Also, by not hitting the ground, the deliberateness of the action is emphasised.

1: Bluegum

There were still a few hours to sunrise and being unable to sleep, I sat up in my sleeping bag[16] and tried to keep warm. The situation was made much worse because my sleeping bag had become very wet along one side[17] from the condensation dripping off the inside of the tent end. I heard nothing more, not even the biped's departure. A few hours later, light began to shine through the door flap.

The discussion immediately began when we got out of bed. None of us had an explanation for what had happened and some of the scouts headed over to the main campsite to learn of their experience. Most of the other patrols had not heard anything significant, although a few individuals had heard some strange noises early in the evening. I remember that our patrol met with Guraki, seeking his wisdom. He had not heard anything and only repeated the dingo rationale[18] from the night before. As we would later be told by Elders of the Parramatta Aboriginal Land Council in 1993, the light from a campfire or any light source attracts these animals and also repels them. The attraction is curiosity in finding out what is there and the repulsion is the light which blinds their excellent night vision.

---

[16] My sleeping bag, like my brother's, was made from kapok inside a cotton covering, which our father bought from an army surplus store. They were not very warm, so from previous camping experience George had mum make a flannel sleeve insert. I wish that she had made two.

[17] The sleeping bag had to be dried alongside a fire, but was still damp the next night. It also took on a smoky aroma.

[18] If a pack of dingoes had been on the cliff top the night before, they could not have safely descended to the valley floor in the short time available, without going the long way around. Many encounter reports to the 'Octopus' (community database and support) describe the rapid vertical movement of 'marsupial hominoids' in both directions.

While the others did what they did, Rodney[19] and I walked down to the river and began looking for any indication of what had visited our campsite. We walked along the alluvial sand bars of the floodplain for a considerable distance, looking for animal tracks. We mostly found kangaroo and wallaby trails, where they had come to the river for a drink or had crossed over the river at that point. I remember that we reached a very distant part of the eastern river that had thicker vegetation along the banks and was noticeably darker. We both felt very creepy, so we immediately turned back.[20] After returning to the campsite, we continued in the opposite direction for a much shorter distance. We had been specifically looking for dingo tracks but found none.

Back at the campsite, I had three urgent duties to perform. I had to dry my sleeping bag next to our fire. I also went into the bush and cut a large number of fern fronds with my scout knife, which I used as insulation and padding underlay for my bed. Most of all, I was extremely hungry. This would have been due to the cold but also because the older and larger 16-year-old boys were disproportionately eating the majority of the food. Consequently, I visited my brother George, who gave me a packet of biscuits from the store, telling me to hide in the bush while I ate them and to return when I needed more.

The remaining three nights of the camp were very uneventful. Even if something had happened, I severely doubt that anybody would have had the stamina or willpower

---

[19] Previously I had been a 'Sixer' in Cubs and Rodney was my 'Seconder.'

[20] This fear response is very commonly reported by encounter witnesses. As researcher Andrew experienced during a camping trip at Acacia Hills in the Northern Territory, "I walked into some sort of virtual wall. I just stopped and had an overwhelming feeling in my gut to turn around and return to camp. Odd, yes, explainable, no."

1: Bluegum

to remain conscious! The extremely cold and unseasonal weather required me to wear as many clothes as I could to bed and make several additional visits to the quartermaster. The ferns greatly enhanced my comfort and warmth, but the sleeping bag remained damp for the remainder of the camp.

WITH MORE THAN fifty years of experience and thirty years of research, certain common aspects of this encounter are evident.

The use of a campfire was, undoubtedly, the primary instigator of this encounter incident. As known by Aboriginal Elders and confirmed through our experience, a light of any kind acts to attract these curious animals, however, a large bonfire casts an even bigger influence. Similarly, the presence of food is a great attractant, as with all animals. In this case, it was most likely the temptation of food being cooked, with many campfires being used to prepare a variety of fragrant, meat meals, that would have travelled with the wind along the valley. Having been drawn to a location, their typical behaviour is to stand behind some bush cover, outside of the illuminated area and observe the human activity for lengthy periods, without moving. From our experience, these intelligent 'marsupial hominoids' like to observe people, particularly children and are known to hold their position for hours. In situations like this, the reflective rhodopsin 'glow' of their night vision eyes is frequently seen from around a campfire and is commonly known as 'Old Red Eyes.' This is the most common form of contact experienced by people.

On the other side of the coin, a campfire also has a repulsive influence, because the bright light overloads the animal's night vision ability, which blinds them. The brighter the bonfire, the further back this blinding influence extends into the bush. If the campsite layout is considered, it was large, clear and circular. At its centre was the rock-lined fire pit. Each patrol was positioned around this prepared area, except for our patrol, which chose to isolate itself in an uncleared area, about a hundred metres away from the bonfire and the other tents.

With the bonfire burning brightly on the first night, its beacon would have indicated the presence and location of people camping in the valley. On subsequent nights, though less bright, it would still have been very impressive and would have burnt long into the night. From the pattern of movement heard, it seems that the animal first came into contact with the main campsite early in the evening. With the bonfire still burning, it would have been prevented from approaching too close to the centralised tents. However, whether seen or heard, our isolated tent was soon discovered and the animal was not inhibited in its approach by any bright light.

Clearly, the consequences of our patrol leader's decision to self-isolate[21] were very unfortunate. Being too far away from the bonfire, we did not benefit from its protective glow. Pitching our tent in an uncleared patch of bush, allowed the biped to closely approach our tent using the natural bush cover and without being exposed to the bright light. These 'marsupial hominoids' rarely move through open areas, unless they feel very secure.

A COMMON STRATEGY used among Aboriginal tribes was to have a fire burning at night at their campsite to keep these animals at bay.[22] Currently, in some remote areas of Australia, these practices continue. One tribe in the Northern Territory places a ring of drums around their

---

[21] This was edited during the COVID-19 pandemic.

[22] See Aboriginal Elder Merve and his advice, Chapter 3.

1: Bluegum

camp that burn large pieces of timber, called 'all-nighters', to maximise the deterrent.[23]

Consequently, some general advice can be provided to campers and isolated households, located in active areas. For campers, it is advisable to locate the campsite in an open or cleared area. If possible, position the tent in the centre of the clearing, not on its boundary.[24] Similarly, isolated houses with one or more borders fronting the bush should have motion-sensing lights facing outwards.

There is one important aspect remaining that should evolve over the length of this book. What are these animals? From the beginning, it should be understood that they are not wild humans. Nor are they a remnant population of hominins, for example, *Homo erectus*. They are not gorillas or any of the other Great Apes. They are not an Australianised variant of the North American Bigfoot.

Instead, these animals are indigenous to Australia. They are marsupials. They are a remnant group of Australian megafauna.[25] They are most probably a convergent form of biped similar in form to humans that are closely related to macropods, like kangaroos.

THERE IS NO shortage of witness encounters involving Australian 'marsupial hominoids' in the Blue Mountains and elsewhere. The majority of these occur where human settlement or activity meets the bush interface.

---

[23] This term was used by Northern Territory Aborigines and was reported by researcher Andrew.

[24] Gary, a former colleague, located his tent on the edge of the bush on a cleared paddock, without a fire, at Mudgee, NSW. Gary had a similar experience.

[25] Steve Rushton was the first serious researcher to suggest this to us. His alternative paradigm went against the prevailing dogma that continues to dominate, despite the overwhelming evidence to the contrary.

Over many decades, our informal collaborative group, the 'Octopus', received many thousands of encounter reports. Below is a newspaper article from *The Macleay Argus*, which has been used as an independent source example. The encounter occurred at Blackheath in 1954, twelve years before and eight kilometres from our 1966 scouting camp.

> Who Made the Strange Cry[26]
> One night, in the mid-fifties (probably 1954) a teacher at the Blackheath Girls' School was awakened by a loud, mournful call that went on almost at the same pitch for over half an hour.
> The teacher, Miss V. Everingham, now retired and living at Macquarie Street, Frederickton, relates her strange experiences below:
> "It occurred during the winter at about 2 a.m. and whatever it was could be heard calling as it made its way slowly up from Centennial Glen between Paradise Hill and the Blackheath township. As it went it uttered an unforgettable, plaintiff call such as I have never heard before or since. It seemed half human, half animal, but I could not catch sight of the animal making it. It went on for half an hour, giving me an extremely apprehensive feeling. At first, I thought it was cattle calling, but I soon realised it was a completely strange call. It was too loud, even in the distance. It woke me from a sound sleep. All I could hear was this loud sad call that

---

[26] 1976. Who made the strange cry. *The Macleay Argus* (28 September).

came from the bush in the distance. As the animal drew nearer, the intensity of the sound increased, as if it was in great distress. The noise came from heavy timber below the school ground, but when the animal seemed to reach the top of the hill, the sound stopped entirely. There was complete silence. I pinpointed the spot behind the school tennis courts and the next day I went out to see if I could find tracks. There were none. I lived at the school for almost 37 years and that was the only time that I ever heard that noise. It still mystifies me. At the time I thought an animal might have escaped from a circus, but there were no circuses travelling on the mountains at that time."

Miss Everingham said most of the cottages in the vicinity were weekenders and unoccupied. The headmistress came out on the verandah after the calls stopped. She had not heard it, but something had wakened her.

MANY ENCOUNTERS INVOLVING our 'marsupial hominoids' occur as they attempt to secretly navigate around the human and natural barriers that lie in their path—for example, as they contemplate crossing the highway through Blackheath, from one valley to the next. Many of these circumstances bring them into proximity with human settlements where plenty of different food and entertainment opportunities exist, but at the increased risk of discovery. The encounter by Miss V. Everingham was undoubtedly of this type.

During the 1966 camping encounter, the 'marsupial hominoid' vocalised its anger from the cliff top, six hundred and fifty-six metres above and the following day made its way into the untenanted valley floor below by some alternate route. Almost three decades later, when looking for a 'lost passage' through the upper mountains, I went searching for two topographic features that were promisingly named: 'Orang Utan Gully' and 'Orang Utan Pass.' The tantalising reference to the placental primate was suggestive of something similar having been seen in the earlier European history of the Blue Mountains. The reference to a nearby 'Pass' was also encouraging.

To find out more information about the naming origins of these two topographic features, I telephoned the Geographical Names Board of New South Wales and was given the only information that they had in their database in 1993. I was given the name Dorothy (Dot) and her

Fig. 1.7 Three freshly broken saplings on a remote 'game trail' between Perry's Lookdown Road and the cliff at the edge of Orang Utan Pass in 2010. See map of Blue Gum Forest. (Photo: Neil Frost)

1: Bluegum

telephone number. I called Dot and we had a long and pleasant conversation.

Dot, like many other people relaxing on weekends between the two World Wars, regularly went bushing. As would later become known as the 'standard question', I asked her, like anyone with the potential to harbour an interesting encounter story, if she ever had an unusual experience in the bush. Asking this question, I think, offers a release, or a rare opportunity for the witness to speak freely about a memorable incident that might have troubled them, within an implied context of safety. The standard question typically has a high success rate amongst those who might have experienced something out of the ordinary, but chose, for whatever reason, to keep the secret. For example bushwalkers, timber cutters, tourist operators, pig hunters, bush photographers, national park workers and many others. Over her long bushwalking life, Dot could not recall anything of importance, so I raised the names of the two topographic features and asked if 'orangutan' had any significance for her. She told me that the boys in their group, some of whom she named, would often perform around the fire at night, imitating orangutans with their dance movements and vocalisations. She thought that this campfire behaviour might have been responsible for the names being associated with these bush locations. I then did something that I normally don't do, I asked her if she had seen a 'Yowie'. She said "No", laughed and then said, "You are being silly." As promised, I returned the phone call to the Geographical Names Board and related the information as received to the archivist who recorded it, even though I was not completely convinced of the depth of Dot's evidence, particularly as a solitary source. I think that probably there may have been much more hidden background to this story, particularly from the boys. In knowing very well how socially taboo this subject is, I think that the reporting prejudices, general apprehension and the personal fear of ridicule in the 1930s, would have been just as great as they are now!

Driving along dirt roads, frequently on the centre or wrong side of the road, is a very quick and efficient method of covering long distances in comfort when looking for promising indicators of Hairy Man movement or evidence of road crossings. It provides rapid inspections at a relatively close range, from out of the driver's window or open door. I had been using this technique for decades, with caution, and clearly, this is not a recommended practice. However, in 2010 when slowly driving along Perry's Lookdown Road, looking for footprints and developing tracks on the side of the graded road, I recognised the subtle indications of a habitual track. These tracks can be very difficult to spot and are best observed when the weather is overcast, or at sunrise and sunset.[27] They are normally not worthy of a photograph. The indicators are difficult to explain and could best be described as a mild flattening of the ground vegetation, that is consistent with infrequent traffic usage, with a dominant vertical absence of plant material, caused by aggressive, manual defoliation. After stopping the car and investigating, the simple track was typical of those that we had found elsewhere in the mountains. It led southward towards the Orang Utan Pass and showed signs of recent and past track maintenance activity, where new growth or hindering vegetation had been removed or severely inhibited. The general area was not thickly bushed and occasionally some of the smaller eucalypt regrowth had been consistently snapped in a line,

---

[27] Satellite imagery, as used on Google Maps and elsewhere, provides an excellent track detection method, particularly for remote and difficult to access areas. Any anomalies can then be examined more closely on the ground using GPS data.

at a height of about four feet, as shown in the above photograph. Some inexperienced observers have attributed this damage to wind gusts, but clearly, it would require two hands to achieve, with one hand being used as a fulcrum at the breaking point. Although known for their exceptional strength, I would prefer to think that this damage could not be achieved by using the fingers only. For me, these affected paths were not simply wallaby trails because of the upper tree breaks and the clear indications of manual track maintenance.

It would seem, therefore, that Orang Utan Pass and Gully might provide a suitable, alternative access route to the Grose Valley at its upper reaches, particularly for these well-adapted indigenous climbers. Such an access point and its associated track may have been the site of many historical sightings by European bushwalkers and campers since the early 1800s when the ignorance of the time would have naïvely linked these beasts with the newly discovered gorilla, orangutan and the other large primates.

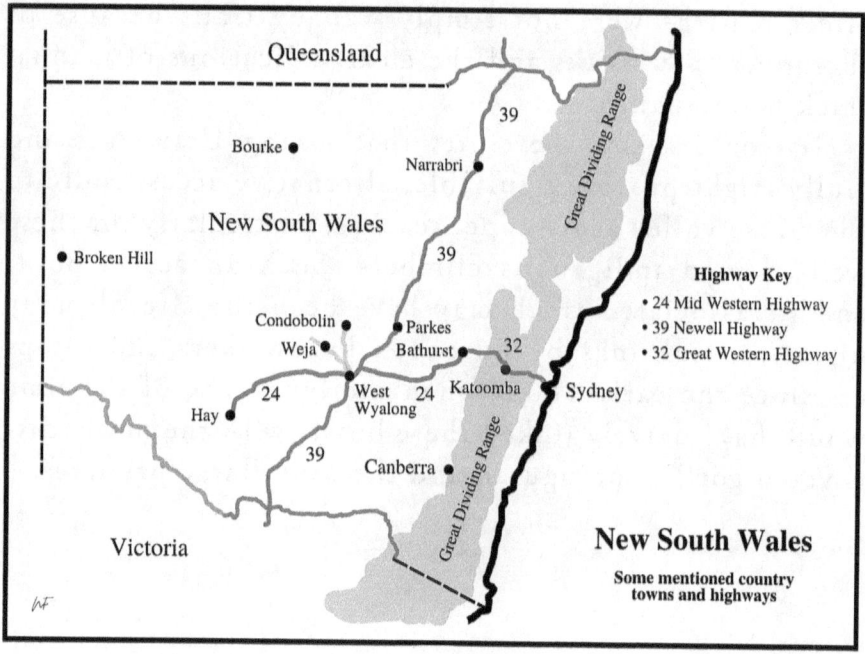

Fig. 2.1 Some country towns and highways mentioned in the chapter. (Map: Neil Frost)

2: Providence

FROM 1978 I was serving as a reserve teacher in Sydney because I was a bonded student and anthropology didn't pay very much. In early 1981 my circumstances involuntarily and substantially changed when I was given my first permanent teaching appointment beyond the 'Western Line',[1] in the remote central west of New South Wales (NSW), a 540-kilometre and six-and-a-half-hour journey by road from Sydney. All teachers were expected to do 'country service' at some time during their careers but the requirement was very frequently exercised as a coercive method of staffing, used to fill teaching positions and other school employment requirements in remote and undesirable locations.

After staying in a very small, one-bed motel room with a separate kitchen for about half the year, my cooking became legendary in a small culinary-deprived country town where the comestible alternative was a meat pie with

---

[1] The 'Western Line' was a reference to the 33° Celsius (91.4° Fahrenheit) and above, average summer temperature isotherm for western New South Wales. It was used to demarcate the boundaries for various working and teaching conditions. West of this line, schools could expect to have air-conditioning; the summer holidays were extended by one week; and teachers were provided with subsidised housing of a reasonable standard or might be paid 'forage' as compensation.

tomato sauce from the pub. My cooking establishment was very attractive to many casual diners because the closest restaurant served basic Australian-Chinese meals and was sixty-three kilometres away. I was deceptively enticed to move into a bedroom on a farm called 'Providence' near Weja, mainly because I could cook a variety of curries, samosas, garlic chapati, raita and poppadoms with rice.

Like most rural properties in the region at that time, the 15,000-acre farm produced sheep and wheat. It had been purchased by farmer Ned to consolidate his local holdings, to take advantage of the economies of scale because the vagaries of the climate were making the land increasingly marginal and many farmers were either leaving the land or merging with other properties. Consequently, the very large farmhouse was vacant, so Ned rented the property to local teachers for nominal rent[2] because inadequate and insufficient accommodation was being provided by the NSW Teacher Housing Authority (THA). As partial compensation, the NSW Department of Education paid some teachers in this predicament 'forage' which, as the name suggests, provided financial assistance in the basic provision of food, transport and shelter. I did not receive forage whilst living in the motel room because the THA regarded my situation as adequate because I was living in a small town with a regional population of four hundred people and could take advantage of its comprehensive range of facilities. When I moved to the farm, I became entitled to share the forage payment but the person who currently received the payment regarded the entitlement and money as his own.

---

[2] Ned was very keen to support teachers and encourage them to stay by only charging ten dollars per week rent, which included electricity and water cartage.

2: Providence

The farmhouse was very old, large, traditional and neglected. It was probably built sometime in the 1930s and reflected the halcyon days when agriculture was highly profitable and much less stressful than in current times. The homestead sat on a knoll in the home paddock. It enjoyed a cooler southerly aspect that overlooked the many paddocks and provided a good view of the surrounding region for tens of kilometres. Both the external and internal walls were made from double red brick, to help moderate the high summer temperatures. The roof was corrugated, galvanised iron that fed a water tank made from similar material next to the southern verandah steps, directly in front of my bedroom and several other metal tanks on the western side. Around the southern and eastern sides were the original, wide verandahs without railings. On the western side, the verandah had been filled in to provide a winter sunroom. There were two main entrances to the house. On the southern side, the door entered directly into the kitchen and immediately to the left was my bedroom. The eastern entrance opened into a corridor that ran through the house to a western verandah door. There were other connecting doors off the main corridor which allowed exits from other parts of the house and permitted cool breezes to pass through the building from any prevailing direction. The internal ceilings were generally fourteen feet high which acted as heat wells to assist with cooling. For winter heating, the kitchen and living room both had very large open fireplaces. Some of the bedrooms also had fireplaces, except mine. My bedroom had two doors, one opening onto the kitchen and the other onto the enclosed verandah with a western, external door. As is typical of a country home, there were no locks on any doors or windows. My bedroom window and most of the other windows had no curtains. My bed was a mattress on the floor without a base, underneath the window.

The services to the house were rudimentary. Water was usually trucked in from town because the local rainfall was well below average during the time that I lived there and despite every effort being made to conserve it. To save water on the farm I frequently showered at school, which also provided the luxury of having hot water. The electricity supply traversed many kilometres over paddocks from the nearest road and after several voltage step-ups, there was very little current remaining for cooking or any other useful purpose.[3] The telephone line was a party network. Every farmhouse sharing the common landline had a unique Morse Code ring call so that farmhouses could unmistakably identify their private telephone messages and avoid any accidental pickups. Sometimes there were errors made with two or more parties picking up in response to the same coded ring. Sometimes one of these parties would forget to hang up. As telephone connections were manually connected, to maintain a full service, it was necessary during the evening for a student from the school to sleep at the telephone exchange in town. Consequently, personal telephone calls tended to become public domain.

On the northern and western sides of the homestead was natural bush. It was one of only a few areas on the farm that still had remnant native vegetation. The bush generally approached about ten metres from the house and was mostly light, increasing to moderate density as you moved up the hill. At that time, the bushland to the north was fragmented but still made a loose connection with more extensive bushland surrounding Condobolin and the Lachlan River, about 70 kilometres away. About 60 kilometres to the east was the notorious Newell Highway,

---

[3] Spike Milligan defined a 'small town' as a place where the street lights dim when you plug in your electric toothbrush.

2: Providence

or A39.[4] The 'Newell' is a major inland highway, over a thousand kilometres in length that carries a large proportion of heavy vehicles from northern Victoria to southern Queensland, through inland New South Wales.

Even though I spent many Christmas holidays as a child staying on several farms in northern Victoria and southwestern NSW, this was not the same experience for a variety of reasons. Since my grandfather had a long family history associated with the farming district that he was living in, we also tended to be acknowledged and accepted by the local population, without needing to individually serve time. As a teacher blow-in, it was generally regarded that unless you married locally soon after arrival, your stay was only going to be as long as it took to earn a general transfer back home or after having earned sufficient hardship bonus points to gain a priority teaching transfer to the idyllic location of your dreams.[5] I didn't want to marry a local country girl and I didn't have the thirty or more years required to become a recognised semi-local. After completing the minimum of two years of country service and acquiring the maximum number of sixteen transfer points during that time, I moved back home.[6]

---

[4] The Newell Highway #39 runs from Goondiwindi in Queensland to Tocumwal in Victoria, a total distance of 1058 kms.

[5] That was the naïve and romantic belief held by many teachers who stayed far too long until the inevitable loss of connection with home sealed their plight.

[6] Teaching in the Western Division earned 4 points and 4 points for working in a small Central School, resulting in a maximum of 8 points per year or 16 points for my two-year country service. Normally teachers earn 1 point per year. Regardless of having maximum transfer points in this new transfer programme, I was appointed to my 110th priority school on a list of 120. Doesn't seem right somehow? The new school was similarly undesirable, providing 4 points per year because of its disadvantage.

Living on a remote farm, far from the nearest isolated hamlet meant that, outside of work, there was "nothing to do and all day to do it."[7] Staying in the area during holidays seemed to be an insane proposition, however, after a while, some did this because it was apparent that they had lost touch with their roots and had no one to visit. Even if you wanted to, watching regional analogue television was difficult under the best of atmospheric conditions and you also needed to have an acute interest in stock and grain prices to fulfil your enjoyment ambitions. Consequently, to bring some entertainment to the area, I started a travelling movie theatre that we called the 'Humbug Film Cooperative.' Using the school's Eiki film projector, my stereo hifi and after purchasing a cinemascope lens, I hired 16mm film movies from Village Films in George Street, Sydney, which were delivered by rail to the school. As this setup was very portable, I showed these films around 'nearby' schools during the week which also raised money for them. It wasn't destined to last, since the video home system (VHS) was gaining rapid acceptance.

Other than this, there was sport. There were several choices, including darts, tennis, cricket, basketball, rugby league and the main interest in the region, Australian Football, or AFL. As this region was the ancestral home of the very extended Daniher family, everyone was expected to support the Essendon Football Club, by default. This was a good thing because everyone assumed that I was a supporter and interested in the sport, which avoided any additional alienation. Some of the sporting activities conveniently took place during the week but many were held over the weekend. For me, weekends were a time for travelling back

---

[7] One of my father's favourite quotes, inspired by years of service in very remote Australian Air Force bases during World War II.

2: Providence

home and after breaking several of my ribs during a staff versus student football match, getting seriously injured was no longer attractive.

Consequently, I desperately needed to find alternative activities and since the night sky was outstandingly dark, I bought myself a Celestron Schmidt–Cassegrain telescope and restarted my hobby in astronomy. In addition to the superb astronomical viewing and limitless horizons, I fortuitously managed to see the Aurora Australis from this relatively low but dark latitude. I also had an interest in radio, so I bought a medium spend, short, medium and long-wave radio receiver from a shop at West Wyalong, that knew a needy market when it saw it. Since the farm had many kilometres of north-south and east-west wire fencing, they made excellent and selectable radio antennas, like a radio telescope, that enabled listening to radio stations from anywhere around the world. Finally, I bought some pigeons that I intended to breed from. This was an experiment to see if breeding and selling squab might serve as an alternative future occupation to teaching. As the water tank outside my bedroom was on tall timber stumps, I enclosed the space underneath with some bird wire and made a coop.

Surrounding the farmhouse was the home paddock. This was a fenced-off area that was primarily used to keep the sheep from neighbouring paddocks away from the house. As this fence was slightly higher and more robust than others on the property, it was also optimistically used to discourage kangaroos from entering the area and consuming the few remaining drought-tolerant garden plants or from seeking water. Containing kangaroos is a very difficult task. If a kangaroo can't easily jump over a fence, it might try forcing its way by 'torpedoing' through any gap in the wire. However, the fence was more than adequate in stopping the sheep from entering the home paddock and eating anything there, but more importantly, it prevented the sheep from making any noise that might disturb the sleeping residents. Although, occasionally Ned or one of his many sons would allow the sheep into the home paddock to clean up the weed growth. When they did this, the sheep had the habit of climbing the stairs and walking about the verandah, particularly during a moonlit evening for some reason. Worse still was when a sheep died, either inside the home paddock or close by the fence on the other side. When this happened and the body could not be easily removed, I would wait for the maggots to do the majority of their work and then burn off the carcass with petrol in an attempt to reduce the number of potential flies. This occurred more frequently during the lambing season, during winter and spring.

    During the breeding season, some ewes would succumb to milk fever or lambing sickness, due mainly to poor pasture nutrition or a lack of supplements in the later stages of pregnancy. As Peter and I lived at Providence with Thomo, we often shared a journey to school and back as well. We would sometimes drive past ewes in the outer paddocks on the way to the main gate, that were recently dead or dying. Occasionally we would come across a ewe that was experiencing a difficult labour, so we would stop and assist, if impossible. As Peter was the agriculture teacher, he would usually fish around within the lower, internal workings of the ewe, while I would pull from the front end, applying Newton's Third Law of Motion. It would seem that Ned or, most large-scale graziers, would not know how many ewes or lambs they had, until later in the year when shearing was completed and an account was made. It would be common, therefore, for overall stock numbers to be vague, with any breakdown of losses due to sickness, predation or other sources remaining unknown with their causes being very difficult to determine.

2: Providence

With three terms in the 1981 NSW school year, we would have seven weeks off over Christmas and two weeks of vacation between terms. Teachers and students got one additional week over the Christmas break if they were located west of Isotherm 33, where temperatures in January would typically exceed an average of 33 degrees Celsius for long periods. Over one weekend, we experienced more than 50 degrees Celsius.[8] During this heatwave, a mob of about a hundred or more Eastern Grey Kangaroos *(Macropus giganteus)* closely approached the western side of the farmhouse to drink from the unmaintained swimming pool. They were unperturbed by my approaches and lowered the level of the pool by about an inch in an hour and then laid down under any available shade for the bulk of the day, trying to keep cool by licking their forearms or panting.

Before leaving for my end-of-term two break in 1981, I filled the bulk water and seed hoppers for the pigeons and left their small bird door open.

WHEN I RETURNED to the farm two weeks later at the beginning of spring in September, I found that the water tank stand had collapsed, drowning and crushing the majority of birds. A few survived and were living on the rafters of the garage. When driving around the farm

---

[8] Early that Saturday morning, Peter and I drove to school, unbolted a refrigerated air conditioner from a window at school and stopped off at the bottom pub to stock up with a few cases of beer. We set up the air conditioner in the lounge room, stayed there for most of the weekend and excitedly watched the changing stock and grain prices on the poorly received regional television. From this experience, I more fully appreciated my father's wartime service as a Flight Sergeant at the remote dispersal facility at Charters Towers airfield in outback Queensland, during similarly quiet times. In addition to supporting elements of Australia's Army and Air Force, it was also home to the U.S. Army's Fifth Air Force.

on the first Monday of term, I saw Ned and told him what had happened. The next day I went to school. When I returned home, I was shocked to see that Ned had removed the damaged water tank, stumps and collapsed pigeon cage with a bobcat. In their place was a new concrete water tank, sitting neatly on the ground and full of water that he had trucked in from town. The ground around the new water tank had been churned up by the bobcat and was very wet and muddy.

That night I had been listening to ABC Radio Sydney, as was my habit. I was home alone as Peter was staying the night with his girlfriend in town. I went to bed early as it was the beginning of a new term. Sometime later, before midnight, I woke for some reason and lay awake. Soon a noise outside the window gained my attention and I listened intently for a few minutes. After thinking about it I thought that the noise must have been quite loud because the walls of the house were double brick. I heard another sound and thought that I could hear walking. Thinking of the whereabouts of the sheep, I knew that none were in the home paddock. It wasn't a kangaroo either. A few minutes later, I heard the walking again. It sounded very much like a person. The next time that I heard the noise, it sounded like very slow walking, but this time the movement was on the verandah at the far western end and seemed to be approaching my window. Whoever it was must have been very heavy because the normally quiet floorboards were creaking noticeably. As there were no curtains on my window, I slid closer to the wall and I felt sure that someone was standing at the window. Very slowly I sat up in bed. Looking out through the bottom left-hand corner of the window, I saw nothing, but there was an immediate reaction! Someone accelerated to high speed, noisily jumped off the eastern verandah edge and could be briefly heard running away.

2: Providence

Cautiously, I sat up in bed, got to my feet and turned on as many lights as I could easily assess. Going to the kitchen, I turned on the light and armed myself with two large cook's knives, one for each hand. Then, moving to the centre of the house where the telephone was, I turned on more lights and kicked open the doors at each end of the corridor. I rang the exchange.

Identifying myself to the known student I told him that this was an emergency. I thought about requesting the presence of the local policeman but thought about the details of the situation becoming known around town. Instead, I asked him to call a colleague who was staying in town. After explaining the urgency of the situation to Peter, he optimistically told me that several teachers would be at the farm in about fifteen minutes. Standing in the corridor opposite several potential entry points, I thought that it might be safer to stand outside in the open. Turning on the verandah lights, I moved to an open space away from the house in the home paddock with a knife in each hand and looked eagerly towards the south, the direction that the cars should approach from. Fortunately, there was sufficient moonlight, once my eyes became dark-adapted, to see anyone who might approach me out of the shadows.

After a nervous start, I could finally see very distant pinpoint glows of several car headlights travelling along the main road, about ten kilometres away. Following the winding path of the headlights along the dirt backroads, the cars turned right and stopped at the main farm gate. Constantly scanning the area around the paddock where I was standing, there was no sign of the intruder and after waiting for the vehicles to pass through several paddock gates, two cars arrived at the farmhouse. With me holding two knives as I spoke, I think that the seriousness of the situation was much more quickly appreciated. I got the drivers to position their high beams at my bedroom window on the southern verandah and then we moved towards the house. From the southwestern corner of the verandah, there were a series of very wet, muddy footprints that crossed the verandah boards towards my bedroom window and then, diagonally away to the southeast. The muddy footprints first appeared in the churned-up area on the western side of the new water tank, but there was no obvious track continuing from where the individual had jumped off the verandah. The footprints were noticeably large, but they were also confused and ill-defined, especially around the floorboards outside my bedroom window.

There had been an intruder at the house and the incident was very worrying. Although the other teachers stayed at the farmhouse that night, understandably I didn't sleep. After school the next day I went to the general store in town and bought four sliding bolts, some screws, curtain fittings and a long length of cloth. On the inside of each door, I attached a top and bottom bolt, screwed the window closed and installed a curtain across the window. Suburbia had come to the country!

In town, there were two hotels, the top and the bottom pubs. The clientele of the top pub typically consisted of the majority of cockys,[9] stock and station agents, banking staff[10] and other aspirational types. The bottom pub was regarded as the worker's hotel, mostly supported by truck drivers, teachers, farm workers and other inspirational types. When seeking advice, the best range of opinions was immediately available from the bar at the local pub. When I mentioned the incident with the intruder a few nights

---

[9] Graziers or land owners.

[10] This was a time when banks still played an active role in small country towns and maintained staffed branches.

2: Providence

before, I remember that there was much discussion along the bar, mainly centred upon the potential seriousness of the situation and the wider implications for the local community. The discussion continued for several hours with the consensus explanation suggesting that it was a 'hermit' or a homeless person who was responsible. Several local people recounted historical stories of 'sundowners' who would arrive at a farmhouse late in the afternoon, requesting work in exchange for a hot meal and a place to sleep. Other accounts were much more serious, with several other people telling stories of 'hermits' who would raid isolated farmsteads at night looking for food and terrorising the inhabitants. The local police officer, who frequented the bottom pub until closing time required him to leave, similarly agreed with this opinion. When asked, it was interesting that no one was able to identify a responsible hermit, though the excuse made was that these homeless people are transitory, constantly on the move and therefore, untraceable.[11]

When talking to the truck drivers, they had a different perspective. I was friendly with several of them and they told me that there was "a lot of weird stuff" happening on the bush-lined roadside at night. It wasn't just the interstate drivers who had strong opinions and drew upon examples from colleagues. The local drivers also had stories of tall dark figures standing in the bush, or making fast dashes across the road in front of the truck's driving lights, in the middle of nowhere. They knew that I did plenty of long-distance driving and asked me if I ever

---

[11] During a Blue Mountains police investigation into our situation in 1993, the default explanation was also the activities of hermits. Homeless people living in the bush are an unfortunate reality, but as far as I am aware, none are seven feet tall, have large reflective red eyes, or can run through thick bush at nighttime at speeds approaching 40 km/hr.

travelled along the Newell, particularly at night. I told them that I regularly travelled a thirty-five-kilometre section, in both directions, along the Newell from West Wyalong to the turn-off for the Mid Western Highway,[12] mostly late at night. They told me to avoid using the Newell if I possibly could and that night was the worst time to travel on it. They particularly advised me to never stop on the roadside or at truck stops. They said that the most notorious sections of the highway were the heavily bushed and hilly districts and that many truck drivers refused to stop in these well-known areas after some drivers had been severely harassed within their vehicles and in the bush near their rigs. One story that they recounted involved a truck driver turning around after urinating, to find a large beast standing behind him. On returning to the cabin, the large animal shook the truck before he could drive off. There were other stories that I can't fully remember, however, common experiences involved sightings of very large, dark figures standing beside the road or crossing it.

THIS INCIDENT WAS not an open-and-shut case. At the time I was ambivalent about most explanations. It seemed like an intruder may have visited the farmhouse, possibly to rob the house after I had gone to work in the morning. Although, this explanation seemed unlikely since there was only one way to travel to the house by vehicle and no other car had been seen or heard of. If the intruder had walked in, it would have been a long journey through vast amounts of nothing to get here. The front gate alone was more than three kilometres away. However, the incident did have a strange but familiar feeling about it. The

---

[12] The Mid Western Highway #24 runs from Bathurst in NSW to Hay in NSW, a total distance of 516 kms.

2: Providence

intruder was heavy and the muddy footprints were large and unusual. This did make me reflect on the encounter with a strange biped on a scouting camp fifteen years earlier, but there wasn't that much to compare!

WITH HINDSIGHT AND the benefit of experience, this incident was undoubtedly an encounter with a Dooligahl,[13] a medium-sized, seven-foot-plus tall 'marsupial hominoid' and part of a remnant population of Australia's marsupial megafauna. Typically these bipeds have a large body mass that weighs upwards of several hundred kilograms. Such a mass would have been sufficient to cause any floorboard to creak, even the very solidly constructed verandah decking on our homestead. Similarly, peeping through windows is another very common behaviour that was very frequently recorded by the Octopus.

From experience, it would seem very likely that it was the loud and intense construction activities of Ned that were responsible for attracting the attention of a Dooligahl from a distance. People-watching, particularly involving workers or young children, is a common behaviour. Having been drawn to the farmhouse by the unusual sounds, it would have watched the work of Ned from afar, probably from the bushy knoll to the north of the homestead and then approached closer to the house after dark to observe my activities. It would have completed its final approach to my bedroom across the open muddy area after I had turned off the light and gone to bed. However, it was early spring and I doubt that this was the first and only time that this biped had visited the farmhouse and the adjacent animal paddocks in recent days or weeks.[14]

---

[13] Dooligahl is a south coast New South Wales Aboriginal name for this medium-sized 'marsupial hominoid'.

WINTER AND SPRING are the seasons in which ewes give birth to lambs. For any observant and intelligent predator, the lambing season would be an easy opportunity to get a substantial meal over a sustained period and provide sufficient food to meet the needs of hungry, growing joeys.[15] Lambs and ewes are very docile animals, particularly in comparison to the indigenous kangaroo population which are dangerous and require much more effort to take down.[16] During many decades of research, I have read numerous reports and interviewed witnesses involving the disappearance of lambs and sometimes ewes, by something other than wild dogs[17] or wedge-tailed eagles.[18] Occasionally, there are reports of cattle and calves disappearing,[19] or being mutilated as well.[20] Although the commonly perceived predators are responsible for many of the losses, it is good to have reliable eyewitness accounts where something else was responsible.

---

[14] It would seem to be very worthwhile to set up an infrared camera at night on a hilltop overlooking a valley with lambing ewes.

[15] As these bipeds are marsupial and probably descendants of a macropod progenitor, then the correct term for their young is 'joey'.

[16] Ian Price was a very knowledgeable Australian biologist. He said that all animals tend to take the easier option to conserve energy and reduce food demands. This was particularly the case in Australia because the environment is extremely harsh and marginal.

[17] A wild dog is any undomesticated or feral dog, including dingoes and their hybrid offspring.

[18] If the base of a wedge-tailed eagle nest is inspected, it is usually possible to find many National Livestock Identification System ear tags on the ground nearby and other indicators of their lamb predation, including fleece and body parts.

[19] From the Nerrigundah region during the 1980s.

2: Providence

I spoke for more than two hours by telephone to a retired NSW Police Senior Detective in 2014[21] about his encounter. His name was Ray and he had an unforgettable experience with his detective partner, now deceased, on 7th June 1990. Ray said that they both received a large amount of ridicule over the incident and like most encounter witnesses, they quickly learnt to keep quiet. However, they both made a pact for the surviving partner to publicly reveal their story after the other had died.

Ray said that they received a call from a sheep farmer near Bungonia National Park, not far from Goulburn, NSW. The farmer said that over the previous three weeks, during the commencement of the lambing season, he was regularly losing several of his lambs each night.[22] The lambs weren't been taken away but were butchered in the paddocks instead. When the farmer found the bodies, they seemed to have been professionally butchered in the field, with only the rear quarters and backstraps taken. The farmer was mystified by this flagrant waste of the lamb but even more so as the remainder of the animal was usually rolled up into a bundle using the animal's fleece, which made the lamb's slaughter blatantly obvious. Ray said that this was not like any other professional animal theft from the area, where the lambs are taken alive and butchered elsewhere. They noted that the lambs had been butchered using an incredibly sharp knife or scalpel. Like the farmer, they couldn't understand why the whole animal was not butchered and utilised or, simply sold. As this problem

---

[20] There is also indisputable evidence involving the predation by other cryptids, such as panthers, by researchers Mike Williams and Rebecca Lang.

[21] Original contact through AYR.

[22] Such nocturnal predation excludes the role of wedge-tailed eagles.

was ongoing, they decided to have a stakeout overlooking the paddock as soon as possible.

After visiting the paddock they chose a position behind a small gully and amongst some large rocks that would provide some visual cover and physical protection. In preparation, they chose several different firearms, munitions and other devices, so that they could deal with the perpetrators under a variety of circumstances. They were very heavily armed. After settling in for a night of surveillance and freezing temperatures, nothing happened for a long time. Then around 11:30 P.M., they began to hear noise coming from the edge of the paddock. Following a short period of inactivity, they heard a crashing sound coming straight towards them. The animal was initially very cautious, which allowed it to detect the police presence within the group of rocks. As it was almost a full moon they could identify a dark animal coming straight towards them on two legs. It ran to within twenty-five metres of their position, roared and threw several branches at them before returning noisily back to where it had first appeared. It then began to circle their position and then repeated the cycle of events. It maintained this activity until about 1:30 A.M., after which it ran off. Whatever it was, it was very aggressive and angry, although Ray thought that it meant them no harm and simply wanted them to leave so that it could feed its young. Ray also thought that the animal would have been very difficult to stop with the rifles, had they decided to use them!

It was very fortunate that Ray was a skilled and well-trained police officer, who was capable of accurately recording his observations and analysing them. From what he saw, he determined that the biped was seven to eight feet tall and that the shoulders were the same height as a tree branch he later measured at six feet. The hair on its body was dark and blotchy. It was a very sure-footed and fast biped that took very long strides, able to move easily and fluidly across the irregular and rocky surface in the dark. For this reason, Ray thought that the biped had excellent night vision. Ray said that its feet or 'pads' made a loud thumping noise as it ran and it also grunted in time with its gait.[23]

When speaking at length to Ray it was obvious that he had been paying very close attention during this intense ordeal. The information that he provided was accurate and demonstrated detailed information acquired over relatively short periods of observation. It was also highly consistent with our observed knowledge obtained from many decades of research and personal experience. Typically when conducting an encounter interview I would listen for any one of several well-established, but discrete behaviours or physical traits, to establish the veracity of the account and the accuracy of the descriptions. The numerous specific examples and depth of detail that he provided gave him instant credibility from my perspective.

It was interesting that he experienced the roaring, as this behaviour is not commonly heard. This is a reaction that tends to be heard only in response to anger or perhaps frustration. In 1993, Flatfoot roared in my face because I had discovered her hiding place and shone a torch above her head. Also, people who have observed a Dooligahl running typically comment on the loud thumping noise that their feet make, which is similar to, but not the same sound made by a kangaroo hopping.[24] Furthermore, suggesting that this biped might have had pads on its feet

---

[23] These very specific observations are only rarely reported and greatly increase the veracity of the witness.

[24] "Not quite right," was the first comment made by Ian Price when describing the sound of Fatfoot's walking and running.

reinforces the connection with macropods. For macropods, or 'big feet', these pads are used to increase grip and provide a high-impact cushion when hopping. He also mentioned that the running was extremely fast, covering a lot of ground in a short time using very long strides. This is similar to kangaroos which achieve their high speeds by using their large leg tendons, which store the energy of landing for reuse on the rebound. More significantly, however, kangaroos can achieve higher speeds by increasing their stride length, rather than increasing their stride frequency. This also gives the eyewitness impression that the Dooligahl is 'gliding along' the ground as it moves along. Many witnesses have commented on this. This efficiency of locomotion is very typical of Australian marsupials living in this harsh environment. He also mentioned grunting which is something that we have heard when these bipeds run hard. It seems to be an involuntary vocalisation associated with the movement of air in and out of the lungs, in response to their large diaphragm moving in sympathy with their stride and shifting body mass. Kangaroos partly do this to make breathing and hopping compatible by exhaling when landing and inhaling when rebounding. Dr. John,[25] a trauma neurosurgeon who has heard the breathing of a Dooligahl at his home, estimated their lung capacity to be 5 litres, or twice that of humans. Once again, this aerobic efficiency in kangaroos seems to be shared with these 'marsupial hominoids' and would account for their greater physical abilities and increased stamina or endurance. Ray was also impressed by the animal's night vision ability. This is because the biped was able to move confidently about in the low light.

---

[25] Dr. John is a pseudonym for a highly trained trauma neurosurgeon who accompanied us many times during our research. He also had a Dooligahl in the valley where he lived, which he reported on.

2: Providence

Perhaps the greatest revelation was the apparent use of a knife or other cutting device. It is known that these animals use tools. In our valley, Fatfoot used a stick to remove fixed lights from trees. She also used sticks to interfere with cameras, flood lights, passive infrared (PIR) detectors and other devices. Our neighbour, Ian Price found a termite mound that had been broken open with a stick still in place, which seemed to have been used to extricate the ants—something that Jane Goodall had documented many years earlier in Africa with chimpanzees and their tool usage. So, it would seem that this intelligent biped was using something sharp as a cutting tool to butcher the lambs. What is unknown from this encounter is whether the cuts were made by a hand-held tool or were achieved because of a biological adaption or modification. It is known that these bipeds have long, black claws on their hands and feet and there have been Aboriginal reports where they have been observed sharpening a claw on a rock. Apart from this, having black claws[26] instead of flat pink nails rules out any possibility that these animals are biologically related to the Great Apes! In other words, our Australian 'hominoids' are not hominins like *Homo erectus*, nor are they apes, like gorillas. It also further opens up the possibility that these bipeds are not placental.

Finally, Ray's intuition was that the animal was attempting to feed its joeys. The two police officers both had this gut feeling because it wanted them gone, rather than leaving, which suggested some sense of immediacy involving the food. The reasoning was that alternatively, it would need to move the entire family to a different

---

[26] "Claw" and other root derivatives are mentioned 29 times in: Tony Healy and Paul Cropper. 2006. *The Yowie: In Search of Australia's Bigfoot*. Sydney: Strange Nation.

Fig. 2.2 Rear black claws on a Swamp Wallaby, *Wallabia bicolor*. (Photo: Neil Frost)

2: Providence

location and begin a new search for lambs on another farm. This conclusion was similar to our realisation, three years later, that Fatfoot was the matriarch of a large family group, with her joeys maintaining a safe distance away from us in the bush. Consequently, this speculation could be extended to suggest that this dominant biped was female also. Additionally, Ray was able to accurately confirm that the animal's stature was six feet to the shoulder, with a possibility of having a total height of seven to eight feet. With a total height most likely being around seven feet, the same height as Fatfoot, this would further suggest that the biped was female because males seem to be taller and heavier (sexual dimorphism), where males need to be able to compete for mating opportunities.

Fig. 3.1 The author standing in bush near the house during the daytime. (Photo: Sandy Frost)

Fig. 3.2 Extremely thick swamp vegetation—tunnel visible, lower left of frame. Main north-south swamp track. (Photo: Neil Frost)

3: Home

AFTER 11:15 P.M. on Sunday, 21 February 1993, everything changed! At the time, I was extremely confused by what had just happened. What I heard in the bush across from our home, was a combination of sounds made by a very heavy and powerful animal moving and the vegetation in its path being destroyed. However, it was also immediately obvious that this animal was walking, and then running on two legs. It wasn't Skippy!

Without hearing any further movement, I decided to go inside the house and wake Sandy who had been asleep for several hours. On waking her I simply said, "You need to come outside and hear something!" As she got out of bed she asked, "What is it?" In reply, I said, "I don't know! You need to listen to this!" After haphazardly throwing some rocks at the last known location of the animal, we both heard three slow, heavy steps moving away from the house. Without hesitation or any discussion, Sandy said "I'm going to phone the police!" I replied, "What are you going to tell them?" Sandy said, "I don't know. That there is something big in the bush!" Fearing that she had come to the same inexplicable conclusion that I had, that is, that whatever it was, it wasn't human and extremely unusual, I cautiously advised that she should tell the police that we had a "prowler." In agreement, Sandy phoned the police.

About thirty minutes later, we heard the sound of a police car speeding down our road and shortly after we could see the light from two torches. The Senior Constable began by asking, "Why do you think there is a prowler on your property?" I replied, "Because we heard him running in the bush." She asked, "Did you see him?" I replied, "No. I only heard him." She then asked us where we had heard the running, the duration and direction of travel, how long ago and whether we had heard anything since. After we answered her questions, she commented that "someone being in the bush in such a remote location late at night, must be assumed to be up to no good." She then said that they would search the area for us.

As the two police officers started to move into the bush, I became concerned and stopped them. I said to the Senior Constable, "I think that I should come with you!" Grateful for my offer of assistance, the Senior Constable said, "Sir, it would be better that you stay here and allow us to search the bush." Fearing for their safety, it seemed that I would need to explain my concerns more fully. I then said, "This thing sounded big." The Senior Constable said, "You seem very concerned. How big?" I replied, "Like an elephant on two legs, wearing size twenty boots!" She said, "Are you suggesting that this thing is a Yowie?" I replied, "You said that not me! All I know is that this thing is bloody big and powerful!" Turning to her partner she said, "Perhaps we won't go into the bush. I think that we will drive down to the end of the road and see if we can find anyone walking around. Phone the Station again if you hear anything else." The police left. For the first time, we locked the doors of the house!

THE FOLLOWING MORNING we were still highly confused and stressed. Reflecting upon recent events in light of our new experiences, I remembered an unusual incident that occurred the previous week.

Fig. 3.3 Main north-south swamp track. East-west barbed wire and dog mesh fence, 30 m north of our land. (Photo: Neil Frost)

I was washing our car in the driveway around 11 A.M., with our daughter Avril. We were both washing the same front guard when Avril said, "Daddy! There's a man down the back." Not quite catching what she had said, I asked her to repeat it. Thinking about what Avril had said, I thought that her statement was highly unusual and perhaps inaccurate because we rarely saw anyone on our property over the previous decade. I looked down towards the swamp but saw nothing. After telling Avril to keep washing the car, I called our female collie, Bess, and together we walked down to the swamp. Bess immediately began sniffing back and forth along the well-established game trail, backed up, growled into the swamp and then ran back to the house. I had a good look around, saw nothing suspicious and returned to the car.

On Monday morning, I packed the children into the car after Sandy had left for work, and drove off. Speaking to Avril in the rear seat, I asked if she remembered seeing the "man down the back"? She replied, "Yes." In an attempt to test the veracity of what Avril was saying, I asked her some misleading questions, including checking out the activities of the Cookie Monster and Beaker. I asked Avril, "Did the man have blue hair?" She said, "No." I asked, "Did the man have orange hair?" She answered, "No." Then Avril suddenly offered her description. She said, "The man had long, yucky hair, Daddy. It needed cutting!" After stopping at our preschool destination, I felt highly perturbed!

THIS INITIAL PERIOD of uncertainty was becoming increasingly difficult for us, as Sandy and I continued to process what we had experienced the night before, plus assimilate the new and old information. When Sandy had gone to work on that first Monday, I spent the morning exploring the bush surrounding our house after I returned home from Avril's preschool, looking for anything that might be telling. After Sandy got home late from work, we discussed the options that we had both considered. By sundown, we had rejected the notion of an animal being responsible and began to believe that we were dealing with a person—a very large person! Consequently, I decided to spend the remaining twilight hours walking around the house perimeter carrying an air rifle, thinking that for any person present in the dark, it wouldn't be recognised as being anything but a formidable weapon! Later that evening, I heard branches breaking and walking to the north. In response, I yelled challenges into the bush, with copious amounts of supplementary swearing to help raise the machismo. Then, I caught a glimpse of the red eyes. The same red eyes that we had frequently seen during the past decade. This called for another re-think.

The next day our belief had swung back the other way. We started to think, like the police, that we were dealing with a Yowie, something that we had heard about in the Mountains for many years, but had not given any thought to. Sandy suggested that I should talk to Barry, an Aboriginal consultant for the NSW Department of Education, whom she knew well and recommended highly. When I rang him, I told him that we were having problems with a Yowie. He laughed! He said, "What makes you think that you have a Yowie?" I replied, "Because I had one run past me and I have seen its eyes!" Barry then asked, "What sounds does it make?" I told him, "It sounds like an elephant on two legs, wearing size 20 boots!" He finally asked, "What are the colour of the eyes?" I told him, "Red or yellow." Barry told me that he had a meeting at the Parramatta Land Council that night and that he would ask the Elders some questions on our behalf and get back to me tomorrow. Two days later Barry telephoned me. He said, "Half of the Elders suggested turning all of our lights

Fig. 3.4 Track passing through low-density section of southern swamp. (Photo: Neil Frost)

on and the other half suggested turning all of our lights off."[1] I laughed.

After about the fourth or fifth night, with terrifying things still happening outside, Sandy and I started to think that we might be on our own. However, our attention instinctively turned towards our neighbour, Ian Price.

Ian was a powerfully built, bearded and heavily tattooed biker. He had a presence that gave him authority and the seeming ability to deal with any situation, no matter how difficult or dangerous. He joined the 'Life and Death' motorcycle club at the early age of 17, shortly after gaining his driver's licence. This initiated his involvement in a wide spectrum of crimes, involving assault, drugs, guns, prostitution and more. Indeed, as a former convict and category 'A' prisoner, Ian had experienced more extreme circumstances than most people could ever imagine and had survived the ordeal of prison life in many high-security establishments, including the notorious Long Bay Gaol. Ian also survived an attempt on his life involving stabbing by other inmates, after being incorrectly placed in the minimum security gaol at Oberon. He often boasted, apart from other credentials, that he had been in most goals throughout New South Wales and that the famous Australian criminal, 'Chopper Read', was a pushover.[2]

There was another side to Ian. The person that we knew was also highly intelligent and thoughtful. From the age of three, Ian started reading *National Geographic* magazines. He had a very strong interest in science and particularly biology, reading everything available to him. Unfortunately,

---

[1] From later experience, both Elder groups were correct. Lights will attract these animals in an attempt to see what is happening and lights also repel them because of their acute sensitivity to light (NV). Barry also spoke to Merve, a very knowledgable Elder who would later provide excellent advice.

[2] A euphemism.

Ian suffered from autophobia, the fear of being alone. This craving for human contact meant that his mother needed to be near him when reading and was a significant problem for him in gaol. Although his condition improved with age, we would frequently find that Ian had arrived at our house and was sitting on our verandah on many occasions during the week, reading my *New Scientist* magazines, because his wife Cheryl was not home. On hearing Ian outside, we would unconsciously deliver him a strong white coffee with no sugar, to which he would invariably say, "Thank you, neighbour."

Ian attended Epping Boys High School. Due to various issues in class, Ian spent much of his time in a one-to-one teaching arrangement with the Principal, something which he manufactured and enjoyed. Even when Ian ran away from home several times, he would always be found at school. Ian liked school, and although he left after Year 10, he completed his Higher School Certificate by correspondence many years later, achieving an outstanding result in Biology, a subject that he loved!

Ian's first job was working at Macquarie University as a biology laboratory assistant in 1975. Ian loved this job, aspects of which he would reference years later when we were dealing with these hominoids. After this, around 1979, Ian worked for the Commonwealth Scientific and Industrial Research Organisation (CSIRO) which he also enjoyed immensely. It was here that Ian obtained his dog Toby, where "undisclosed research" had been taking place.

In 1989, the anti-social neighbours living next door to us put their house up for sale. This was a significant improvement for the community, with the property remaining unsold for more than a year and a half because of their unreasonable vendor demands. Then in 1990, Ian and Cheryl bought the property and in no time, they both became very good friends. Ian and Cheryl both worked several jobs. Ian drove concrete trucks and worked for a local landscape supply business. He bought a trike and provided customised tourist trips around the Blue Mountains, raced motorcycle sidecars and also maintained his membership of the bike club, where he held the position of Sergeant-at-Arms, or as Ian would simply say, "maintaining discipline". Cheryl also worked various jobs, particularly cooking meals for the patients and staff at the Queen Victoria Memorial Hospital, an institution that provided Cheryl with many contacts who reported encounters across the upper Blue Mountains.

WITH SOMETHING LARGE and powerful moving around the bush, about fifteen metres from our house, I spoke to Sandy and we decided to seek our neighbour's help. It was sometime after nine o'clock and looking up our driveway, I could not see any lights on next door, though I already knew that Ian and Cheryl would be in bed because they both started work early in the morning. After climbing over the fence, I tapped on their bedroom window and Ian answered, "Who's there?" We had a brief discussion as I attempted to explain that there was something in the bush opposite our house and we would appreciate his assistance. As we later found out, Ian said to Cheryl, "I don't know what drugs the neighbour is on, but I want some!" A few minutes later, Ian appeared outside and summoned his dog, Toby. We walked to the bottom of our driveway and faced the bush.

Standing there, I attempted to outline the inexplicable events of the past week to Ian. Then suddenly, we heard heavy movement in the bush immediately in front of us, a short distance away. It seemed to be a deliberate attempt to gain our attention. It was intimidatory and it wasn't human! Turning to Toby, who was standing a few metres behind us, Ian commanded his dog to attack. Toby didn't respond, so Ian repeated the command. Ian said to me that Toby had never refused him before and that this meant that Toby was afraid. As if honour was at stake, Ian began to roll up his flannelette sleeves and said, "I'm going in there! Whatever it is, I'm going to crash-tackle it! You're either going to find a clump of hair in my hand or a lump of flesh in my mouth. I need you to promise me one thing. I want you to name it after me, 'post-humorously,'—*Homo priceii*." I said to Ian that this thing sounded very big and powerful, that it could probably rip him to pieces! Ian replied, "It's probably going to hurt but not for very long!" I tried to talk Ian out of tackling it, but he couldn't be dissuaded.

With his sleeves rolled up, Ian stepped into the bush. After only a few steps, Ian stopped and said to me that he could see something in front of him and that he was going to tackle it. As Ian charged forward, the animal ran away at high speed. The only sound heard was the loud thumping of the biped as it ran off towards the northwest.

After a few minutes, Ian and the animal were off somewhere distant. Standing with Toby, I thought that I should attempt to follow him. After entering the bush the same way that Ian had gone, I pushed my way through some dense vegetation until I came upon the east-west track on our northern boundary. It is funny how acute your senses become in situations like this! I followed this track to the swamp and then along another 'game trail' to the north. Moving quietly and very slowly, I came upon Ian who was standing atop a corner fence post, with both hands cupped around his ears. Asking where it was, Ian pointed to the northwest and said, "In the swamp." After a few minutes, Toby arrived. Ian was not very happy with his dog and I was very glad that Toby had arrived after me!

The following night, Ian and I walked down to the swamp. We soon heard heavy movement in the thick bush to our north, followed by an extremely loud thump. Ian said, "That's a threat display! Gorillas do this to warn you!" I said, "Sounded like a very large rock or log being hurled at the ground!" Ian replied, "Or, the beast is thumping its chest." Without warning, Ian started to move towards the house saying, "You stay here and keep track of where it is. I'm going back to the house and getting a torch. See you shortly." As the situation didn't seem to be negotiable, I stood opposite the animal, fully expecting it to charge me at any second! Using my newly heightened senses, I could hear everything around me, including possums in nearby trees and numerous insects. I could also hear the bipedal animal moving slowly towards the swamp, carefully placing its foot on the detritus and very slowly applying its weight. It seemed to be circling me. At the same time, I could hear two people arguing inside their house, several hundred metres to the north, on the other side of the swamp, down towards the fork in the valley. I could follow some of the muffled arguments and wished that I was there instead! After a lengthy wait, that seemed longer than it probably was, Ian returned to where I was standing holding a torch. He asked me where it was located and said, "I'm going to turn on the torch and we are going to charge it." I said, "Do you think that this is a good idea, Ian?" Ian replied, "The worst that it can do is kill us!"[3] With that, Ian turned on his large quartz-iodine spotlight and fortunately, the biped noisily ran away at very high speed to the north, leaving us standing where we were!

---

[3] Something that Ian told me, undoubtedly from his experience of prison life in maximum security gaols, was that, "There is nothing more dangerous than someone who doesn't care!" However, in this situation, I think that that statement is trumped by, "There is nothing more dangerous than something that is two to three times your body mass."

Over the following nights, Ian and I spent time engaging with the biped, frequently up to and beyond midnight. Consequently, Ian decided to limit his involvement to Friday nights because he was finding it difficult to function at work. However, Ian would arrive on our verandah most afternoons for a cup of coffee to discuss the weekly events and plan our strategies for the upcoming Friday night.

SOMETIME AFTER IAN had reduced his nightly involvement with pursuits, I received two telephone calls from Merve, a tribal Elder from Mogo. He said that Barry had spoken to him at the tribal council meeting, telling him that we were experiencing difficulties with a Dooligahl.

From the beginning of our conversation, Merve asked a series of inquiring and knowledgeable questions, based on his experience. He asked how long we had been living at this location and if we had recently built a house there. Concerning the land, he asked if we had cleared it, fenced it, if there were any tracks across it and how far from the swamp our house was situated. He then asked some very troubling questions. He wanted to know if we had any children and their ages and sexes. His tone changed when I told him we had two young children, two-and-a-half and one-year-old. He then said, "Now this is very important! Have you ever lost any children's clothing, like off the clothesline?" I said, "No. But I have found some!" Merve replied, "In the swamp?" I said, "Yes." Merve exclaimed, "That's it! You've got a female!"

Talking to Merve in more detail, I told him that I found a yellow '000' baby's jumpsuit in the swamp, that did not belong to us. He reiterated that a female Dooligahl would be interested in the scent of the baby. He told me to never leave our children outside, day or night, without Sandy or myself being present. He said that there was a real danger that the female Dooligahl would snatch the child. I said to

3: Home

Merve, "What! to eat?" He replied, "Put it this way. You probably wouldn't see your child again!"[4]

A nineteenth-century account of Devil-devils taking children in New South Wales was written by Mrs Charles Meredith during the 1840s.

> ". . . but they have an evil spirit, which causes them great terror, whom they call 'Yahoo' or 'Devil-devil': he lives in the tops of the steepest and rockiest mountains, which are totally inaccessible to all human beings, and comes down at night to seize and run away with men, women, or children, whom he eats up, children being his favourite food; and this superstition is used doubtless as a cloak to many a horrid and revolting crime committed by the wretched and unnatural mothers, who nearly always when their infants disappear, say 'Yahoo' took them."[5]

I was deeply concerned about Merve's warning, so I asked him how he knew this. He told me that during the "Stolen Generations" period, the Elders took the children into the bush to hide them from the government agents who would have placed them in orphanages, mission stations and domestic schools, not allowing contact with their families and country. One evening, the mob had a large fire burning and the children were playing around it. A female Dooligahl went through the campsite, picked

---

[4] John would draw a comparison with Lindy Chamberlain and what Merve warned of, on Thursday July 1, 1993. See log.

[5] Mrs. Charles Meredith. 1973. *Notes and Sketches of New South Wales: During a Residence in that Colony from 1839 to 1844.* Sydney: Ure Smith. P. 95.

up one of the children and ran off into the bush. He said that all the men grabbed their weapons and took pursuit. They followed the Dooligahl and child until dawn. The Dooligahl took up a position across a large clearing, while the men formed a line and raised their weapons. The Dooligahl released the child and walked off into the bush. He said, "That's how I know!"

Merve also mentioned other things. He said that within his tribe, it was considered 'good luck' for a young girl or woman to be touched by a Dooligahl and that this would usually occur when the mob was moving through a swamp. Researcher Andrew told me of a recent incident in the Northern Territory where a pregnant woman, sleeping with her husband, had her hair stroked by a Dooligahl after it reached its arm through a cavity in a wall where an air conditioner had once been. Also, children typically slept in the centre of rooms, followed by a circle of women, with men on the outside. Sleeping quarters are frequently surrounded by drums full of burning logs, called 'all-nighters'.

Additionally, Merve mentioned that Dooligahl dislike cleared land, fences and anything that prevents them from moving freely through an area. They are regarded as "protectors of the environment where he lives." He also said, "They live in the trees." He strongly advised, "Do not feed the local animals!"

With Merve's warning in mind, I thought that I would undertake a door knock in the neighbourhood to find out who owned the '000' jumpsuit. I could only think of two close neighbours who currently had young children. After speaking to Anthony and Megan, only the renter next door to them remained a possibility. When Wilhelmina answered her door, she immediately recognised the clothing as hers, asking me where I had found it. Thinking that this would be difficult to explain, Wilhelmina told me that the

3: Home

jumpsuit had disappeared from her clothesline one evening and was amazed that I had found it in the swamp below, more than two hundred metres away. Playing it safe, I asked her if she had experienced anything unusual around her house recently. To my greater surprise, Wilhelmina told me that early one morning, around 2 A.M., she woke to someone moving about in the kitchen.[6] She got out of bed and was expecting to see her eldest son, who frequently sleepwalked. Instead, she saw a large figure that rapidly exited through the back door, ran down the wooden verandah steps and beat a noisy path down her backyard and onto our land. I then told her what Merve had said.

BY NOW, SEVERAL weeks had passed and we had become 'Yowie-aware', however, the frightening activity outside continued nightly and seemed to be escalating, probably in response to our stimulatory involvement. As Ian reasoned, these hominoids are intelligent and after hunger has been sated, the next need is entertainment and we fulfilled that role! As Ian said, "They don't have satellite back in the cave!"

The local Aboriginal Elders had been helpful too, allowing us to partially understand the situation and provide a basis for dealing with the problem. One of the first changes involved the installation of an outside light, at the end of the pathway from our front door to the driveway. This could be left on overnight if we felt that we needed to create a Yowie-free bubble or provide security for someone arriving home late and wanting to avoid being ambushed.

---

[6] Home invasions and attempted intrusions are quite common! More frequent violations occur: under houses; on verandahs; at windows; and around the perimeter of houses.

Based on the Aboriginal advice, we found that the light would attract these hominoids to the area, as they were interested in observing human activity, but they were also prevented from approaching closer because it would blind them. In further following the recommendations, we started to thin out the thick vegetation around the house so that there would be less cover, however, this 'Yowie-free zone' was not fully achieved for several more years. Also, we stopped feeding the local carnivorous birds. We had started feeding the birds about four months earlier, in November 1992, so that our children could enjoy feeding them. The ritual would begin late each afternoon in the kitchen, as we cut up the raw meat. Over time, more and more birds arrived and as a group, they showed up earlier each day, keenly anticipating our preparations within the house. Afterwards, we would move outside to the eastern verandah and the birds would follow, fighting for position along the railings and nearby trees. At the height of this insanity, in February 1993, the scene each afternoon reminded Sandy and me of Alfred Hitchcock's movie, *The Birds*. At the time, we were amazed at how far these birds seemed to be travelling and the variety of routes taken, to get fed. To any observant carnivorous predator, seeing such a large and regular assembly of kookaburras, currawongs, magpies and butcherbirds, would have alerted it to the presence of meat and an easy meal.[7] Similarly, both Bess and Tobys' food bowls must have been an easy acquisition, although it would be many more months until we realised that!

---

[7] Not surprisingly, food, particularly an easy meal, is always a very strong attractant to any animal. These 'marsupial hominoids' are no different, although it seems that, from encounter interviews and personal experience, their favourite meal types are barbecued and roasted meat, cats, dog food and fish.

3: Home

Thanks to the local Aboriginal community, we had slightly better knowledge of what was happening around us and we better understood most of the major risks. If there was an urgent incident, we knew that we had the immediate support of Ian, without any need to explain ourselves or wait for a response. Although our initial contact with the police suggested an underlying awareness of the Yowie phenomenon, it wasn't something we thought we could generally rely upon, which proved true. Consequently, in the beginning, when we telephoned the police in response to a late-night situation, we avoided any suggestion of it being anything but a prowler.

From the start, we were advised by the NSW Police to telephone the local police station every time we had an incident, to establish a record of events. It was also suggested that we keep a diary. The police said that a record of events around our house could be useful in determining patterns of behaviour and could be used as evidence in court against an offender. During this early stage of police involvement, the natural assumption was that we were dealing with an intruder. We avoided using the 'Y' word for the first few weeks, hoping that the extremely unusual events would become self-evident. Two intruder profiles were proposed, a "homeless person" and "someone interested in our young children".

The police said that several homeless people lived in the bush throughout the Blue Mountains, which we were also aware of and had occasionally encountered ourselves. Over the time that the log was kept, they suggested several strategies. As they became increasingly aware of our efforts, a Senior Constable told us to leave a wine glass prominently displayed at eye height in a tree, in the hope that it might be handled and fingerprints obtained. Even though we thought that this would achieve little, we did it anyway. During the time that the glass was left outside, nothing touched it. Ironically, however, on September 9 and 10, Fatfoot touched the external microphone that was hung high up in vegetation along a swamp track, sometime after the police had finished their investigation. Similarly, we were asked to secure a wine cask bladder in the bush on July 22, four days before the police stakeout would commence. This was a final attempt to avoid the alternative conclusion. On September 22, the wine cask bladder was found on the ground, with two puncture holes, 60 mm apart and bitten through the middle, leaving an impression of a jaw. This occurred about two months after the stakeout.

There was also the very scary prospect that someone was interested in our young children. This was partially raised during the police response to our first call for assistance at midnight on February 21, when the Senior Constable said, "With someone being in the bush in such a remote location late at night, it must be assumed that they are up to no good!" This possibility was more clearly articulated when raised during early discussions with police officers, where various motivations for these intrusions were examined. It was further compounded by the exclamation made by Senior Detective Sergeant Graham[8] on July 16, who said to Sandy, "What are you doing? Don't you have two young children? Get inside and look after them!" However, the most powerful comment of all was the warning given by Merve, an Aboriginal Elder from the NSW South Coast town of Mogo, who told us to watch our children day and night because they could be taken by a Dooligahl.

---

[8] See log entry for July 16, Chapter 4.

4: Log

THE PRELIMINARY ENTRIES for this log were mostly contemporaneous notes collected during the day and night on loose sheets of A4 paper. The intent behind this record-keeping was to generate potential criminal evidence that could be used by the police to support the arrest and prosecution of an offender. For us, its purpose was to help detect any hidden patterns of Dooligahl behaviour that weren't obvious in the short term.

During the night, I would keep paper and a pencil on my bedroom side table, scribbling entries as they happened and referencing the time displayed on the clock radio. Other important synchronising events—like the time a distant coal train passed through, a twin-engined propeller plane flew overhead, or a particularly noisy truck applying its exhaust brakes on the highway—were helpful in identifying moments during a tape recording and were often written down. Also included were comments and actions of individuals involved. These raw entries would then be transferred to the "Activity Log" as they were originally noted, along with any other empirical information for that day. The diary used for this purpose was a Teacher's Notebook for the year 1993.

FRIDAY, JUNE 11
Robert and Kerry[1] stayed the night. Robert brought cameras, video and lights.[2] Investigated back of swamp around 8:00 P.M. Heard footsteps on Ian's block about 10 metres in from border at edge of swamp. Heard loud noises shortly after to north. Investigated—found large bird or owl in tree. Talked to Basim,[3] he said that the dogs were barking all night from 10:00 P.M. He investigated by torch but heard and saw nothing. Basim lives up other valley from valley fork.

SATURDAY, JUNE 12
Heard nothing. Walked with Robert down valley about 500 metres into the bush—very dense. Robert videoed the whole journey. Found very old footprint. Looked like it was made jumping off a tree that was across the fence[4]—photographed. Size about 16" long. Found a bush 'hide'[5] next to a log near a waterhole, as if made by lying down next to log—photographed. Found chicken wing nearby

---

[1] Robert and Kerry were friends. Kerry was a former staff member in Sandy's History faculty.

[2] Robert and I had an encounter together on Saturday, June 5. After sitting in the bush for some time, we advanced towards Fatfoot, turned on the spotlight. She stood up, lent forward and roared into my face. I kept the spotlight trained on Fatfoot's head, we could see the head turn, the red eyes and heard grunting. Understandably, Robert was greatly affected by the encounter.

[3] Basim was a colleague who taught Biology and Science. He was a trained marine biologist. He lived in the next valley to the west of our house.

[4] Ian and I found a number of fences along the valley that had medium-sized trees laying across them, bridging the tracks.

[5] It seemed to us that this 'hide' was being used to ambush birds and animals at arm's length, coming to the water source. Feathers and miscellaneous body parts found around the hide. Since this time, we found frequent examples of other hides, 'beds' and a 'shelter'.

4: Log

Fig. 4.1 A bush hide, about half a kilometre down into the natural valley, with vegetation purposefully moulded around a fallen log (top, left quadrant of photograph). A water source was located on the far side of the log within easy arm's reach. The volume of the cover was sufficient to keep a very large animal hidden from most viewing angles. (Photo: Neil Frost)

with feathers attached and had been pulled off at the joint. Found footprints through bush alongside creek. Along way found faeces of various types. Some looked human,[6] others definitely animal—wallaby. Found wallaby and dog tracks.

SUNDAY, JUNE 13
Heard nothing.

MONDAY, JUNE 14
Ian came down to look at the swamp late in the afternoon—3:30 P.M. Noted the growth in size and number

---

[6] Many times we found faeces that were suspiciously 'human' looking. Some were dried and frozen to preserve.

of tracks through the swamp. Noted extension of tracks onto Ian's block and onto block to the north. Heard dogs barking to the north of the valley fork. Barking continues up valley, very loud at around 5:20 p.m. Heard noises alongside northern side of house near the track. Bess barked. Investigated immediately. Heard footsteps. Fired 2 photographs from near septic tank—camera held overhead. Heard footsteps moving away to the northwest—stopped 3-4 steps, then moved again 3 or 4 steps. Ran inside and called Ian. Waited outside. Ian and I approached the septic tank. Toby[7] growled and became very excited, hesitantly charged into the bush. Ian entered bush above house and I walked down bush track on northern side of house. I went to the bottom of the swamp. Heard footsteps moving constantly for 30 seconds in the swamp, moving south. Heard large dog on this side of valley about 200 m away to the south barking frantically as if it was on a lead and choking. Dogs barked on further up valley—south. Woke up at 1:00 a.m.—Sandy heard noise. Took photographs from out of second storey window to the south of the house—2 photographs.

Tuesday, June 15

Dogs barked down at valley fork at around 5:30 p.m. Few dogs barking up valley by 6:40 p.m. Good Dog only gave a few barks. Heard thump, one only at 6:45 p.m. Followed by Bess barking constantly for 30 seconds—went quiet then disappeared. Dressed, camera charged and torch, opened lounge room door—instant reaction, heard footsteps

---

[7] Toby was Ian's elderly 'anti-personnel' German shepherd. Many years earlier, when Ian was being held in the Taree Lockup, Toby was placed in a separate gaol cell because he wanted to eat the police. Ian had to feed Toby because no one could get near him!

moving rapidly through reeds across from driveway. Approached sound and fired camera held above head from edge of driveway—reaction, heard footsteps moving away to north, then towards swamp—west. Dogs on far side of valley barking. Went inside. Came back out with camera and torch and stayed still on grassed area on western side of house. Moved down to the bottom of the swamp near northern border—heard footsteps moving away to north into swamp. Came back up to the grass clearing. Stayed 25 minutes—heard nothing more—returned inside. placed tape recorder on table on NE side of house—tape stopped recording after 1 minute—nothing. Woke up in Avril's bed at 5:45 A.M. after hearing thump. Dressed and went outside—heard and saw nothing.

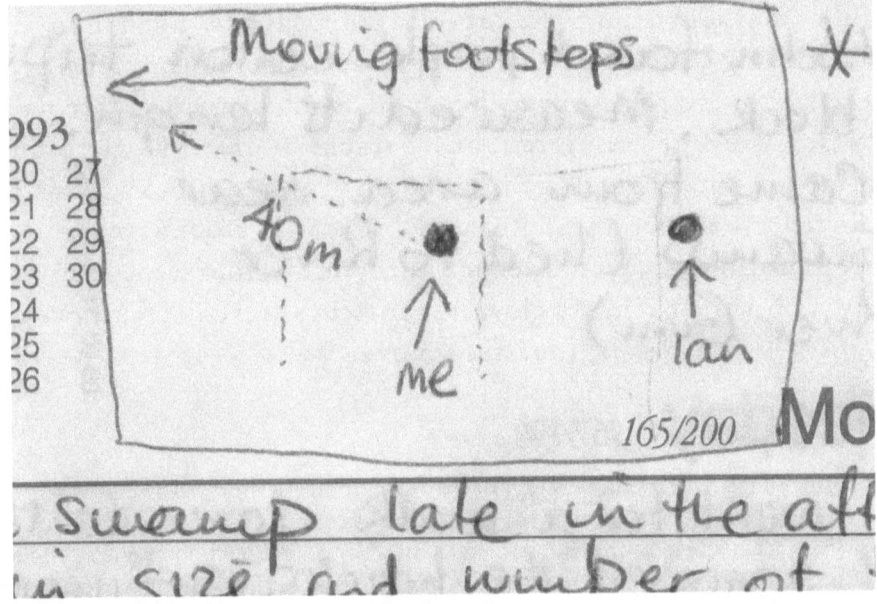

Fig. 4.2 Sketch from 'Teacher's Notebook' showing our movement along main north-south swamp track—14 June 1993 (Image: Neil Frost)

Wednesday, June 16

John[8] and I went for a walk down to the swamp at around 3:45 p.m. Noted that some of the tracks seemed to be enlarged further. Saw that the banana was missing. John said that it was missing at least by last Sunday. I had only noted its disappearance earlier at 1:30 p.m when I also noted that the grass was flattened near the bush house and ferns across the driveway were also disturbed and flattened. John found purple cotton tripwire 20 metres down track on Ian's block. Measured its length. Came from area near swamp, tied to river red gum. Heard the dogs bark at around 6:30 p.m. Gordon's[9] dogs barked constantly. The Good Dog only barked a number of times. Heard thumping repeatedly on northern side of house by about 6:40 p.m. Bess barked constantly for more than a minute and ran away. Thumping continued. Dressed, charged camera and took light, left house by door on southwestern side (main bedroom). Carefully approached bush house. Thumping continued but, stopped as I entered the clearing. Heard footsteps. Approached and fired camera (extremely scary event). Entered bush track, footsteps mirrored mine around bushed area. Footsteps rotated clockwise as I moved clockwise also. Rotated back anti-clockwise, footsteps followed. Went deeper into bush, footsteps moved away to the north. At closest, footsteps were 20 feet away. 7:00 p.m. left bush, footsteps moved away. 7:45 p.m. heard

---

[8] John was a work colleague. His son David, together with a group of friends had had an encounter at the top of our land seven years earlier when a very large bipedal animal grunted and approached them at high speed from behind, along the edge of the road. It then swept back down into the valley, continuing to grunt as it ran off.

[9] Gordon was a neighbour on the far side of the swamp, to our northwest. Ian and I met Gordon during a nighttime encounter in the swamp. He had several dogs, some caged birds and a number of other animals.

dogs bark in valley again. Bess and Toby stayed around house without barking.

Thursday, June 17
Woke up at 6:15 a.m. Heard general but suspicious noises in bush in front of house (east). Bess started barking for 30 seconds, then Toby. Got up at 6:20 a.m. went outside with camera only. Heard dogs in valley to west barking. Heard nothing but, noticed large flattened area near driveway southern edge. Contacted police at 10:30 a.m. at local station. Suggested that I ring them every time that there was a disturbance in order to keep a record. Police visited at 12:45 p.m. and inspected site. Had the photographs from the night before developed at Kmart. Showed two red eyes. Measured from photograph—distance to eyes = 13.5 metres. Heard dogs bark at 5:30 p.m. Bess unsettled at 5:45 p.m. John and I went outside from bedroom door. Heard possible—vague noises in bush to the north. Heard dogs barking on other side of valley. Moved to swamp area. I heard footsteps in swamp moving south. Dogs barked, particularly Good Dog, who barked constantly and loudly, owner came out several times to quieten him. John and I heard long, deep growl from swamp, fluctuated in tone. I heard movement to south. Dogs barked. Returned to house. Dogs barked occasionally until 10:00 p.m. Went to bed at 11:10 p.m.—quiet. Woke at 2:45 a.m. (Friday). Heard general noises east of house, thumps, no barking by Bess—lasted 30 minutes approximately—went to sleep.

Friday, June 18
Heard dogs barking down valley at about 5:30 p.m. Dogs barked less than usual. Barked again at 6:45 p.m. Waited inside house with Robert and Kerry who arrived at 6:30 p.m. At 7:30 p.m. after hearing dogs earlier, I opened lounge room door and heard branches braking across from

driveway to the northeast. Left house by bedroom door at 7:40 P.M. with Robert. Sat on grass near first line of trees and listened. Heard noises to the north and south.[10] Heard crashing through the bush about 10 metres behind neighbour's house and again about one minute later. Heard noise to the north again. Heard footsteps and the breaking of branches to the south. Heard noises and footsteps for several minutes. Heard a bird make a noise for several minutes.[11] Listened to sound move into swamp area. (See separate detailed entry)[12] Decided to stir things up. Robert and I moved to the bottom swamp area and turned on torches. I entered the swamp near Ian's border. Heard growling and panting.[13] Came out of swamp and ran to the centre of our block—heard growling and panting. Saw two yellow eyes in swamp for half a second. Went to higher ground (8:50 P.M.) shone torches to other side of swamp—heard grunting and growling and running along far fence line—dogs barked loudly on far side. Rang police—talked to Mark.[14]

---

[10] Strongly suggests that there was more than one individual.

[11] The bird was a currawong—identified by Robert and myself by its call. Bird was shaken violently off its perch in the tree, called as it fell to the ground, flapping its wings, then became quiet. Presumed taken as food. Confirmation of Fatfoot's excellent night vision by being able to identify the black bird in the upper tree canopy.

[12] I wrote out a separate and detailed account of this incident because of insufficient log space.

[13] Kangaroos begin to rapidly pant after they stop exerting themselves, in order to cool down. They also sweat and lick their forearms to control their temperature. However, it is important to realise that the biped in our swamp was not a kangaroo.

[14] Local police officer. Senior Constable.

4: Log

SATURDAY, JUNE 19
Drove to Canberra at 9:50 A.M. (To meet and speak with Dr. Helmut Loofs-Wissowa from the ANU)

SUNDAY, JUNE 20
Arrived home from Canberra at 4:00 P.M. Spoke to Ian. Contacted Matt[15]—said he would try to get there by 6:00 P.M. Heard dogs bark 4:15 P.M. to 6:00 P.M. Heard thumps 6:10 P.M. Bess barked. Contacted Matt. Came to house via Ian's block. Ian went down northern side track. Matt and I went to swamp. Heard noises in swamp. Heard footsteps traveling south. Used torch, heard growl. Dogs barked on far side of swamp. Matt left after 7:30 P.M. Heard dogs bark until 8:00 P.M. 9:00 P.M. heard thumps on northern side of house. Investigated with Sandy. Noticed grass flattening on northern border. Went inside. Heard thumps on southern side of house at 9:15 P.M. Went to bed at 9:30 P.M.—heard nothing.

MONDAY, JUNE 21
Basim said that dogs in his valley went mad from 4:00 A.M. Heard dogs in our valley bark from 5:50 P.M. Went outside at 6:00 P.M. to get bottle from car—quiet. Dogs barked again at 6:20 P.M. At 6:27 P.M. heard noise while putting Drew to sleep. Sounded very near to northern wall of house and at an angle below our floor level (also heard in same place night before). At 6:30 P.M., Sandy placed her ear to the northern wall and heard a general noise and a sound of shuffling. At 6:55 P.M. I went outside lounge room door and heard footsteps directly north of house. At 7:00 P.M., heard dogs barking. Rang police at 7:05 P.M. and reported above. At 7:30 P.M. heard dogs barking.

---

[15] Local police officer. Senior Constable.

Basim arrived at 7:35 P.M. Basim said that photograph of eyes did not look like a mammal (phosphorescent eyes?). Eyes were not offset but, forward facing. Commented on broken branches at the back of his house and tracks in remote and difficult places in the bush.[16] Basim leaves at 8:15 P.M. Dogs bark again shortly after. Sandy has result of measurement of eye spacing from Ralph:[17] range 54.2 mm to 77.8 mm (mean = 66 mm). Talked to Barry[18] earlier at 4:30 P.M.

Tuesday, June 22

Peter[19] arrived at 5:45 P.M. to discuss the wiring needed for the camera. Dogs barked from 5:45 to 6:15 P.M. Then quiet. Heard thump at 6:30 P.M. to the northeast of house. Heard other vague noises to 6:50 P.M. Toby barked from compost heap in direction of our driveway for 3 minutes. Investigated—heard a few noises such as twigs braking. Rang police—talked to Katoomba. Heard thumps to the north. Sandy thought to the south. Quite loud! Checked southern neighbour's house—no sign of people near car—

---

[16] Basim said that a steep incline at the back of his house had multiple trees broken at waist height which trailed downhill along a well-worn track. He thought that the trees could serve as handrails for dexterous 'anthropoids' making the ascent and descent easier.

[17] Ralph was a professional photographer who made calculations of the eye separation from the 35 mm negative and lent us equipment.

[18] Barry was an Aboriginal Consultant with the NSW Department of Education. He spoke to Aboriginal Elders on our behalf, seeking advice and attempting to answer our questions.

[19] Peter was a local electrician who provided a 240-volt relay needed for the automatic film camera that I was making. Peter also reported on activity in his valley to the east, together with Basim who reported on the west which, together with our own central records, established that activity in the three valleys was limited to any one at given time—there was one group that worked the length of a number of adjacent valleys. This led to a search for short cuts between valleys.

approximately 7:30 P.M. Heard very loud thump to northwest of house at 7:47 P.M. Heard dogs on far side of valley barking loudly (4 dogs) at 7:50 P.M.

WEDNESDAY, JUNE 23
Woke up at 3:45 A.M. to the sound of running alongside the house, followed shortly by a loud thump. Got up and turned on the outside lights—saw nothing. Very windy outside. Went back to bed—heard thump at 3:55 A.M. Heard 'drumming' noise near bedroom corner at about 4:05 A.M., followed a few minutes later by a 'scrapping' noise and a thump. Got out of bed and turned lights on—saw nothing (noises mainly occurred during still periods between wind gusts).[20] Turned kitchen lights on and wrote notes—listened—nothing. Went to bed at 4:30 A.M. About 4:35 A.M, heard loud footsteps—scrapping noise—footsteps, followed by a very heavy object hitting the roof (once). Rang Katoomba police about 4:45 A.M., arrived about 5:10 A.M. Heard and saw nothing. Showed them the swamp and driveway area. Police left at about 6:00 A.M. Heard dogs barking across valley at about 6:10 A.M. Still very windy. Went to bed at 6:10 A.M. Prepared an area NE of driveway, ready for attempt to photograph him on Saturday. Dogs barking noisily at 4:30 P.M. Heard thump to NNW of house at 5:36 P.M. and again to the NNE at 5:37 P.M. Bess started barking at 5:40 P.M. for 30 seconds and disappeared. Dogs barking noisily at 5:50 P.M. and at 6:05 P.M. Went to bed at 9:30 P.M. Woke up in Avril's room at 11:30 P.M. Bess, Toby and two other dogs barking very loudly at area near driveway for about 5 minutes. Heard

---

[20] Wind noise was either used to mask movement or gaps between gusts were used to allow certain demonstrative behaviours like thumping, to be more clearly heard.

scrapping noise near house and possibly bush being disturbed (?). Heard thumps (gentle) to the N/NE of house. Dogs barked again shortly after.

THURSDAY, JUNE 24

Jo[21] said that the dogs barked loudly and constantly at midnight and 1:00 A.M on June 23. Extremely quiet in the valley from early evening to 9:30 P.M., a few dogs barked at 9:30 P.M. Heard 2 thumps (very loud) as we were going to sleep at 9:50 P.M. to the NE of house. Heard thump that woke me sometime after this and before 2:15 A.M. Heard loud thump at 2:15 A.M. Heard second loud thump at 2:20 A.M. Bess barked and ran into the bush to the NE. Heard thump at 3:25 A.M. to NE (turned on outside lights at 2:20 A.M. and 3:25 A.M.). Sometime before dawn heard 6 more heavy footsteps running across eastern side of house.

Basim said the dogs were barking at his house at 11:50 P.M. (June 24) to 12:05 A.M. (June 25).

FRIDAY, JUNE 25

Bought more equipment for camera. Dogs barked 5:55 P.M. to 6:05 P.M., then 6:10 P.M. to 6:20 P.M., then 6:25 P.M. to 6:30 P.M. Basim said that the dogs are not barking at his house at 6:50 P.M. 7:55 P.M.—dogs barking—Bess barking down in swamp area. Went to bed at 8:00 P.M. Woke in Avril's room to the sound of thumps to the NE at 10:25 P.M. Woke again sometime during the night—heard a loud noise. Sandy heard (?) something at 12:30 A.M. Swamp area extremely cold and covered in frost. Basim reported that the dogs barked constantly and noisily at his house after about midnight (June 26). Investigated, heard and saw nothing.

---

[21] Basim's wife.

Saturday, June 26

Prepared camera and automatic firing mechanism. Talked to Ian. Robert and Iain[22] arrived, setup camera and lights in bush NE of house.[23] Basim arrived. Tried to find footprints. Found footprint—damaged. Heard dogs barking around 6:00 P.M.—previously very quiet. Heard thumps at 6:20 P.M., 6:25 P.M., 6:45 P.M.—until 8:00 P.M., thump and at 9:15 P.M. Iain and Robert prepared to leave at 10:15 P.M.[24] Heard noise to N of house, then at 10:16 P.M. to NNW. At 10:17 P.M. Iain heard four footsteps moving west towards swamp and saw briefly a reflective disk of an eye for a fraction of a second, as if the head turned towards him. 10:20 P.M. heard noise in swamp—no dogs. Bed at about 10:50 P.M. Five minutes later heard 3 heavy thumps to NE of house—camera didn't fire that night.

Sunday, June 27

Heard dogs barking at around 6:15 P.M. Heard nothing else all night except, Sandy heard a very loud thump to the NE of house before dawn.

Monday, June 28

Checked camera at 8:00 A.M. and found that it had taken a photograph during the night. Developed photograph at Kmart. Picked up at 1:00 P.M. Found image of human head at the eastern edge of the infrared scan limit. Heard some dogs barking at 6:15 P.M. Barked for about 15 minutes. Robert arrived to look at photograph from night before.

---

[22] Robert's friend, not Ian Price.

[23] One of the lights that Robert and Iain had secured to the trees was removed by Fatfoot using a simple tool—see Wednesday July 14 entry and photographs.

[24] Deliberately giving away her position was a common behaviour used by Fatfoot to gain attention and play games.

Bess barked timidly for about 30 seconds in direction of bush opposite house to the north. Robert left at 7:00 p.m. No noise. I heard howl, long and eerie for about 15 seconds at about 7:45 p.m. Phoned Katherine[25] and Ralph.[26] Sandy heard a growl, long and eerie for about 15 seconds to the north at about 8:50 p.m. Rang Lawson Police, talked to Matt. He said that he would come down the next time that we heard it. Ian came to the house after 9:00 p.m. after Matt phoned him[27] to see if he heard anything. Sandy also heard thumps around the time of the howl at 8:50 p.m.

Tuesday, June 29
Sandy, John and I took independent measurements of face from photograph and camera position. Distance to face was 16.7 metres. Height to top of head = 1570 mm (5′2″). Checked measurement of photograph of eyes[28] with John. Height in the range of 1550 mm to 1600 mm—very difficult to measure accurately. Found evidence of broken wood at position were face was photographed. Dogs barked at 6:15 p.m. in valley. Heard thump at 6:30 p.m., NW of house. Heard dog bark 100-150 metres SW of house bark at about 6:45 p.m. Heard nothing to 9:00 p.m., then thump. Bess barked for one minute under house. John came to house

---

[25] The editor of a computing textbook that I was writing.

[26] Professional photographer who was assisting us with technical advice, printing photographs and providing equipment.

[27] When Ian arrived at the house, he told me that Matt rang him to find out if we were "nutters" or "drug addicts". Ian told Matt that we were both "boring teachers".

[28] The top of the head was difficult to accurately determine. Used the position of the eyes on the face as a datum or reference point. Believe that eye measurement made with Sandy using staff at 1850 mm was more accurate because the eroded trough on bottom of track had not been taken into account.

at 5:45 P.M. till 7:15 P.M. Ian arrived at 6:30 P.M., left at 10:15 P.M. Dogs barked at 10:35 P.M. Ian and Cheryl said that Toby barked and ran to the compost heap at about 10:30 P.M.

Wednesday, June 30

8:15 A.M., went to Ralph's house to do enlargements of photographs—face, swamp. Lent me a Winder 2 [electronic motor drive] for camera. Set camera using old equipment. Whilst setting camera heard dogs opposite valley barking madly—about 5 dogs. Very loud at 5:03 P.M. Heard dogs barking about 100-150 metres southwest of house, up valley at 5:10 P.M. Bess barked at 5:30 P.M.—under house, as I went to Ian's house. Dogs opposite valley barked. Heard thump SW of house and dogs at 6:05 P.M. and again at 6:10 P.M.— 2 thumps, SW. Heard various other noises between 7:00 P.M. and 8:00 P.M., thumps and a noise like drumming— though this noise could be explained by rapid hammering at this time of night, to the NE of house? Some dogs barking during this time. Went to bed at 9:00 P.M.—very quiet —no dogs or any noise at all. Woke during the night— heard thumps to the east of the house several times. Woke again sometime during the night, heard thump and the sound of bush being parted to the east of the house.

Thursday, July 1

Developed film from the night before. Matt came to the house just as we were leaving to get the film developed. Went for a walk—showed him log near face and tracks— found footprint in the sand[29] to the north of house. Set

---

[29] I had wheelbarrowed some white builder's sand and placed it along the east-west track, many months earlier, with no success. The track widened around the sand. The sand by now was contaminated with dark material and tannins.

up camera with motor drive—works brilliantly! Cast footprint. John arrived at house at about 5:30 P.M.—sat on path to the NE of house for an hour—heard gentle footsteps, twigs breaking to the NE. Came inside and related story at 6:30 P.M.[30]—went back outside at 6:35 P.M.—John and I could hear clear footsteps, braking twigs and the parting of bush very clearly and loudly, moving from the NE towards Anthony's[31] shed. Called Lawson police and spoke to Mark—came down. Called Ian. Investigated bush—heard and saw nothing. Removed cast of footprint and cleaned it. Mark left. Heard noises in bush to the right of driveway when returning home. Sandy heard 3 thumps earlier. Heard thump again to the NE of house. Toby barked at bush near camera. Rang Mark at Lawson again—8:45 P.M.—came down—found nothing. Camera fired at 9:40 P.M.—rang Ian—could have been Toby? Dogs on far side of valley barked at 9:50 P.M.; 9:55 P.M.; and 9:57 P.M. and again at 10:15 P.M.

FRIDAY, JULY 2
Repositioned camera pointing ESE along driveway. Sensor at about three and a half feet, camera above at about 8 feet. Heard dogs barking about 500 metres to north at valley neck at 4:00 P.M.—very loud, many dogs. Again at 5:00 P.M.—close to house. John arrived at the house about 5:30 P.M. Sat near driveway. Joined John near driveway at 6:15 P.M.—I heard bush parted and movement to the NW

---

[30] John was excited by what he had heard. He said, "I now totally believe that these things exist." He also told Sandy and I that, "If the Aborigines are right and this thing takes one of your children, no one is going to believe you. Like Lindy Chamberlain, they are going to lock you up and throw away the key!"

[31] Anthony, Megan and their six children lived in the house on our northeasterly boundary.

in swamp area. Heard gentle movement to the north for some time at about 6:25 P.M. Heard movement away from house at about 6:35 P.M.—moving NW at about 60-70 metres. John and I returned to the house at 6:46 P.M.—heard heavy and loud movement to NNW—both of us heard what sounded like deliberate movement, bush being thrashed about. Dogs barked across the valley at 7:00 P.M. Dogs barked a few times between 7:00 P.M. and 8:45 P.M. Heard dogs barking at 9:00 P.M. Heard 2 thumps to the NNW at 9:03 P.M. Rang Mark at Lawson to report. Heard dogs barking at 9:05 P.M., 9:10 P.M.—Bess under house, 9:13 P.M., 9:23 P.M. Went to bed at 9:25 P.M. Woke up at various times during the night after hearing thumps to the NE and E of house. Camera took 2 photographs.

SATURDAY, JULY 3
Developed more photographs—no face—found nothing. Dogs barked at 6:50 P.M., 6:55 P.M. and very loudly and constantly at 6:57 P.M. Barking subsided, occasional barks at 7:05 P.M., 7:10 P.M. "Good Dog"[32] barked—woofed—constantly every 5 seconds from 7:10 P.M. to 8:45 P.M. Woke up during the night sometime—heard noise to the NE—thump. Sandy heard one very loud thump[33] that woke

---

[32] The 'Good Dog' was given this name by Sandy because of his constrained barking, that only occurred after all other dogs had commenced their fearful bark and other factors were suggesting that a predator was present.

[33] This loud thump was most probably the result of 'rock smashing', or 'rock thumping', where a large rock is forcefully slammed onto another rock or the ground. Can be identified from sandstone rock fragments scattered around an obvious, central impact point on a rock shelf or other large rock. Very loud, subterranean sound, where two separate arrivals of sound can be heard travelling through the different media of air and ground, over a distance. See also, Mike Williams (2009) 'Yowies talking', https://www.youtube.com/watch?v=eFSHLaON6Yw, 2:20 to 3:15 minutes (accessed: 13th June 2020).

her and our daughter Avril (?) to the north, then to the south—very close—scraping noise, followed by thump.

SUNDAY, JULY 4
Very quiet in valley until 6:20 P.M. Heard Good Dog and one or two others barking briefly—3 minutes and not very seriously. Heard thump at about 6:30 P.M. and again at 6:37 P.M.—Sandy and I. Sandy heard two thumps to the NE of the house at 12:30 A.M. and woke me. Avril woke briefly at the same time. I dressed and investigated. Wasn't sure if I heard something, because of the noise of water drops falling from the trees. Returned home after 15 minutes. Sandy and I woke again at 4:45 A.M. I heard several thumps, Sandy heard 8 thumps, every 2 seconds. I also heard something moving on the verandah east side—sounded like a step, followed by a drag. Avril woke—didn't investigate until a few minutes later—saw nothing. Sandy heard possums at 5:15 A.M. Rained lightly during the night.

MONDAY, JULY 5
Gordon's dogs barked frantically at 6:40 P.M. and briefly at 6:45 P.M. Raining constantly but medium light. Thump to the NNW at 6:49 P.M. Heard thump at 7:00 P.M. and 30 seconds later—difficult to be sure, due to the light rain falling on roof. Heard Gordon's dog at 8:40 P.M. and again at 9:00 P.M. Heard nothing else all night. Rained constantly most of the night.

TUESDAY, JULY 6
Heard 2 thumps to the NW at 6:27 P.M. and another thump at 6:20 P.M.—no dogs barked before or during this time. Dogs barking at 6:53 P.M. Heard dogs barking at neck of the valley about 500 metres north at 7:30 P.M. Gordon's dogs barked intermittently from 8:00 P.M. to 9:05 P.M.

4: Log

and Good Dog from 8:00 P.M. to 8:10 P.M.[34] Fed Bess at 9:10 P.M. Dogs woofed across valley, Good Dog and Gordon's. Outside light on. Bess very wary about eating her chicken carcass. Kept looking to the NW of driveway. I moved between her and the driveway, Bess started eating,[35] heard bush being parted and clear footsteps in swamp area at the bottom of the block. Noise seemed to be deliberate—wanted me to know she was there? Good Dog barking at 9:15 P.M.

Wednesday, July 7

Matt and Mick[36] came down at 1:30 P.M. Gordon's dog barked at: 6:34 P.M.; 6:38 P.M.; 6:42 P.M.; 6:55 P.M.; 7:00 P.M.; 7:08 P.M.; 7:13 P.M.; 7:28 P.M. Good Dog barked about 8:00 P.M. Heard 2 possible thumps at 8:02 P.M. Heard very heavy thump to the north or NE at 8:23 P.M.[37] Heard Good Dog bark at 8:25 P.M. Gordon's dogs joined the Good Dog barking at 8:27 P.M. Turned on outside SW light to see if anything happens—8:30 P.M. Dogs stopped barking at 8:31 P.M. Gordon's dogs started barking: 8:32 P.M. to 8:35 P.M.; 8:40 P.M.; 8:42 P.M.; 8:45 P.M.; 8:57 P.M. Several dogs, including Good Dog and Gordon's dogs barked noisily at 9:08 P.M. Good Dog woofs at 9:15 P.M. Turned on outside lights at 9:00 P.M. to 9:36 P.M.—

---

[34] This was the point where we started to notice that Fatfoot was approaching our house from a different route, most probably to prevent the dogs on the western side of the valley from detecting her presence and giving away her position. Ian and I confirmed this new, evolving strategy during many of the following Friday night pursuits.

[35] This was the point were we started to suspect that Fatfoot was "standing over" Bess each night and stealing her food.

[36] Local police officer. Mick spent many hours with me in the bush, until he had an encounter.

[37] Most probably rock or stump smashing.

heard 2 very loud thumps!!!—NW.[38] Dogs quiet in valley!!! Went outside to investigate at 9:55 P.M. 10:00 P.M. Sandy thought that she felt something to the NNE of house—saw orange eyes. Investigated with torch, saw a flash of two yellow eyes at approximately the same place as the photograph of eyes, went along track, heard very gentle noise moving away to the NE. Sandy thought that she saw eyes ahead of me to the NE. Heard thumps several times around 10:30 P.M. Went to bed at 10:50 P.M., set tape recorder on table facing NE—quiet.

THURSDAY, JULY 8
Raining moderately light. Heard dogs barking at about 6:30 P.M., otherwise quiet. Heard thump at 7:00 P.M. NE. Heard 2 thumps at 7:15 P.M.—heavy, NE and another heavy thump at 7:17 P.M. to the NE. Raining lightly. Two gentle thumps at 7:25 P.M. Heard loud thump at 8:05 P.M. Medium rain. Went to bed at 10:10 P.M. Moderate rain most of the night.

FRIDAY, JULY 9
Dogs started barking at 6:15 P.M. Heard thump at 6:22 P.M.—think that there was an earlier one as well(?) and at 6:27 P.M. Dogs started barking again at 6:30 P.M. Gordon's dogs barking loudly at 6:35 P.M. Good Dog, Gordon's dogs

---

[38] This was the point when we expanded our attempts to communicate with Fatfoot using light signals. Earlier in 1993, we had used Ian's powerful halogen spotlight to call in Fatfoot from a distance by shining the beam into the sky, similar to the 'Bat Signal'. Turning a light on inside the house (even from a microwave oven) in the middle of the night typically encouraged a round of foot thumping! Even in 2020, we still turn the outside light on and off in response to thumps made by Fatfoot or 'Son of Fatfoot'—meaning, "We have heard your thump and though we acknowledge it, we are not coming outside!" Failure to do so usually means that the thumping persists until we flick the light on and off!

and one other bark very noisily at 6:52 P.M. Dogs barked on occasions between 7:00 P.M. and 8:30 P.M. Generally quiet. Heard Good Dog woof several times at 8:45 P.M. Heard suspicious noise at 8:53 P.M.—sounded like wood being struck with another piece of wood, several times rapidly, to the west or SW? No lights on at the southern neighbour's house, or across valley?[39] Gordon's dogs barked at 9:00 P.M. Went to bed. Woke sometime before 2:00 A.M.—heard thump to the WSW several times. At 2:00 A.M. heard the sounds "mook . . . mook . . . mook" very loudly to the WSW.[40] Sandy heard 2 thumps around 10:00 P.M. to the east of the house.

SATURDAY, JULY 10
Dogs started barking at 5:40 P.M. Heard several thumps to the north at 5:45 P.M. Investigated at 5:50 P.M. after Bess barked. Heard suspicious noise—twig, to the north about 30 metres. Dogs barking across the valley 6:10 P.M. Dogs barking—Good Dog 6:30 P.M. Dogs barking noisily 6:35 P.M.-6:55 P.M. Good Dog barked *very* noisily 7:15 P.M. to 7:25 P.M. Called Lawson. Matt and Greg[41] came down. Heard nothing. Left at 8:30 P.M. Heard extremely loud thump at 8:40 P.M. Went to bed at 9:30 P.M. Heard 2 thumps around 9:40 P.M. Heard nothing else all night.

---

[39] Wasn't certain about this noise. I checked to see if anyone was home nearby because it sounded like hammering or wood chopping—maybe to start a fire?

[40] Heard this "mook . . . mook . . . mook" another time with Cheryl in our driveway. I repeated the "phrase" back to Fatfoot and she replied the same, but much louder. Did this cycle one more time with even greater volume. Cheryl asked me to walk her home and said, "I think that it likes you!" I wasn't so sure!

[41] Local police officer.

Sunday, July 11

Placed tape recorder in bush next door where swamp paths finish—flattened area.[42] Gordon's dogs barked at 6:05 P.M. very loudly. Heard 2 *extremely* loud thumps at 6:15 P.M. Dogs quiet. Dogs barked at 6:20 P.M. to 6:40 P.M. Heard 4 *extremely* loud thumps at 6:43 P.M. and another at 6:44 P.M. Dogs quiet. Gordon's Dogs barked at 6:47 P.M. Generally quiet.[43] Collected tape recorder at about 9:00 P.M.—letterbox caused it to record the sound of its own motor—nothing.[44] Heard suspicious noises after 9:00 P.M. Heard thumps at 10:30 P.M. Investigated for half an hour. Found newly flattened area across from driveway—opposite where photo of eyes. Heard movement away to the NNW. Bess barked at 10:30 P.M. Toby barked from 10:30 P.M. to 11:00 P.M. Reset tape recorder on table NE corner of house. Bess and Toby barked during the night.

Monday, July 12

Heard dogs barking down valley about 200 m at about 5:15 P.M. Good Dog and Gordon's dogs barking at 5:30 P.M. Placed tape recorder in tree fork, 8 m in from north border near swamp at 5:30 P.M. Dogs barked furiously for next 10 minutes. Returned to tape recorder to rewind tape

---

[42] These flattened areas usually covered a large area in the swamp, some 100 square metres and would appear overnight, generally after much activity.

[43] The dogs rarely went this quiet during the middle of activity—suggests that the dogs were more than fearful.

[44] VOR tape recorders were loud enough for me to hear the motor when recording at a short distance of about 200 mm. To dampen this sound so that it wouldn't be found and to protect the battery from the cold, I initially placed the recorder in an old metal letterbox. Later modifications had the recorder wrapped in cloth inside the letterbox. Final version was wrapped in cling wrap (waterproofing) and cloth (sound proofing and insulation) which was then camouflaged.

at 5:40 P.M. Dogs still barking and Gordon's cage birds very noisy and distressed.[45] Set Cheryl's tape recorder on steps near tap on NE corner of house at 5:50 P.M. Heard very loud thump at 6:15 P.M. and again at 6:23 P.M. Sandy and I retrieved tape recorders at about 7:30 P.M. Played tapes. Tape from swamp has sound of twigs breaking and other suspicious noises for 5 minutes prior to footsteps. Heard approach to tape recorder—walking and touch etc tape recorder. When I retrieved this tape recorder, the machine was not working—thought that it had reached the end of the tape![46]

Tuesday, July 13

Unusually quiet early in the evening. Dogs barked at 6:23 P.M. Heard 7 or 8 gentle thumps to NW at 6:40 P.M., followed shortly after by dogs barking. Dogs barking at 7:50 P.M. Good Dog and Gordon's dogs barking at 8:00 P.M.—variously until 9:10 P.M. Heard thump at 9:15 P.M. Barked noisily from 8:55 P.M. to 9:05 P.M. Good Dog and Gordon's dogs barked at 9:17 P.M. Heard very loud barking—not Gordon's or Good Dog at 9:30 P.M. when getting tape recorder. Sandy heard very loud thump to the NE at 9:54 P.M. and again at 9:57 P.M. and 10:05 P.M.—NE. Bess barked at 10:43 P.M. to NE. Bess barked sometime

---

[45] Gordon's birds were in a large cage located at the SE corner of his fenced property. At a later date, we found evidence of bird seed on the outside of the cage floor being swept up and dragged into a pile.

[46] On this occasion I hung the recorder by it's strap on a short branch about 1.8 metres from the ground at the edge of the swamp. The very small, red LED activity light was left uncovered and the recorder was unprotected. The recording suggested that the recorder was being stalked, multiple twig snaps, followed by a loud bipedal rush towards it and the sound of pressure being applied across the many buttons, which stopped the recording. I sent this tape to Steve Rushton because he had access to a professional audio technician who could improve the quality.

Fig. 4.3 Tree with stick tool in bark and light on ground. (Photo: Neil Frost)

earlier—about an hour and a half before—investigated to see what Bess was barking at—Bess was looking alertly to the NNE.

WEDNESDAY, JULY 14
Mick came to the house around 4:15 P.M. Had earlier found branch thrust into bark up tree and flood light on ground. Photographed stick in tree.

Mick looked around and decided to stay until 7:00 P.M. Mick, John and I stayed behind bush hide[47] down block. Heard *very* quiet movement up from swamp alongside block next door.[48] Investigated at about 6:45 P.M. I went into bush and heard movement away to the NW quietly. Ran to bottom of swamp, heard nothing more. Mick seemed to be convinced that it was "an intelligent animal."[49]

FRIDAY, JULY 15
Took more photographs of tool and light on ground and new swamp track. Dogs barked noisily from 6:02 P.M. Heard several thumps around 6:05 P.M. Heard 2 thumps at 6:09 P.M., 2 thumps at 6:10 P.M. and 3 thumps at 6:12

---

[47] Made the bush hide earlier in the morning with vegetation from another part of the block, so that the disturbance would be less obvious. Also, made hide very quickly to minimise the chance of being observed during the day. By now we were starting to use more camouflage because it was obvious that Fatfoot's night vision was too good! We worked on the basis that, "If it is difficult for us to see during the day, then it should be difficult for Fatfoot to see during the night!"

[48] Spoke quietly into Mick's ear. Suggested that he should come with me. He replied, "No way! And I've got a gun!"

[49] Mick said that if anything significant happened, he would like to be involved. He also said that many of the police involved in this case were being mocked at Katoomba Police Station, having cartoons and drawings left on their desks.

Fig. 4.4 Floodlight on ground and flattened grass around tree base. (Photo: Neil Frost)

Fig. 4.5 Stick tool used to lift floodlight from nail in tree. (Photo: Neil Frost)

Fig. 4.6 Stick tool. Bottom of stick eight feet from the ground. (Photo: Neil Frost)

P.M. to the NW—quite loud and spaced at one second intervals. Tape recorder is hidden up tree about 20 m in from border next to trail near swamp. Quiet in valley from 6:20 P.M.

Friday, July 16

Basim heard dogs early in the evening.[50] Heard Bess and noise to the east, investigated—8:00 P.M. and found police coming quietly down driveway.[51] Talked to police—Matt and Graham.[52] Showed them the swamp, branch up tree, photos, tapes of walking etc. They went at 10:25 P.M. I walked them up to the car, up on road. Graham saw light in tree tops across the road. I investigated, thought it was Ian—wasn't, called to Sandy. Sandy thought that I was nearer. Sandy called out that something was walking across driveway, above east of house.[53] Ran up to police, they drove down driveway very fast with lights on, jumped out and we investigated bush around[54]—found/heard

---

[50] Basim telephoned me early in the evening to let me know that the dogs in his valley to the west were barking loudly.

[51] A police officer that I didn't know, was extremely surprised. He said, "How did you know that we were here? I replied, "Because we are very used to reading what's happening outside!" Although it was dark, I think that he was impressed!

[52] Graham was the Senior Detective Sergeant at Katoomba.

[53] Spent two and a half hours speaking to Graham and walking around the block. He asked many questions. I walked with them up to their 4WD which was parked 30 metres down the road, a total of about 150 metres from our house. As they were preparing to leave, Graham noticed light in the treetops opposite our driveway. I walked down to our driveway and could see torchlight. Thinking that it was Ian, I called out, "Is that you Ian?"and Sandy replied, "Is that you Neil?" I replied, "Yes." Sandy said, "If you're up there, then what is down here with me?"

[54] By now the police were driving up our road, however, I ran after them and managed to slap the back of their vehicle. I said to Graham, "He's

nothing. Police left at about 11:00 P.M. Dogs barked to south house. 11:05 P.M. dogs barked across valley—west. I heard a "growl—woof" noise—shortly after 11:05 P.M. Dogs barked at 11:15 P.M., 11:25 P.M. Light rain falling at 11:30 P.M. Heard nothing all night.

Entry by Sandy: "Footsteps were extremely loud. I thought that Neil had heavy shoes on. They began and ended with the same volume on the driveway. Stopped when I shone the torch in direction of steps."

Saturday, July 17

Working upstairs with Fie on tax. Heard dogs at 6:00 P.M., 6:40 P.M., 6:47 P.M., 6:53 P.M.—dogs up valley to south about 200 m or less. Dogs at 8:00 P.M. Thump at 8:57 P.M., dogs at 9:00 P.M. Thumps 2 at 9:10 P.M. to the NW. Heard 2 thumps to the NE at 10:15 P.M. Bed—heard nothing all night. Murdock[55] heard footsteps about midnight to the south along concrete path.

---

down there now!" They spun the 4WD around, turned on all of their driving and spot lights and drove at high speed down our driveway. As they slid to a stop, Matt and Graham jumped from the vehicle. Matt ran straight into the bush. Graham stopped and yelled out to Sandy, saying "What are you doing? Don't you have two young children? Get inside and look after them!" After Graham had searched around the house, he spoke to Sandy again. He asked her recount what had happened. Sandy added to the audio description by saying that, "They were really heavy footsteps. I thought that it was Neil returning and he was disappointed because they were slow and heavy. The closest sound to what I heard would be the noise made by a car tyre moving across a gravel road. I had walked up the path towards the noise because I wanted to light up the driveway for Neil."

[55] Murdock and Julie were our neighbours on the southern boundary with Ian and Cheryl. Fatfoot's red eyes would regularly be seen in bush across the road when Murdock returned home from work around midnight. Fatfoot would also go underneath their house and spend time beneath the main bedroom. Son Jeremy had a face-to-face encounter with Fatfoot.

Sunday, July 18

Set camera and tape recorder in bush next door near swamp. Also placed walkie talkie[56] in centre of camera field of view. Set tape recorder taping at 5:45 p.m.—quiet, on return to house Ian and I heard footsteps on receiver. Dogs barked at 6:55 p.m. and noisily at 7:05 p.m. to 7:10 p.m. Dogs at 7:27 p.m., 8:30 p.m., 8:35 p.m. Thought that I heard a thump to Nth at 8:48 p.m. Dogs at 8:55 p.m., 9:05 p.m., 9:10 p.m., 9:13 p.m., 9:28 p.m. Woke sometime before 4:15 a.m to noises and a thump, went to sleep, woke again to thump. Camera didn't take a photo.[57]

Ian and Cheryl walked down valley before lunch. Found uneaten remains of currawong,[58] tracks, track from our place to Gordon's, track up embankment at bottom of valley.[59]

Monday, July 19

Generally, extremely quiet, though had heard dogs barking at about 4:15 p.m. Heard few dogs barking during the night. Tape recording proved that Fatfoot was there anyway—breathing, twigs breaking. Set camera and tape recorder near swamp, 8 m from the border with glass on branch of same tree to obtain 'finger prints'[60]—at about 5:00 p.m.

---

[56] Used electrical tape to keep transmit button down.

[57] Like floodlights, certain that Fatfoot can recognise a camera and flash. Avoids anything with a reflector.

[58] Robert and I previously heard Fatfoot shake a currawong from a tree.

[59] Ian and Cheryl followed the tracks down the valley and established their paths to the north.

[60] The police requested that I place a glass on a branch so that they could obtain fingerprints. Wasn't going to argue! They were still entertaining the possibility that we were dealing with homeless person. Had heard reports of homeless living in the bush.

4: Log

TUESDAY, JULY 20

Ian and I walked down valley at 4:15 P.M. about 700 metres. Ian showed me some new tracks, especially the track up the embankment that led N to another road in the next valley. Walked about valley for about one and a half hours, saw and heard nothing. On returning home, we split up. Ian went into the swamp and I went on dry land on eastern side of valley. Heard dogs barking as we moved through the "cleared" land about 100 m from home. Heard dogs barking—Gordon's and Good Dog more seriously as we approached home. Found branch across track on next door's block (track adjacent to where tape recording of walking up to tape recorder) had been pushed into swamp, reeds also pushed or combed into swamp—branch had been in place at 7:00 A.M. Heard nothing else. Went to bed at

Fig. 4.7 Photograph of flattened swamp taken at 4:59 P.M. on 19 July 1993— 19 4 59—see bottom right of photograph. The top left-hand corner of the film photograph shows where light partially entered when removing the film cassette—it is not a 'supernatural artefact'. (Photo: Neil Frost)

about 9:30 P.M. Heard 2 thumps very shortly after. Woke during the night to the sound of bush being disturbed, things being moved(?)—very difficult sound to describe, don't know what time—probably before midnight, to the SW of bedroom, about 10 to 20 metres away. Woke again sometime during the night to thump near bedroom.

Saw Dave a few hundred metres up our road—has heard footsteps at night.[61]

WEDNESDAY, JULY 21
Set camera at 4:00 P.M. in tree near swamp, 8 m from border. Tape recorder set east of camera about 5 m, with external microphone in tree at about 5 foot level at 5:30 P.M. Heard dogs at 5:35 P.M. and at various times to 6:00 P.M. Dogs barked very noisily from 6:00 P.M. to 6:15 P.M. and again at 6:25 P.M. to 6:30 P.M. Some barking from dogs until 7:30 P.M. Good Dog woofed from 7:30 P.M. to 7:40 P.M. Heard possible thump at around 7:15 P.M. Graham came down at 2:00 P.M. to drop off tent for "Stakeout" on Sunday—6:00 P.M. to 2:00 A.M.[62] Dogs at 9:50 P.M.

---

[61] A contact from our local club told me to see Dave and another local resident. Dave told me that they frequently heard someone outside their bedroom at night. When this happened, his wife would slide the outside door open and Dave would run out swinging his baseball bat. Dave said, "I can never catch the bugger, he is too fast!" Peter was the other local resident who lived in another street at the head of our valley. Peter said that someone regularly walks through their vegetable garden at night, ripping up the plants and leaving large footprints.

[62] Graham had previously suggested a stake out if he could get approval for overtime for two police over three nights. After getting approval he dropped off his personal tent and asked me to assemble it at a location that he chose, about five metres from the western foundation wall of the house, next to our bedroom.

4: Log

Thursday, July 22

Heard dogs at 3:15 P.M. Set up camera in new position.[63] Started at about 3:30 P.M., finished basic setup by 4:30 P.M. Placed wine cask bladder[64] in bush to right of centre of camera frame. Set tape recorder in bush to the NE of camera about 5 m distant at 5:45 P.M. Lights came on at 6:50 P.M.—4 times over about 10 minutes. Investigated with Ian, heard and saw nothing, except for twig breaking to the north—from camera position. Went up to area near driveway, heard twig(?) Went inside to listen to tape

---

[63] Each time the camera was moved, it required the 240-volt power supply to be re-routed, as much as 100 metres—together with a charged 12-volt battery.

[64] Police thought that a homeless person might be tempted by the wine cask.

Fig. 4.8 Glass on small branch on left tree. Wine cask taped to bush, right of centre; daytime photograph. Well-established, worn 'old main track' visible, winding through centre of photograph. New swamp track located five metres left of the frame. (Photo: Neil Frost)

retrieved from near camera. Lights came on several more times—10 X. Listened to tape, heard suspicious noises. No lights from about 7:50 P.M. to 9:45 P.M. Dogs barked at 9:30 P.M. to 9:45 P.M., dogs 9:50 P.M., 10:00 P.M. I went outside at 9:59 P.M. to check on sound heard on tape—paint can. Good Dog 10:10 P.M. Bess barked at 10:12 P.M., dogs 10:33 P.M.—Gordon's. Placed two tape recorders, Robert's on SE corner and Cheryl's on SW corner of verandah at 10:30 P.M. Went to bed at 10:45 P.M. Woke at 3:05 A.M. to loud thumps (?) listened, heard 2 more but gentle thumps at 3:47 A.M.—SE. Heard various noises. At 5:10 A.M. heard movement on SE corner of verandah. Heard 2 loud thumps at 5:18 A.M. to the SE. Heard various noises until about 5:30 A.M.

Friday, July 23
Photos from previous night showed nothing. Set up camera in same position as last night—on border up scribbly gum—no tape recorder. Camera turned to the NNE instead of N because Sandy saw eyes up track when lights were on the previous night. Dogs barked noisily at 6:00 P.M. to 6:04 P.M., 6:15 P.M., 6:22 P.M., 6:25 P.M., 6:27 P.M. Very noisy from 7:15 P.M. to 7:20 P.M. with Bess barking also from under the house. Flashed torch in bush around house—dogs across valley barked after turning the light off. Dogs at 8:00 P.M., 8:10 P.M. Went outside with torch 8:15 P.M.—dogs. Heard nothing else all night.

Saturday, July 24
Dogs barking at around 4:30 P.M. Noisy at 5:00 P.M., 5:20 P.M. and variously between 5:30 P.M. and 6:30 P.M. Heard 4 thumps at 6:50 P.M. after returning home from shops—put high beam on down driveway at 6:30 P.M. Heard thump at 7:10 P.M. to NE. Went outside to investigate, heard twigs to the NE—movement away along high side of

4: Log

valley to NE, dog about 100 m to the NE barked. Dogs at 7:40 p.m., very noisy at 7:55 p.m. Sandy heard 2 thumps around 8:30 p.m. and Bess ran into bush barking to NE at 8:45 p.m. Dogs barking noisily at 9:05 p.m. to 9:10 p.m. across valley. Went to bed at 9:30 p.m. Woke at 4:50 a.m. to 3 thumps—very loud to the SE.

Sunday, July 25
Set up tent in morning. Light to medium rain from 10:00 a.m. to 1:30 p.m. Went to see Robert and Kerry. Came home at 5:15 p.m. Dogs barked at 5:15 p.m. noisily and at various times—every 5 minutes until 6:00 p.m. Good Dog at 6:10 p.m.

Monday, July 26
Police stakeout.

Tuesday, July 27
Police stakeout.

Wednesday, July 28
Police stakeout.

THE POLICE STAKEOUT

Monday, July 26; Tuesday, July 27; Wednesday, July 28
The three-day stakeout was the culmination of five months of police involvement. At the beginning of this period, Sandy and I had no idea of what we were dealing with, although we knew that the circumstances were very strange and terrifying. The first time that the police responded to our situation, the Senior Constable surprised us by suggesting that we were dealing with a Yowie after I was concerned for their safety and having described the intruder as "an elephant on two legs, wearing size twenty boots!"

Over the following months, the police began to realise that they weren't responding to the activities of an ordinary criminal intruder. Some constables accepted the new situation quicker than others, although Mick told me that the reality back at the police station was that Yowie caricatures and cartoons were being left on their desks and hanging around the office.

In the final approach to the stakeout, it seemed to us that there was a reasonable amount of acceptance amongst the police, that a Yowie was responsible for the strange activity, however, there were some very understandable, residual doubts. Consequently, requests were made of us to leave a glass and a wine cask bladder in the bush, in the last-minute hope of obtaining the traceable fingerprints of a homeless person. The police also conducted door-to-door interviews with residents to determine our veracity, with many confirming their involvement as part of an active community. As part of their specialised procedure in investigating bush-related incidents, the police engaged the services of Steve Crofts,[65] a professional shooter, bush expert and local resident, asking him to investigate our property to evaluate possible explanations for the activity we were experiencing. Fortunately, Steve was also Ian's boss. Steve told Ian that the police said to him, "These people think that they have a Yowie on their property!" Steve said that he took great pleasure in confirming to the police that he believed us, telling them that he had also seen and encountered a Yowie several times himself. He also told them that there were many game trails throughout the valley, with no evidence of pigs or other feral

---

[65] Steve is the father of Brad, who had an encounter about seven years later, which led to our contact with Jerry and Sue O'Connor and their experiences with a Quinkan.

animals, although there were swamp wallabies. When I spoke to Steve, he told me about his encounter with Doris at a local waterhole. He said that this powerful bipedal animal had rapidly descended an impossibly steep cliff,[66] stopping opposite the pool where they were camping, before running off at high speed. Like us, he also mentioned seeing the red eyes change to yellow as the gaze shifted. He also confirmed Sandy's discovery that Dooligahl eat tree larvae[67] and expressed disappointment at having the game that he had shot, sometimes disappear before he could retrieve it. On these occasions, Steve said, "The whole valley would go quiet"—which was acknowledging the presence of a felt predator that was actively hunting in the area. One night when shooting feral animals on a farmer's property in the Megalong Valley, Steve said that he trained his rifle scope onto some red eyes, but didn't shoot because the outline was very human looking!

On the evening of Monday, July 26, two police officers arrived at our house to commence the three-night stakeout. As they entered our house, they bought in two large zippered bags. Placing them on our kitchen table, the bags were opened, revealing two formidable rifles, several boxes of ammunition, some packets of potato chips and a plastic

---

[66] Steve told me about the location of their encounter and when I visited the waterfall site, it had a vertical drop of about fifty metres. This account made me reflect upon the 600 metre descent that the 'marsupial hominoid' must have made during our scout camp at Blue Gum Forest in 1966.

[67] Sandy had found that Fatfoot would bite the trees at night, to obtain the tree larvae. When we showed Steve an example of a treebite, he confirmed that something with powerful jaws and canines had done this, rather than simply a yellow-tailed black cockatoo, which also rip open the trees but in a different manner. He said that he would sometimes eat the larvae himself but it took a lot of effort to extract them from the heartwood with a hand axe.

box of other food. Looking at the rifles and then at each other, Sandy and I could read the other's minds. The police officers started to load their rifle magazines with hollow-tipped bullets. Sandy said to the police officers, "Get those guns out of our house!" At this point, we were told, "This is what is going to happen! We are going to shoot this thing tonight. We figure that a body will be worth about a million dollars and we want half!" I said to the officers, "What happened to your homeless person theory? Are you going to shoot a homeless person?" In reply, we were told, "I think that we are well past that theory!" I said to Sandy, "Let's go to the bedroom and talk."

In the bedroom, Sandy and I were very shaken. We questioned whether we had the NSW Police in our house or a pair of criminals. We realised that the situation had evolved well past the point where we could count on the police to assist, should it become necessary.

Returning to our kitchen, we questioned the morality of killing this intelligent hominoid and expressed our regret. We were then told, "Don't you want this problem fixed?" The police then moved outside and we continued to discuss the new situation. We concluded that Fatfoot could easily look after herself. After all, we had more than enough trouble trying to shoot her with a camera and the police would have no hope because they had no idea what they were dealing with!

The stakeout achieved nothing. The police spent a lot of time in their tent eating potato chips. When they got bored, they went for walks through the valley and door knocked the neighbours. In the end, the police commented on the high level of support that we received from our valley community.

4: Log

Summary of the Most Significant Log Events
August to October, 1993

Interest in maintaining the log ceased for about a week afterwards. From then on, I didn't log every encounter incident, as it wasn't needed by the police investigators. Instead, I only recorded events that contained significant information

Friday, August 6

... Heard noise in bush opposite driveway. Went inside, got camera, torch—chased Fatfoot through bush—was about 20 m in front of me. Met Michael[68] and he got Phil[69] to come outside. Michael heard more in bush, Phil heard branch break. Heard Fatfoot again at 6:50 P.M.—rang Ian, who waited in ambush.[70] Fatfoot went down into swamp instead. Went outside with Ian at 9:00 P.M., heard movement in bush to the S of Ian's land—moving into swamp—investigated swamp area—heard movement to the south about 20 to 30 m distant. Phil contacted me on Saturday morning. Heard and saw other Yowie SW of his house—saw yellow eyes for 30 seconds. Saw eyes turn for a few seconds away, then returned. We have 2 Yowies![71]

---

[68] Michael is the youngest song of our neighbours, Phil and Helen.

[69] Phil was our neighbour to the northeast of our house. Phil and Helen had a long history of experience with these hominoids, beginning in the early 1980s when they were building their house and living onsite in a metal shed. They would frequently hear walking and heavy breathing on the other side of the thin sheet metal shed walls.

[70] Ian would frequently lie in wait, sometimes with a camera, sometimes without. He would tell me to drive Fatfoot towards him so that he could "tackle the beastie". As if I was somehow capable of achieving that, I would as frequently, tell Ian that he was crazy, to which Ian would reply, "The worst it can do is kill me!"

[71] Phil came to our house on the Saturday morning to tell me that Ian and I had missed chasing Fatfoot. I told Phil that we were close behind Fatfoot

Friday, August 13

... Heard gentle noise of movement to north above humpy. 9:20 P.M. heard very loud thump to SE of house—thought that Avril had fallen out of bed! Sandy heard thump shortly after. Both of us heard thump 2 minutes later to SE again. Called Phil, came down with Tom.[72] Investigated area. Phil heard "growl" in bush, followed by second "growl"[73]—wasn't sure of direction—found nothing. Dogs barked in valley shortly after. Quiet in valley 10:30 P.M.

Sunday, August 22

... Sandy woke at about 10:00 P.M.—heard thumps, banging on the Sth brick wall and long "growl" ...

Sunday, August 28

... Met Ian and Cheryl who were investigating noises. Toby's dog bowl missing from verandah, full of food.[74] Bess barked to E at 8:00 P.M.

(cont.) as she made her way up the valley, until the track became too difficult. Phil said that he was looking at the outline of a hominoid standing beside a tree on our northern neighbour's land. Phil was standing on his western verandah with the outside light on. He said that it seemed to be looking at the two of us as we moved down the driveway and headed into the bush towards the swamp. It turned to face him and Phil saw the yellow eyes for thirty seconds. It then looked away before looking at him again. He also said that the eyes were at a height of about six feet. Although we had suspected that we were dealing with more than one individual, this was substantial proof that we had, at least, two individuals. Over the following months, we paid more attention to the location of sounds and thought that we had a minimum of three individuals, moving as a family group.

[72] Tom, Phil and Helens' eldest son, would assist me with tracking and searching for evidence. Tom was responsible for finding three orange hair strands on the barbs of their swamp fence.

[73] Ian, Sandy and I had heard 'growls' on other occasions.

Fig. 4.9 New swamp track (middle) formed on evening of 13 August 1993. Photograph taken next morning (bottom, right—93 8 14). Previous swamp entry and exit tracks, left and right of new swamp track. (Photo: Neil Frost)

[74] We compared our experiences with dog food and bowls. Bess would only eat if I stayed next to her or between her and the bush. We lost many dog bowls over the years, including the sloping sided bowls, designed to prevent a dog picking up the bowl with the mouth. Ended up using disposable containers. Sometime later we found a collection of Bess's food bowls under a large, dense bush across from the driveway, on our neighbour's land. Toby's bowl had been lifted off the seven foot high, raised verandah and was not found. From then on, Ian used a double handled aluminium pot that was short chained to the verandah post, as his food bowl. This partially worked, although the bowl was occasionally found hanging by the chain. Also, in an obvious attempt to gain Ian's attention sometime later, one of the ceramic pot plants was pushed off the verandah and smashed. Similarly, Fatfoot would stand behind a tree during summer, a few metres from the back door, and provoke Ian, usually with 'growls' and sometimes by deliberately giving away her presence.

A few years later, I interviewed a woman a few kilometres away who, on turning on the light in response to a disturbance, saw a hairy arm removing her dog's food bowl through the 'doggy door'.

Tuesday, August 31

Heard thumps to north of house. Investigated, walked down east-west track towards swamp with Bess.[75] Bess ran away—heard noises of movement to the north, about 30 foot away. Investigated—too thick to go in. Woke sometime before dawn—sound of *extremely* heavy thump next to bedroom, verandah SW, followed by footsteps. branches, twigs braking about 10 seconds later.[76]

Wednesday, September 8

Heard noise to the southeast of house at about 9:45 P.M. Investigated—heard crashing through the bush—chased for a few metres—very dense—circled around area—heard further noises. Bess barked at 10:25 P.M.—went outside, heard top strand of fence wire 'twang'[77] to the east of house, near sewer clearing, noise moved across to Megan and Anthonys' land, Bess followed barking. Placed tape recorder in swamp at beginning of main track near wooden stakes. Tape recorded Fatfoot touching the microphone, rest of tape has dogs barking and rain falling on microphone.

Friday, September 10

Placed tape recorder with external microphone in swamp, about 10 metres in on the main track—Fatfoot touched it for about 90 seconds.[78] Heard Bess barking to north at about 7:30 P.M. Investigated and heard movement away to the

---

[75] Very unusual for Bess to do this.

[76] A bit unnerving.

[77] Fatfoot was capable of clearing the top of 1200 mm wire fences in stride.

[78] An amazing aspect of this recording was Fatfoot handling the microphone and then pausing while a twin propeller aircraft flew overhead for a minute and a half, then continuing to handle the microphone. This reminded me of the 'Cargo Cult' that I studied when in the Papua New Guinea Highlands in 1975, where airplanes passing overhead were

NNW. Went to bed at about 11:30 P.M.—heard 6 very loud thumps to east of house as I was turning off the lights, then 3 more thumps as I was in the toilet. Heard more noises when in bed. Sandy was kept awake throughout the night—sounds of walking, thumping and hitting the wall. Sandy thought that the noises sounded like more than one Yowie.

Saturday, September 11

Placed tape recorder on edge of swamp near entrance on next door block, 8′ up in branch of tea tree. Margot[79] visited. Heard thump to the NE at about 7:30 P.M. On leaving, Margot and I heard movement through bush to NNW for about 30 seconds. I heard movement in bush and saw yellow eyes earlier at about 6:30 P.M. to west of house on swamp edge.

Murdock and Jeffrey visited at 6:00 P.M. Jeffrey saw Dooligahl in bush the night before at about 7:30 P.M.—tape from Friday night had Yowie on it, also at 7:30 P.M.—plane passed over at 7:30 P.M.[80] Proof of 2 Yowies.[81]

---

worshiped in an attempt to get them to land or parachute their cargo, for the benefit of the local natives.

We started to use these audio landmarks: especially scheduled airplane flights, like this 7:30 P.M. flyover, passenger train services, and regular coal trains to establish the time frame of VOR tape recordings. Usually scribbled a note on paper next to my bed, at the time when coal trains passed through.

[79] Margot was a colleague who would later rent the property on our northern boundary. Margot had an encounter with a seven- to eight-foot-tall Dooligahl (possibly a Quinkan) on 15 August 2004. Margot was also a New South Wales State Examiner. She subsequently included a question in a senior English exam paper on Yowies that Drew and many other Blue Mountain students appreciated.

[80] For some time we had been using certain landmark sounds to determine the time of VOR recordings. A regular twin engine commuter flight passed overhead around 7:30 P.M. each week night, flying to Bathurst.

[81] The recording of the 7:30 P.M. commuter flight passing overhead and the sound of Bess barking to the north at 7:30 P.M. was further

Jeffrey told of his encounter. Was at the Green's house—stone house up road, on other side. Having a BBQ, Jeffrey was playing to north of BBQ with burning sticks. Heard twigs breaking, looked into bush and saw Yowie clearly. Was about 6-foot-plus high and mainly black or dark brown. Noticed hair on face and around head. Noticed patches or coloured areas or whatever on cheeks. Jeffrey ran away, as did Yowie to the north, making noises of bush being flattened.[82]

Ian heard thumps from his house in our direction between 1:00 A.M. and 3:00 A.M.

SATURDAY, SEPTEMBER 18
... About 10 minutes later, heard thump outside bedroom window, followed by various noises and then a long deep growl.

TUESDAY, SEPTEMBER 21
... Sandy looked out of the kitchen window at 6:15 P.M. and saw the yellow eyes turn and look up at the house and then away again. She said that "it seemed as if it was walking north behind the line of trees and the septic trench." I charged up the camera and went outside through the lounge room door and down the southern side of the house. Bess followed behind. Fired camera in middle of grassed area—frame #12. Bess agitated and sniffing the ground where

---

(cont.) confirmation that we had two Dooligahl and when Jeffery's simultaneous encounter is considered, about 150 metres away on the other side of the road, probably a family of three.

[82] Murdock came to our house with his nine-year-old son Jeffrey. Interviewed both of them. Jeffrey described the face as being big and surrounded by a thick layer of black or dark brown hair with deeply wrinkled (even emaciated) skin, being mostly hairless. There were flashes of grey or white hair were the ears should have been.

Sandy saw eyes. Walked around area and then up east-west track—saw heard nothing. Sent Bess ahead up east-west track, Bess disturbed. Came up to house and then immediately back down[83] to where Sandy saw eyes and fired off a second photo—frame #13. Came inside at 6:35 P.M. Dogs barked to 6:40 P.M. Gordon's dogs at 7:00 P.M.

Wednesday, September 22
Noticed that the wine cask bladder was on the ground that evening when setting tape recorder at about 5:50 P.M. Cask was torn from tape and had been 'bitten' through the middle, leaving puncture marks through both sides[84] and one side had depression, as if made by jaw. I think that he/she was angry with me[85] for using the flash the night before?

Thursday, September 23
Dogs barked *extremely* noisily at about 6:00 P.M. to 6:30 P.M. Heard a thump about 9:30 P.M. while talking to Steve Rushton. Heard a very loud thump to the east of the house at 10:25 P.M. followed immediately by Toby barking. Went outside and investigated around house. Went down towards swamp and spoke to him/her and it answered with a 'growl-grunt' (?)—long combination. Good Dog barked from 10:40 P.M. to 10:55 P.M. Gordon's dogs joined in at 10:55 P.M.

---

[83] I was pretending to have returned inside, trying to catch the Dooligahl off guard.

[84] The police had requested taping the wine bladder to a bush on Thursday July 22. The puncture marks seemed to have been made by canines. The separation was 60 mm. Appeared to be an angry response to something?

[85] Ian and I noticed that Fatfoot would retaliate after an incident that had provoked or annoyed her. On one early occasion, Fatfoot threw a tree root ball at me after firing the flash in her direction. There were other examples. We simply called this behaviour 'payback'.

Tuesday, September 28
Irris Makler from Australian Broadcasting Corporation *7:30 Report* rang.[86]

Thursday, September 30
Heard thumps to NE of house. Bess and Toby barking. Went outside, Bess ran up driveway and into Megan and Anthonys' block and up into Wilhelmina's block as if shadowing something. Heard and saw nothing.

Friday, October 1
Heard thumps to east of house at 7:00 p.m. to 8:30 p.m., several times. Cheryl rang at 8:30 p.m.—heard thumps to NW of their house, other noises, plastic pot falling from verandah. Ian, Cheryl and Debbie[87] went outside, heard further noises, turned on outside light, heard growling near tallowwood tree on our block and noise of running away to the NW, towards fence near sewer main. Dogs barking afterwards in valley. I investigated—heard and saw nothing.

Sunday, October 3
Ended Log here.[88] P.S.—The activity continued in a variety of forms until February 2002 when we believed that Fatfoot most probably died, after about two decades of

---

[86] Irris Makler telephoned regarding possible interview for the ABC. I didn't want to be interviewed.

[87] Debbie was a relative of Ian. She lived on a farm in central New South Wales. She had had similar experiences with strange activity on the farm. The work dogs would sometimes show extreme fear at night and hide under the house.

[88] It was becoming very tiring trying to record every event. Apart from this, the police had completed their stakeout and left the scene.

4: Log

interaction. However, Fatfoot would intermittently disappear for extended periods over the following decades. This was likely for a variety of reasons, including the loss of vegetation cover due to bushfires and turf wars with a 'mischief' of Junjudee. There was one other special entry.

Friday, November 26

Talked to Russell's[89] father, who told me of Roy who saw a Yowie when cutting wood with his chainsaw in the 1950s. There was heavy mist at the time and Roy saw a shadowy figure in the bush. It remained in that position for several minutes before moving away very quietly. Roy was so terrified of the event that he abandoned the chainsaw and returned home. He was very shaken up and remained so for several days. Roy finally went and returned the chainsaw about a week later. Roy died of lung cancer a week earlier, on Friday, 19 November 1993. Very unfortunate that I didn't get to speak to Roy!

---

[89] Russell was a drinking mate at the club. We regularly spoke about the research and Russell would pass on encounter contacts that he came across.

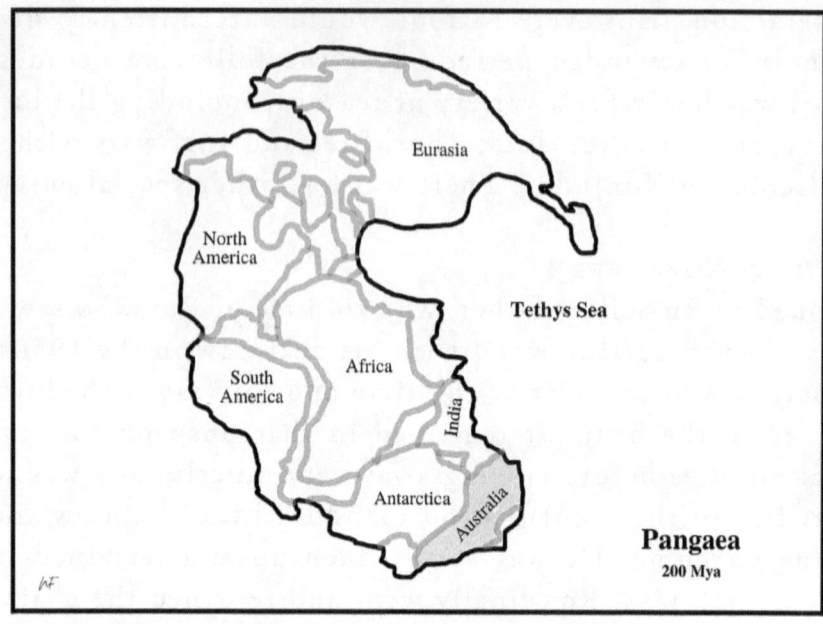

Fig. 5.1 Supercontinent Pangaea, 200 mya (Map: Neil Frost)

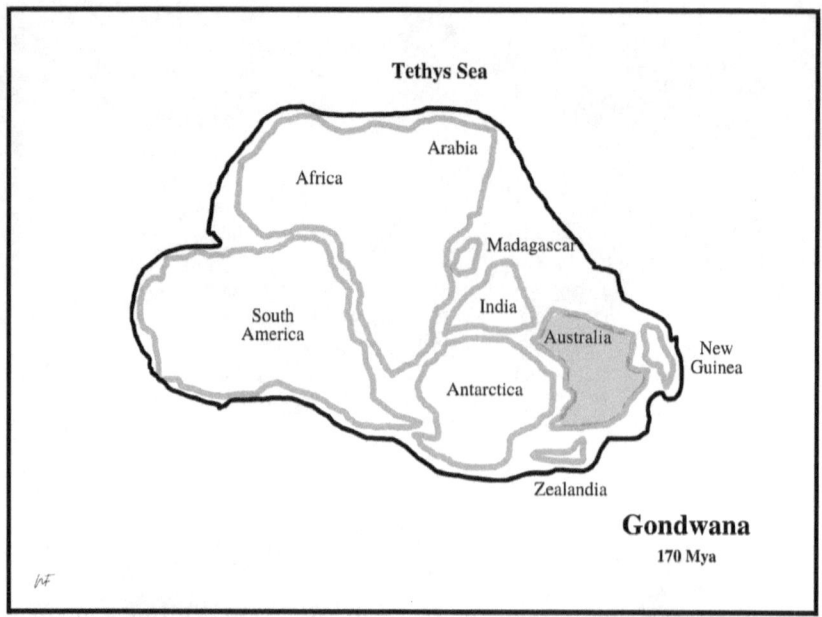

Fig. 5.2 Breakaway supercontinent Gondwana, 170 mya (Map: Neil Frost)

5: Gondwana

THE GEOLOGY OF early Australia played a very important role in determining the evolutionary path taken by the flora and fauna of the newly developing southern continent. At certain times the movement of the Earth's crust provided open corridors between the future continents for plants and animals to freely move between and on other occasions it acted more like closed gateways, particularly those surrounding the emerging Australian region. Changing sea levels also played a similar role. The plate tectonics of southern Gondwana also affected the climate of these future continents, which further influenced the evolution of life in this region.

The early supercontinent of Pangaea was a massive land mass that existed before the Jurassic Period, more than 200 mya. Being a very large and solitary landmass meant that many plants and animals were widely distributed across it, for example, cycads and dinosaurs. When Pangaea rifted apart, it split into two lesser supercontinents. Laurasia was the northern half and Gondwana was mostly the southern part. Following this geological breakup, life on the two supercontinents began to segregate and go their separate ways, although there was still some interchange between them. With the rifting apart of the land masses and the associated volcanic activity, the changing distribution of land and sea also brought about a transformation in regional and global climate, as well as other effects, such as altering the atmospheric composition of the planet, that life needed to adjust to.

Gondwana was the southern supercontinent that was composed of several existing and soon-to-become smaller tectonic plates. These new plates were made up of land mass fragments plus some existing and newly formed seas and oceans that emerged in the gaps between the divergent boundaries. Although now widely distributed across the Earth's surface, Gondwana's earlier existence can be reassembled from current continental and modern country names, like pieces in a giant jigsaw puzzle. Gondwana previously consisted of Antarctica; South America; Africa; Arabia; Madagascar; Australia; New Guinea; New Zealand; and India.[1]

Gondwana began to break apart about 180 to 170 mya with the combined drift of Africa and South America. This was followed by the separation of Africa from South America, but with the latter still maintaining a land connection with Gondwana, through Antarctica. Madagascar and India broke away to the north. Between 130 and 80 mya, New Zealand separated from Antarctica and Australia. The global mass extinction event brought about by the Chicxulub and Nadir asteroid impacts, occurred 65 mya[2] and dramatically ended the potential evolutionary advance of

---

[1] Some of these countries share a common tectonic plate, for example, Australia and New Guinea (Papua New Guinea and Irian Jaya).

[2] Three herbivorous and two carnivorous species of dinosaurs have been found on the Chatham Islands of New Zealand, suggesting that dinosaurs lasted for a million years after the Chicxulub and Nadir asteroid impacts. Stilwell, J., et al. 2006. Dinosaur sanctuary on the Chatham Islands, Southwest Pacific: First record of theropods from the K–T boundary Takatika Grit. *Palaeogeography, Palaeoclimatology, Palaeoecology* 230(3-4): 243-250.

5: Gondwana

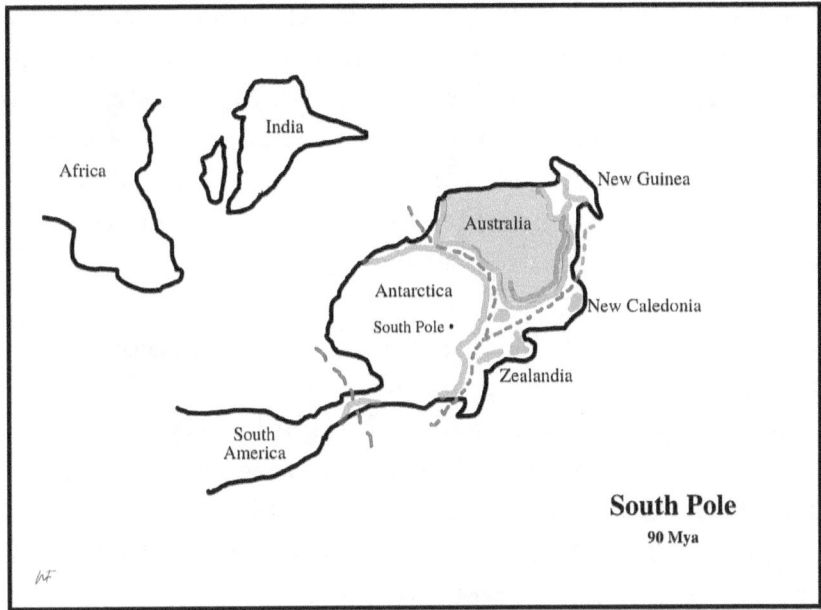

Fig. 5.3 South Pole, 90 mya (Map: Neil Frost)

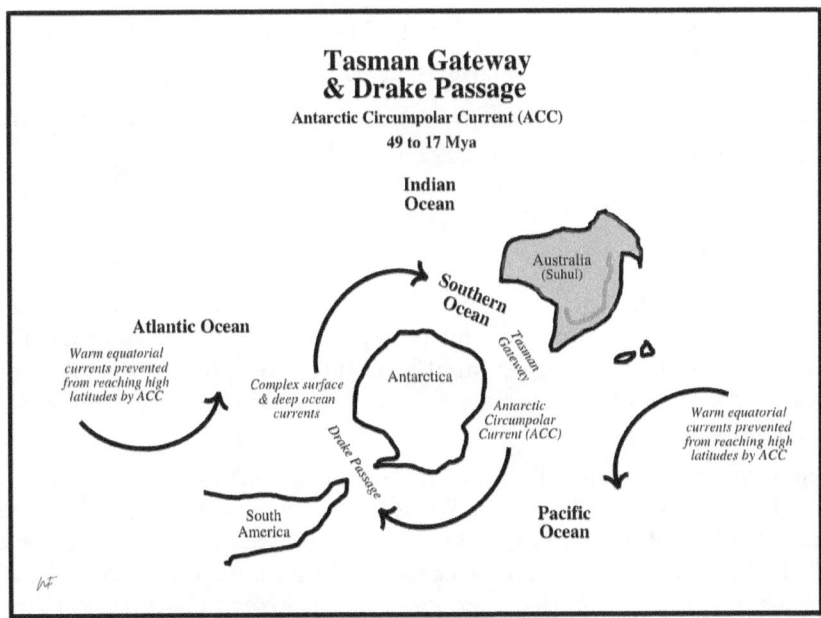

Fig. 5.4 Development of the Antarctic Circumpolar Current (Map: Neil Frost)

intelligent bipedal reptiles.[3] This planetary reset allowed opportunities for mammals to catch up to and eventually supersede the evolutionary head start previously held by reptiles. Australia was the last continental fragment to break away from Antarctica,[4] starting slowly at about 85 mya and becoming completed more rapidly by about 30 mya.[5]

BEFORE THEIR FINAL geological separation about 30 mya, Australia and Antarctica shared similar biota and consequently, their fossil records, though largely incomplete and unexplored, contain common plants and animals, including cycads, ferns, podocarps, amphibians, dinosaurs and marsupials. Despite its high polar latitude, the climate of Antarctica was wet and warm due to many factors, including a powerful global greenhouse effect caused by volcanism and because warm ocean currents were being channelled, without intervening terrestrial obstructions, from the equator to the pole. However, after separation from Australia and South America, the climate of Antarctica became drier and colder because of the developing Antarctic Circumpolar Current, which was brought about

---

[3] Descended from reptiles, many birds are highly intelligent, with advanced problem solving and many other abilities. Having evolved in Australia, parrots are highly intelligent and have spread throughout the world. An Australian example is the sulphur-crested cockatoo and a New Zealand example is the kea.

[4] The separation of Australia from Antarctica was most likely expedited by a very large asteroid impact at Wilkes Land in East Antarctica approximately 250 million years ago. The asteroid created a crater much larger than the Chicxulub impact. It caused a thinning of the crust and produced a mantle plume that eventually rifted the two continents apart.

[5] https://www.antarctica.gov.au/about-antarctica/geography-and-geology/geology/

5: Gondwana

Fig. 5.5 Remnant Gondwana Rainforests of Australia (Map: Neil Frost)

by the opening of the Tasman Gateway[6] and the Drake Passage, falling atmospheric carbon dioxide levels, and other factors.

Before the final Gondwanan breakup, forests of pine and southern beech were common on both continents. After the formation of a polar ice cap, the temperate Gondwanan rainforests in Antarctica were buried and compressed under ice, to become coal deposits. In Australia, the Gondwanan Rainforests continue to survive in recent times, with substantial pockets still in existence along the elevated eastern Australian coastline and collectively known as the 'Gondwana Rainforests of Australia'.[7] The oldest rainforest in the world is the Daintree Rainforest in tropical, Far North Queensland, with a low latitude of 16 degrees south and an age of 180 myr.[8] As a result of its long biological isolation, it maintains the most ancient and diverse biota in the world and preserves the major stages of the Earth's evolutionary history. The oldest and

---

[6] The Tasman Gateway is the oceanic rift that opened between Tasmania and Antarctica as the plates moved apart during the final breakup of Gondwana. Together with the opening of the Drake Passage at the tip of South America, they allowed for the establishment of the Antarctic Circumpolar Current that cut off and insulated Antarctica from the warm northern oceanic currents. This resulted in the development of the polar ice cap and lower temperatures at higher latitudes.

[7] Gondwana rainforest originally covered the entire Australian continent after separation until the climate changed and became drier. Remnant pockets of the Gondwana rainforest still exist along the east coast of Australia, surrounded by fire-dependent eucalypt forest, on what is generally known as the Great Diving Range.

[8] The Daintree began its Gondwanan existence at a high polar latitude and during its 180 myr existence, has been transported north to the low latitude of 16 degrees by the Indo-Australian Tectonic Plate.

(then) rarest tree in the world was the Wollemi pine[9] at 200 mya, which dates to around the time that Gondwana and Laurasia were rifting from Pangaea. Not surprisingly, this ancient Gondwanan pine can survive temperatures that range from -5° C to 45° C and can tolerate low-light environments.[10]

Rainforests around the world are renowned for their biological experimentation resulting in their extremely rich biodiversity. They are home to the majority of the Earth's plant and animal species, with every available niche typically being occupied by more than one competitive species. Trees are an essential part of these dynamic ecologies which provide other challenging habitats for many species of plants and animals.

Meanwhile, on the African continent, the early placental primates evolved in the rainforest trees at about 55 mya and adaptively radiated. It is believed that the East African Rift System and its complex arboreal environment helped to develop many early human traits in these animals, such as colour vision and manual dexterity, leading to the development of a larger, problem-solving brain. With increasing aridification along the rift valley, survival in the trees gave way to a terrestrial lifestyle on the savanna floor and the development of bipedalism, freeing the hands for other purposes and further developing the brain through eye-hand coordination. Eventually, Man evolved during the last 8 myr or so. It would seem very reasonable,

---

[9] Several small groves of Wollemi Pine *(Wollemia nobilis)* were discovered in a deep, wet and dark rainforest pocket about 100 kilometres northwest of Sydney in 1994. As Ian Price said at the time of the discovery, "It's no big deal. The most that the tree can do to avoid detection is to fall over! The Hairy Man can run away and hide!"

[10] http://www.wollemipine.com/aboutwp.php

therefore, to assume that a similar and convergent evolutionary process involving the quarantined tree-dwelling marsupials may have occurred in the Gondwanan Rainforests of this Australian 'Lost World', during a similarly long period of opportunity and experimentation. Like the East African Rift System, many marsupial animals on the Australian continent adapted to the available niches in the treetops. (For example, tree-kangaroos are arboreal macropods.) Similarly, some moved to the ground as the increasing aridification of Australia, brought about by the formation of the Antarctic Circumpolar Current and climate change, caused the rainforests to be replaced with more open, dry sclerophyllous tree communities. As Australia's 'marsupial hominoids' still maintain a connection with the trees, it would seem that this transitional phase is incomplete.

ALSO IN EXISTENCE when dinosaurs were present on Gondwana were early mammals. Various genetic studies suggest that the origins of mammalian lifeforms were derived from reptiles and birds. Perhaps the best example of this early reptilian-mammalian cross is the Platypus *(Ornithorhynchus anatinus)* which is an egg-laying, milk-feeding mammal or monotreme, that diverged from the common lineage about 166 mya. Platypus appear to have evolved in South America, which was then part of Gondwana, as was Australia. Similarly, the four species of echidna (Tachyglossidae), the only other known monotremes, also share these basic traits, in addition to having shoulders that are closely related in form to those of reptiles. Consequently, echidnas move in a waddling manner, similar to reptiles.

Marsupials seem to have evolved after monotremes but before placentals. They are thought to have first appeared in North America, in what was then Laurasia about 125 mya and probably migrated by a land bridge through to

Fig. 5.6 Short-beaked echidna, a monotreme. (Photo: Patrick Cavanagh, CC BY 2.0, https://creativecommons.org/licenses/by/2.0/)

Fig. 5.7 Marsupial thylacines in a zoo. (Photo: *Smithsonian Institution Annual Report,* 1903)

South America and similarly on to Antarctica and Australia. Placentals seem to have evolved later, with fossil records suggesting that they appeared around the time of the Chicxulub and Nadir asteroid impacts, however, genetic evidence suggests that they were present much earlier, at about 100 mya. Based simply on physical evidence, monotremes and marsupials were present in Australia at the time of separation from Antarctica, whereas placentals continued to dominate the rest of the world.[11]

With monotreme, marsupial and placental mammals all sharing an unknown common ancestor, their essential difference is how each subclass reproduces. Monotremes have a cloaca[12] or 'single opening' from which their genetic name is derived. They lay eggs like reptiles and birds and feed their young secreted milk. Echidna share some skeletal similarities with reptiles, which can be seen in how they walk. Marsupials give birth to live, underdeveloped young that are raised attached to a milk teat in a pouch or marsupium, the reproductive feature from which the class is named. Placentals internally nourish young using the mother's blood supply through a womb using a specialised, disposable embryonic organ known as a placenta, the reproductive structure from which the class is named. After giving birth to mature young, the placental offspring are fed milk from mammary glands.

From the time of final separation, being in geographical and biological isolation meant that the Australian mammalian progenitors, monotreme and marsupial, could potentially expand to fill every biological niche on the new continent, with marsupials being more successful and

---

[11] Some marsupials were present in North and South America.

[12] Like birds, monotremes have a cloaca (Latin for 'sewer') which is a common outlet into which the intestinal, urinary, and genital tracts vent.

5: Gondwana   143

Fig. 5.8 Lumholtz's tree-kangaroo. (Photo: © Frank Fichtmueller)

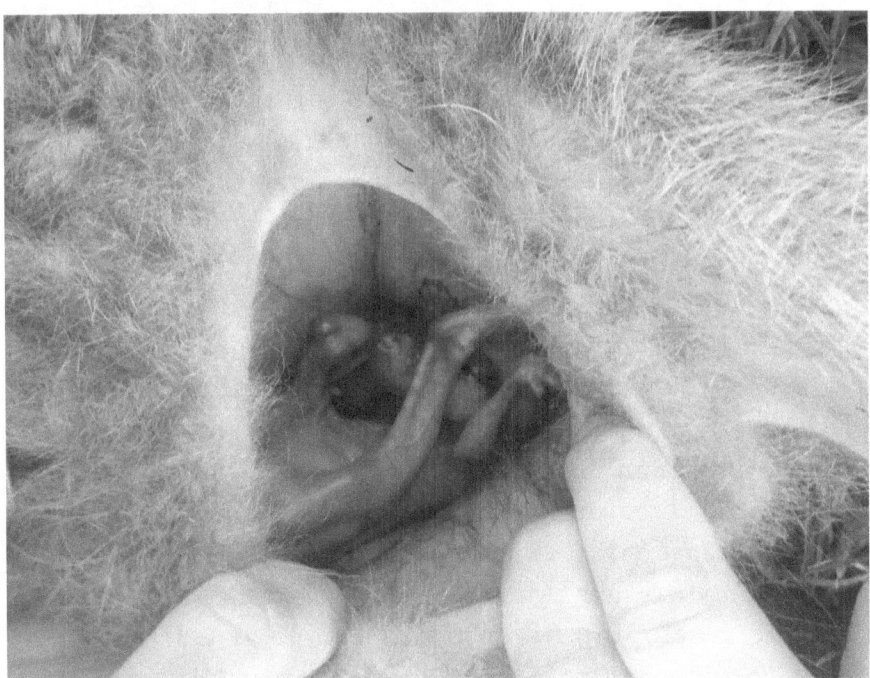

Fig. 5.9 Joey in the marsupium attached to the teat of a swamp wallaby. (Photo: Neil Frost)

a few monotremes managing to survive to the present. Similarly, on the remainder of the planet, placental mammals had their independent opportunity to do the same. This is like having two experimental worlds, each contained within its isolated bubble and subject to its unique developmental conditions.

IN ADDITION TO its geological and biological attributes, Gondwana was affected significantly more than most other continents because of its high latitude and the obliquity of the planet's orbit. At high latitudes, the incoming solar radiation is attenuated by its longer and denser path through the atmosphere and the surface albedo. The tilt of the Earth's axis is responsible for producing four seasons at medium latitudes, but at these polar extremities, it results in either prolonged periods of light or total darkness, similar to the contrasting wet and dry seasons that are experienced at the other extreme regions near to the Equator. In combination, these two influences should produce environmental extremes of continuously sunlit and cool summers or continuously dark and freezing winters. However, high atmospheric levels of greenhouse gases and heat released through increased volcanism associated with tectonic plate movement warmed the planet generally and resisted the accumulation of reflective ice sheets. Also, warming ocean currents from lower latitudes circulating offshore at that time would have further moderated the local Gondwanan climate, making conditions mostly temperate all year around and much more sustainable than could be reasonably expected for this latitude. After the separation of South America and Australia from Antarctica, the developing Antarctic Circumpolar Current had a major cooling and drying influence on the climate.

The Earth's North Pole would have experienced these same effects, however, Gondwana was advantageously different.

5: Gondwana

In contrast, Gondwana had a concentration of large land masses surrounding its true geographic South Pole, which would heat more quickly than the surrounding ocean. Having a large proportion of land instead of water enabled terrestrial animals to exist and uniquely adapt to unusual conditions. Also, the climate was much better. At the North Pole, there were no large permanent landmasses and the cold northern oceanic currents did not produce a warm moderated climate. Consequently, the extreme southern hemisphere encouraged the adoption of several very unique life adaptations, with broad consequences shared across a wide spectrum of antipodean biota.

Having maintained high southern polar latitudes during its long existence, Gondwana generated a set of uniquely difficult environmental circumstances for plants and animals to overcome and adapt to. As the Earth is tilted on its axis between 22.1° and 24.5° relative to its orbital plane or obliquity as it variously oscillates under the influence of lunar gravity as well as solar system and galactic influences, the length of the day and night within the variable antarctic circle[13] would approach its maximum 23-hour duration,[14] around the time of the dinosaur extinction. For plants, the main issue would be photosynthesis, either overproduction during the long polar summer day or underproduction during the long polar winter night. *Nothofagus gunnii* or the Australian Beech is Australia's only cold-climate, winter-deciduous tree and amongst

---

[13] Due to the variable obliquity of the planet, the antarctic circle latitude changes from 65.5° south through to 67.9° south and back again during a 40,000 year cycle.

[14] The length of the Earth's day was shorter in the past due to the planet's higher rotational velocity. The tidal gravitational influence of the Moon and Sun has slowed rotation and lengthened the day to the current 24 hours.

the oldest species of flowering plants. It is endemic to the Tasmanian highlands, having evolved in the Antarctic Gondwanan rainforests. For animals, the decision would be whether to stay or go and the consequent need to hibernate,[15] migrate or develop some other adaptive response, such as torpor, which is influenced by the ambient temperature and food prevalence.

The remnant Gondwana landmass was home to many animal types. Sometime before their global extinction, southern dinosaurs had adapted to the unique antipodean circumstances of Gondwana and it seems reasonable to assume that other animals did the same. For example, as shown by fossil evidence obtained from Dinosaur Cove in Victoria and elsewhere (Rich and Vickers-Rich, *Dinosaurs of Darkness*, 2020: 72), large-eyed southern dinosaurs like *Leaellynasaura amicagraphica* had larger optic lobes than those living closer to the equator. Having larger eyes and a brain adapted to processing low light levels would be a mandatory necessity for any nocturnal animal, whether hunter or prey, under these polar conditions. The alternatives to hunting or foraging through the dark winter would have to be northerly migration or hibernation.

LIKE AUSTRALIA'S TASMAN Gateway,[16] the northerly advance of South America away from Antarctica about 25 mya resulted in the opening of the Drake Passage, which removed the final obstruction for the completion of the Antarctic Circumpolar Current that would cool the remaining

---

[15] The mountain pygmy-possum is an Australian marsupial that engages in short periods of torpor. This is similar to hibernation but the sleep is not as heavy and is involuntary.

[16] The submarine rift between Tasmania and Antarctica was produced by the northern tectonic movement of the Indo-Australian plate and is thought to have been caused by an asteroid impact at Wilkes Land about 250 mya.

5: Gondwana

polar continent. This led to Antarctic glaciation, accelerated by increased surface albedo. As water vapour was being turned into ice, global sea levels fell which heavily contributed towards widespread aridification.[17]

Meanwhile, Australia and its surrounding seas, oceans and islands, increased their northerly advance towards the equator on their own Indo-Australian Tectonic Plate. For Australia, unlike elsewhere, falls in the sea level caused by glaciation did not give rise to land bridges outside of Sahul, that would allow for any terrestrial migratory exchange with neighbouring continents, because of the very deep marine subduction zones to the north and vast areas of hostile and uninterrupted oceanic water to the south. However, during these glacial periods, the land area of Australia did expand along its shallow continental shelf, linking Tasmania, New Guinea and numerous islands with the mainland. Also around 25 mya, Arabia began its separation from Africa which resulted in the formation of the East African Rift, a favourable geological region that supported the preservation of fossils, including the evolutionary history of early hominins.

Around 8 mya, the genus *Homo* evolved in Africa. During various glacial periods and tectonic interactions, from then until the present, many land bridges would facilitate terrestrial animal movements around the globe, including recent human migrations,[18] but excluding any

---

[17] This marked the beginning of the end of Antartica's and Australia's Gondwanan rainforests. It also brought about the drying of the Australian continent and the transition towards grassland.

[18] The Bering Strait land bridge (Beringia) has variously allowed the migration of animals and humans between Asia and North America in response to glacial fluctuations and changing sea levels. Such connections would have allowed the diaspora of primates from Asia, however, the majority of early primate fossil evidence is found in North America and Europe. Unlike Australia's hominoids, the origins of Bigfoot may be more complicated.

physical connections between Australia and the outside world. For animals and humans, migration to Australia could not simply depend upon a parting of the waves. Consequently, the antipodean continent would remain quarantined, except for any animal that could fly across the oceanic barrier, such as birds and bats.

Early forest and savannah elephants evolved in Africa, from where they were able to distribute themselves across Europe, Asia and North America via land bridges. As they did, these elephants interbred with other species and adapted to their unfamiliar environments, resulting in new forms, like the Sicilian dwarf elephant, mammoths and the Asian elephant. This was possible because many of these continents were physically connected at one time or another or had temporary land bridges formed between them as a result of frequent glacial events and the associated falling sea levels.

Elephants were never able to reach Australia,[19] although *Stegodon* were able to colonise as far as the Indonesian island of Flores, as did *Homo floresiensis*.[20] These elephants and hominins were both affected by insular dwarfism.[21] The island of Flores is located about 500 km north of the farthest-reaching and currently submerged part of Sahul,[22] or the Australian Continental Shelf, with the intermediate island of Timor-Leste maintaining a gap of 200 km. Flores lies above the tectonic subduction zone where the Indo-Australian plate is disappearing beneath the Eurasian plate at the rapid rate of about 70 mm per year. South of

---

[19] "Several mammalian taxa described as fossils from Australia are even more problematical. Undoubted elephants were described by Richard Owen as *Mastodon australis* (Owen 1844) and *Notelephas australis* (Owen 1882). While it is possible (perhaps even probable) that these fossils were not collected in Australia, collection data indicate otherwise which leaves these records as provocative mysteries." M. Archer et al. 1999. The evolutionary history and diversity of Australian mammals. *Australian Mammalogy* 21: 1-45. P. 15.

5: Gondwana

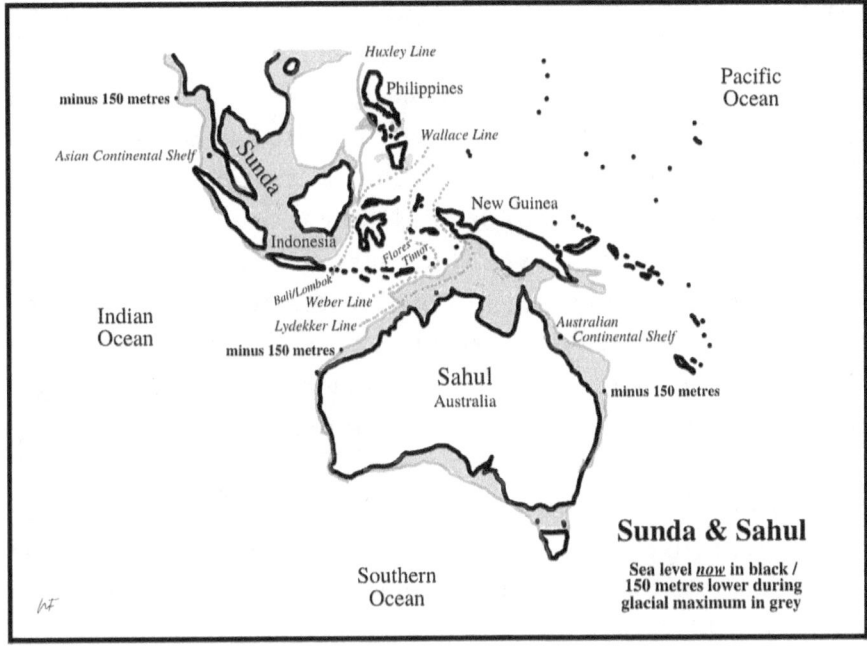

Fig. 5.10 Sea level 21,500 years ago when the height was 150 metres lower than today, during the last glacial maximum. (Map: Neil Frost)

[20] Having corresponded with Dr. Mike Morwood whilst at the UNE about the possibility that *Homo floresiensis* may have reached Australia and were responsible for Junjudee encounters, he thought that it was "possible but, unlikely". Since then it has become obvious the Junjudee are not hominids but are indigenous animals.

[21] Insular (island) dwarfism is a process where large animals are reduced in size because of a reduction in the normal availability of food resources, or their quality, due to the limited size of the physical environment, as for example, was the situation on the small island of Flores and also on Sicily. *Homo floresiensis* are believed to be the dwarf form of *Homo erectus*.

[22] The continental shelf and mainland Australia together with many islands, including New Guinea and Tasmania.

Sunda[23] and Flores, towards Sahul or the Australian continental shelf, is the Sunda Trench, a more than 7000 metres deep, oceanic linear depression. At no time in the recent or distant past has a glacial event resulted in a lowering of the sea level to such an extent that a land bridge between Australia and the Indonesian islands could form across such a bottomless chasm. Furthermore, as Australia has been continually moving northward towards Asia from the open waters of the Southern Ocean for the past 30 myr or more, past opportunities for the formation of land bridges would have become increasingly nonexistent.

FROM THE MOMENT that John Gould and other early European naturalists viewed samples of Australian biota, the uniqueness of these specimens compared to the rest of the world was blatantly apparent. Their uniqueness was so absolute that they beggar belief.[24] With this realisation came the search to identify the very formidable biogeographic barriers that must be capable of constraining any external genetic contamination. From the beginning, the northern and northwestern coastal border regions, between the continental shelves of Asia and Australia, were extensively studied to determine how this biological blockade functioned and the extent of the barrier. The remaining coastlines to the east, south and west, with their endless vistas were obviously regarded as indomitable.

The various biogeographic boundaries between Asian and Australian biota shown on the map Fig. 5.10, have been well studied, variously determined using differing criteria, mapped and debated. The most famous of these

---

[23] The continental shelf of Southeast Asia.

[24] The platypus is a common example of what was thought to be a hoax by eighteenth-century scientists.

5: Gondwana

is the Wallace Line (coined and also modified by Thomas Huxley in 1868) which follows along the edge of the Sunda continental shelf and marks the easterly limit of purely Asian plants and birds. Further to the east of this line begins the mixed appearance of Australasian plants and birds. The Weber Line (named by Paul Pelseneer in 1904) is a mammal and mollusc study that follows a section of the Sunda Trench along part of the Sahul continental shelf, where the combined faunal populations are equally Asian and Australian. It was referred to as a line of 'faunal balance'. The area between the Wallace and Weber Lines was considered to be a transitional faunal region and was named Wallacea by Dickerson (1928). The Lydekker Line (1895) is a boundary that follows the outer profile of the Sahul continental shelf, including the shallow Timor and Arafura Seas that connect Australia with New Guinea and other islands. The Lydekker Line defines the western limit of Australian fauna exclusively. It follows the outer parts of the Sahul shelf which is the underlying geological structure that is above water during falling sea levels, forming the only possible land bridges between the Australian mainland and the associated islands of New Guinea and Tasmania, but not beyond.

In recent times Australia's isolation has not been absolute. For a considerable period, birds and bats have had unrestricted aerial access to Australia. Similarly, insects and plant seeds have been carried by the wind and other means. However, some larger placental animals have reached Australia, including rats, cats, bats and dogs, that may have arrived as foreign immigrants onboard early trading or fishing vessels,[25] or possibly, rafting here from Asia

---

[25] From the beginning of the eighteenth century, fishermen from Makassar Sulawesi sailed to Arnhem Land each year to obtain trepang (sea cucumber) from Aborigines in exchange for metal blades, cloth and tobacco.

after surviving tsunamis generated by local tectonic movements. The success of such chance voyages in maintaining and expanding a viable population upon arrival in Australia would be highly dependent upon multiple arrivals of a significant number of individuals over a short period.

The imagined existence of connecting land bridges on Australia's northwesterly coastline and the multiple rafting voyages of marooned jungle apes propelled by Asian tsunamis are typically used to explain the sustainable arrival of cryptids to the tropical savanna of northern Australia. No evidence exists to support the migratory movements of primates along contiguous land connections from Asia to Australia. Similarly, the more complex possibility of staged multiple journeys involving island hopping during periods of low sea levels, though more appealing over lengthy periods, is still thwarted on the last leg of the journey by the formidable moat that is the Sunda Trench. Best estimates have suggested that this final oceanic barrier may have been at least ninety kilometres wide at the time of the last glacial maximum of 21,500 ya when sea levels were about 150 metres lower than today.[26]

"Anatomically Modern Humans (AMHs) dispersed rapidly through island southeast Asia (Sunda and Wallacea) and into Sahul (Australia, New Guinea and the Aru Islands), before 50,000 years ago. Multiple routes have been proposed for this dispersal and all involve at least one multi-day maritime voyage approaching 100 km."[27]

---

[26] https://en.wikipedia.org/wiki/Prehistory_of_Australia

[27] Michael I. Bird et al. 2018. Palaeogeography and voyage modeling indicates early human colonization of Australia was likely from Timor-Roti. *Quaternary Science Reviews* 191: 431-439.

5: Gondwana

MOST CRYPTID RESEARCHERS and others with a casual interest in the subject, tend to think that our Australian 'hominoids' are hominids, which are primates from the family Hominidae that include humans, their close ancestors and also the great apes, including gorilla, orangutan and chimpanzee.

A frequently mentioned cryptid candidate is *Gigantopithecus blacki*, a flat-nailed pongid closely related to orangutans that lived in the rainforests of southern China during the Pleistocene, that would not have been capable of surviving, let alone adapting to, the Australian environment on arrival. Like elephants, how this very large placental mammal could manage to make the long journey to Australia without the aid of a terrestrial corridor is extremely problematic. Additionally, this very large herbivorous primate was undoubtedly incapable of planning and taking discretionary voyages overseas. Regardless, if this pongid somehow managed to survive an oceanic journey to Australia by good fortune, its joy would quickly turn to disappointment on arrival in Northern Australia because of a lack of rainforest and no partners to sustainably procreate with.

Other proposed cryptid candidates include hominins or archaic species of *Homo* that were ancestral to *Homo sapiens* and are thought to be extinct. These supposedly remnant populations of early humans are typically identified as *Homo erectus, Homo neanderthalensis* and others. Even though most of these early humans were highly sophisticated socially, used tools, had language, culture and possessed other valuable attributes, they may not have been fully capable of making discretionary oceanic voyages from Asia to Australia using manufactured seacraft, as *Homo sapiens* were able to repeatedly achieve during successive migratory waves over the past 65,000 years. Additionally, unlike *Homo sapiens*, there is no archaeological evidence to suggest that these earlier hominins reached Australia.

From an Aboriginal perspective, it has been traditionally said that these 'hominoids' were "here before us and are not the same as us". I agree with this statement because the supporting evidence is conclusive. Of course, many people would dispute the veracity of such an assertion with 65,000 years having passed between then and now. However, there are similar stories that date back many thousands of years, which are culturally and historically accurate accounts, probably because the event and their details were so highly significant that they were worth being retained as a mnemonic story. For example, other tribal memories also exist involving very distant but verifiable events. Some descendants of early settler marriages with local Tasmanian Aboriginal women recounted stories of being able to walk across the Bass Strait from Tasmania to Victoria on mainland Australia. Before scientific investigation, these stories were regarded as 'quaint.' This island separation occurred about twelve thousand years ago as a result of global warming. Unfortunately, as many of us are too well aware, the veracity of oral tradition and testimony are frequently regarded as fanciful.

Although Aboriginal knowledge and history were not recorded in a classical written manner, as with the twelve-thousand-year-old record of rising sea levels, memorised oral narratives were very effectively used to pass on and even update the information to successive generations. These were not the only methods employed and included other practices such as rock art, painting, story-telling and dance. All of these cultural elements were used to store information while keeping the knowledge active and alive in the community.

Many Aboriginal tribes made significant contributions to astronomy that in some regards, predate Western findings.

5: Gondwana

Except for supernovae or stars that violently explode at the end of their stellar lifetime, stars were believed to be fixed in brightness by classical astronomers. It wasn't until the late sixteenth century in Europe that variable stars were first detected and measured. In Aboriginal culture, astronomy was used for navigation, to determine the calendar, seasons, animal migration and a variety of other requirements but as with many other underrated societies, its application has not been fully acknowledged. In the article 'Observations of red-giant variable stars by Aboriginal Australians,' two Aboriginal oral traditions from South Australia were studied which "show that these traditions describe the variability of the pulsating red-giant stars Betelgeuse (Alpha Orionis), Aldebaran (Alpha Tauri), and Antares (Alpha Scorpii), and allude to the relative periodicities in the stars' variability."[28]

Apart from these and other examples of oral history, I am certain that Aboriginal people would have been reminded, almost daily, of the eccentric, non-human nature of these hominoids. If Australia's 'hominoids' were a remnant population of early humans or primates, then the empirical evidence would support it. For example, a defining characteristic of the Great Apes is having flat, pink nails, whereas eyewitness accounts report the presence of black claws. By itself, this should be sufficient proof of the alternative origins of these 'hominoids' but there is much more. Australian 'hominoids' are known to have reflective red eyes. Humans and other Great Apes do not have a tapetum lucidum, a reflective layer behind the retina that assists with night vision and appears at nighttime as a

---

[28] Hamacher, D. W. 2018. Observations of red-giant variable stars by Aboriginal Australians. *The Australian Journal of Anthropology* 29: 89-107.

reflective red-eye disc, similar to that associated with cats and owls. Many more areas of biological difference will be covered during the course of this book.

WITH CURRENT DATING suggesting that Australia's First People[29] arrived in Australia 65,000 years ago[30] and with many, but not all, of the marsupial megafauna disappearing from the environment around 45,000 years ago, this begs the question, "What do the Aborigines know about these 'marsupial hominoid' megafaunas?" The simple answer is, "quite a lot."

A major reason why Australian cryptozoology is crippled by misinformation and many misleading concepts is that we have tended to appropriate all research and theories from North American experience and applied them directly to our situation. We have tended to be blinded by the conclusion that is based upon the syllogism:

> All hominoids are primates
> Dooligahl are hominoids
> Therefore, Dooligahl are primates

The conclusion is only as valid as the propositions. A significant part of the problem is that these Australian

---

[29] 'First People' was an earlier and more accurate term used to honour the original Aboriginal inhabitants of Australia, however, as with Bigfoot research, it seems that we have adopted the North American term 'First Nation', which more legitimately applies to the Algonquin, Iro-quois, Huron, Wampanoag, Mohican, Mohegan, Ojibwa, Ho-chunk, Sauk, Fox, and Illinois people. Ethnographic names are constantly being reviewed in response to shifting cultural and political perceptions and circumstances.

[30] Clarkson, C., et al. 2017. Human occupation of northern Australia by 65,000 years ago. *Nature* 547: 306-310.

mystery animals have common morphologies and consequently, similar behaviours, to primates. They look and behave alike but they aren't the same beastie. Just because both bipeds appear to share common biological traits, does not mean that they are biologically related and capable of interbreeding. True placental hominoids and 'marsupial hominoids' are very distantly related, with their similar traits and behaviours best explained by convergence.

Fig. 5.11 Marsupial thylacine in captivity, demonstrating a wide gape. These Tasmanian 'wolves' exhibit convergence with placental canids.

6: Convergence

JOHN GOULD WAS a nineteenth-century English ornithologist who noted how many finches from the Galápagos Islands had developed beaks with specialised characteristics that affected their form and function. These physical differences were a response to changing environmental factors which allowed the finches to increase and broaden their limited sources of food to include new and unexploited seed types. Guided by Gould's research, Charles Darwin later accepted that the graded series of beak shapes and sizes were progressive modifications that originally came from finches from the South American mainland. This collective work would become part of the theory of natural selection.

Finches can respond and take advantage of beneficial changes in the environment and so can other lifeforms. When environmental change provides new opportunities or produces empty biological niches, organisms attempt to advantageously adapt to fill these vacancies. Often, new environmental niches are separated from each other by distance and/or time. Consequently, the isolated organisms are usually not closely related biologically. Regardless, independent solutions to the problem are found as each organism evolves similar traits as they all converge towards a similar target.

The process of convergence is common throughout evolutionary history. Despite the need to overcome considerable obstacles, similar solutions to environmental problems have been developed throughout biological antiquity. Photosynthesis seems to have independently emerged many times over the past four billion years. An often-quoted example is powered flight, which has independently evolved at least four times. Other less-acknowledged but related examples include bipedalism and high intelligence, which are associated with some early reptiles, anthropomorphic marsupials and hominids.

SUGAR GLIDERS, *PETAURUS BREVICEPS*, are commonly found throughout our valley, however, during our forty years of occupation, their population has not been stable.[1] When we started building the house, there were plenty of sugar gliders living in the hollow, burnt-out trees after the 1977 bushfires.

Sugar gliders are small, nocturnal, omnivorous, arboreal and airborne marsupials. They are very social animals. Like the majority of Australia's native animals, sugar gliders are marsupial, meaning that their offspring are born incompletely developed and are typically carried and suckled on an attached teat, in a marsupium or pouch in the middle of the mother's abdomen, until mature. Females are polyestrous, having two ovaries and two uteri, with males having a bifid penis, allowing up to two joeys per litter and several pregnancies per year.

---

[1] Similarly, the number of bandicoots, *Perameles nasuta*, were very common in our valley during the 1980s before their population crashed by the 1990s. The bandicoot numbers briefly recovered during the early 2000s, before totally disappearing shortly afterwards. We think that the main predators were cats, wild dogs, foxes, owls and 'marsupial hominoids'.

6: Convergence

Sugar gliders can glide, not fly, because they have a thin membrane that is stretched from their abdomen to between their fore and hind limbs. Although this sail acts like a wing, it does not provide powered flight. The power required to glide is obtained by climbing a tree and converting the potential energy into kinetic. Adjusting the tension and elevation of the membrane with their limbs allows sugar gliders, with some assistance from their tail, to steer, adjust height and brake. The maximum distance travelled with each flight depends mostly upon the density of obstructions in the forest and the spacing of suitably tall trees.[2]

Gliders are opportunistic feeders that will eat almost anything. During summer, cicadas are food in oversupply, as are members of the general insect population. Many flowers from a variety of plants are eaten, as well as nectar, buds and seeds. Consequently, the gliders perform a vital pollinating function. During winter many other plants flower which continues to provide nectar for these small marsupials. As an alternative, or if food is scarce, gliders will wound trees to tap their sugary sap and red resinous gum, or kino.[3] They will eat eggs and small birds, lizards, gastropods, fungi, lichen and fruit.

They are very social animals living together in colonies. During cold periods, gliders will group together to conserve heat and will go into a state of torpor, which is a

---

[2] When we observed our sugar gliders at night with an infrared camera, they tended to glide along the outer margins of cleared forest areas in order to travel longer distances, rather than taking a more direct route involving more hops.

[3] The commonly attacked trees are red bloodwoods, which are tapped on a regular basis to keep the wound flowing. It also appears that the gliders have opened shallow larval chambers belonging to many species of wood borers to extract the larvae.

Fig. 6.1 Sugar glider in flight. (Photo: © Anom Harya)

Fig. 6.2 Flying squirrel in flight. (Photo: © Anaredif)

6: Convergence

lowering of their metabolism with a consequent reduction in body heat output.

The genus *Petaurus* is believed to have evolved 18 to 24 mya in New Guinea and then spread to Australia where other species developed. As an evolutionary solution, gliding is considered to be an extremely energy-efficient method of rapid and widespread foraging that does not require excessive climbing and ground exposure. It also allows for the rapid evasion of predators.

THERE ARE MANY species of flying squirrel found across North and Central America, northern and southeast Asia and parts of northeastern Europe. However, the Northern Flying Squirrels (*Glaucomys sabrinus*) are commonly found throughout the northern part of North America.

Northern flying squirrels are small, nocturnal, omnivorous, arboreal and airborne placentals. They are very social animals. Like the other species found outside the New World and mammals generally throughout the rest of the planet, these northern flying squirrels are placental, meaning that their offspring develop with the aid of a disposable placenta, which facilitates the exchange of nutrients and wastes between the blood of the mother and the foetus, finally resulting in the birth of live, mature young. Females are polyestrous, having up to six kits and up to two litters per year.

Northern flying squirrels can glide, not fly, because they have a thin membrane that is stretched from their abdomen to between their fore and hind limbs. Although this sail acts like a wing, it does not provide powered flight. The power required to glide is obtained by climbing a tree and converting the potential energy into kinetic. By adjusting the tension and elevation of the membrane with their limbs allows flying squirrels, with some assistance from their tail, to steer, adjust height and brake.

The maximum distance travelled with each flight depends mostly upon the density of obstructions in the forest and the spacing of suitably tall trees.

Northern flying squirrels are opportunistic feeders that will eat almost anything. During summer, cicadas are food in oversupply, as are members of the general insect population. Many flowers from a variety of plants are eaten, as well as nectar, buds and seeds. Consequently, the squirrels perform a vital pollinating function. During winter other plants may flower which continues to provide some nectar for these small placentals. As an alternative, or if food is scarce, the squirrel will wound trees to tap their sugary sap and resinous gum. They will eat eggs and small birds, lizards, gastropods, fungi, lichen and fruit.

They are very social animals living together in a scurry. During cold periods, squirrels will group together to conserve heat, however, they will not go into a state of torpor, which is a lowering of their metabolism, to conserve heat.

The genus *Glaucomys* is believed to have evolved 18 to 20 mya in North America where other species developed. As an evolutionary solution, gliding is considered to be an extremely energy-efficient method of rapid and widespread foraging that does not require excessive climbing and ground exposure. It also allows for the rapid evasion of predators.

CONVERGENCE CAN BE simply defined as organisms that are not closely related biologically, independently evolving similar physical and behavioural traits, in response to similar environmental circumstances or niches. Convergence is very common in nature. With the above examples, the Australian sugar glider is a marsupial, whereas the North American flying squirrel is placental. Both are mammals, that mainly differ in how they reproduce. Both are very distantly related biologically, geographically and

6: Convergence

temporally, however, they share several fundamental physical and behavioural traits.

Taking another example, powered flight has independently evolved many times in different animals. There are at least four known animal groups that have attained powered flight—insects, pterosaurs, birds and bats. Over time there has been much conformity in the independent design of wing mechanisms in animals, as there are certain principles that are essential for a successful flight that can not be ignored. The key requirements are strength, weight and power.

AUSTRALIA IS AN excellent example of what was a biologically isolated continent, where the dominant pre-placental lifeforms each adaptively radiated in response to the largely vacant environmental niches, over an initial period after separation from Antarctica. Adaptive radiation is a process whereby organisms rapidly diversify from their original species, in response to new, challenging circumstances. However, from rainforest to desert, the wide-ranging environmental conditions experienced in monotreme and marsupial Australia were to eventually lead to the same or very similar evolutionary solutions that were being determined by the rest of the placental world. In other words, as with the sugar glider and the northern flying squirrel, the evolving lifeforms in each of the two worlds were mostly convergent.

The platypus is an overused example that illustrates the uniqueness of Australian animals. Nonetheless, these monotremes might serve one more purpose, by possibly demonstrating convergence, since males have a spur on each of their rear ankles that secretes poison which is similar to a snake. Although poisonous delivery systems are very rare in extant mammals, counterarguments suggest that these spurs may have been inherited instead, from unknown progenitors.

A better, less ambiguous example of convergence in Australian mammals, is the echidna. Continuing with this theme, echidnas are also monotremes. Like the platypus, they are egg-laying mammals with a close relationship to birds and additionally, with reptiles. Echidnas have spines, or modified hairs, which are used for the animal's defence. This is convergently similar to placental hedgehogs and porcupines from the northern hemisphere which occupy similar environmental niches. Additionally, echidnas have powerful fore claws and long, sticky tongues that are used to extract ants from their nests, as do the species of anteater and others.

The thylacine or Tasmanian tiger *(Thylacinus cynocephalus)* is a reclusive marsupial carnivore that once roamed Sahul but is now probably confined to a few, very remote wild areas of Tasmania, in critically endangered numbers. A sign from the thylacine exhibition in the Hobart Museum uses the term 'functionally extinct' to describe their prevalence, meaning that they probably still exist in the wild, but in such low numbers that the species is unsustainable.[4] As such, it remains the fourth largest extant marsupial carnivore in modern Australia, after Junjudee, Dooligahl and Quinkan.

The thylacine is a nocturnal predator with camouflage stripes across its rear back and sides, like a tiger, but morphologically, it looks more like a dog or wolf. By occupying the same environmental niche as dogs, thylacine have evolved analogous structures, particularly in relation to body, cranial and dental features. However, being marsupials they are not closely related to any of the above placentals, but still possess the characteristic pouch,[5] together with a rigid tail, like a kangaroo.

---

[4] An opinion also held by Mike Williams, thylacine expert and author of several books on the topic.

[5] Both sexes have a marsupium. In males it is used to protect the genitals.

6: Convergence

There are numerous examples of extant marsupial species that are convergent with their placental counterparts.[6] Marsupial moles and placental moles. Tasmanian devil and wolverine. Striped possum and aye-aye lemur. Wombat and groundhog. As well as many others.

SINCE THE TIME of the dinosaurs, many different types of large animals have appeared on every continent of the planet. With mammals, gigantism was common to all three groups, although monotremes did not become as large as the other two, with the largest monotreme known, the giant echidna *Zaglossus hacketti* growing to about one metre and only weighing around 30 kg. Large mammals above 45 kg are generally referred to as megafauna. With marsupials and placentals, herbivores generally grow much larger than carnivores. With most mammals, males tend to be larger than females (sexual dimorphism), because males usually need to fight with other males to win the right to mate.

Most placental megafauna are well known because many of them are still around today. Elephants are an example of large placental megafauna that typically weigh between 5 and 10 tonnes. Since they weigh more than 1000 kg, they can be more specifically referred to as megaherbivores. Similarly, placental predators like lions, which weigh in excess of 100 kg, can be more specifically referred to as megacarnivores.

Unlike the placental elephant or blue whale, most marsupial megafauna are lesser-known because very few of them are around today or, as with the case of the 'marsupial hominoids', they have become extremely reclusive. Since the arrival of humans on the Australian continent

---

[6] Koalas have fingerprints which assist them with gripping trees. Possums have opposable thumbs.

around 65,000 years ago, most megafauna were systematically hunted to extinction, whilst others failed to adapt quickly enough to the changing Australian environment, particularly from increasing aridification and land clearing brought about by the Aboriginal use of fire. Additionally, around 42,000 years ago there was a mass extinction of Australian megafauna that could be correlated with a localised geomagnetic event known as a Laschamp Excursion Event. This was a short period of a few hundred years when the Earth's magnetic field temporarily reversed and partially collapsed, which resulted in the localised weakening of the magnetic field in parts of Europe as well as Australia and New Zealand. The greatly reduced power of the magnetosphere above Australia, estimated at only 5% of today's strength, would have been almost like the harsh radiation conditions currently found on Mars. Without this magnetic shield, cosmic and solar radiation would have penetrated the atmosphere and damaged life in these regions. The atmosphere would have become ionised by the high-energy particles bombarding it which would have affected the climate and atmospheric circulation patterns in these regions. The local ozone layer would have become depleted by the intense radiation resulting in a large atmospheric hole, allowing increased ultraviolet radiation to reach the ground in Australia. Interestingly, the impact of this Laschamp Excursion Event would have been minimised for the 'marsupial hominoid' megafauna because of their predominantly nocturnal behaviour, as they would habitually shelter in the dark during the daylight. In doing this, they would be less exposed to high levels of ultraviolet and other high-energy particles from solar radiation. As a general consequence, caused by a combination of these negative factors, Australia has had a much higher proportion of megafaunal extinction compared to the rest of the world.

6: Convergence

Size matters in animals. With terrestrial herbivores, larger animals like elephants can reach higher to obtain food or simply push over trees that are beyond their grasp. Although they may eat more, megaherbivores can eat lower quality food and process it very efficiently, compared to smaller animals that must eat higher quality food that is a bigger percentage of their body mass, thereby requiring more foraging effort. Larger animals have a smaller surface area to volume ratio which affects how they deal with heat and cold, which generally means that they require less food to maintain their metabolism. Larger animals and their offspring tend to be safe from predators. As larger animals tend to be taller, their legs are longer leading to extended walking strides. As stride length and frequency are major determinants of walking and running efficiency, long legs and slow movement reduce the demand for food.

Perhaps the largest marsupial megaherbivore was the *Diprotodon optimum*, a three-metre-long, two-metre-tall, three-tonne browser, which was an animal that mainly ate high-growing vegetation, like saltbush. In size, it was similar to the African rhinoceros, which is a grazer, or an animal that mainly eats low-lying grasses and plants. It was also similar to the African hippopotamus, which is also a grass-eating grazer but is mostly aquatic in nature. As this diprotodon does not have a clear match with the two placental megaherbivores, it is regarded as an example of the uniqueness of Australian megafauna.

Another large-sized megafaunal herbivore was the *Zygomaturus trilobus* which was around two metres in length and weighed about half a tonne. Its fossils have been found in coastal wetlands which suggests that its niche was similar to the hippopotamus. Of similar size and weight was *Palorhestes azeal*, a highly unusual herbivore with curved forearms and large claws. It also had a long thin tongue, similar to a giraffe and is also believed to have had a trunk, which is why it is commonly called a marsupial tapir.

Equally of interest is the range of megafaunal carnivores. Not all of them were marsupial. Perhaps the most intriguing was the megacarnivore *Dromornis planei*, a very large and fast flightless bird that stood at about two and a half metres and weighed about a quarter of a tonne. In appearance, it was similar to an emu but much stockier, with a very powerful beak. It is commonly referred to as the 'Demon Duck of Doom'. Another exceptional megacarnivore was the giant goanna, *Megalania prisca*, which was a reptile about seven metres in length and weighed about half a tonne.

Perhaps the best-known marsupial megacarnivore was *Thylacoleo carnifex*,[7] or the marsupial lion, a two-metre-long, 160 kg, cat-like predator with a very powerful bite force, shearing teeth and a semi-opposable thumb with a retractable, hooked claw. It has been incorrectly suggested that this extinct predator was the largest Australian carnivore, however, it ranked third-largest after the extant 'marsupial hominoids', Dooligahl and Quinkan.

Amongst the marsupials, the Diprotodonts adaptively radiated to become kangaroos, wallabies, possums, koalas and wombats, with most of these being herbivorous, except for the megacarnivorous *Thylacoleo carnifex* and some early kangaroos, as well as the omnivorous possum. *Balbaroo fangaroo* was a basal macropodiform that stood about wallaby-height, had enlarged canine teeth, and galloped instead of hopping.[8]

---

[7] A carnivorous wombat.

[8] As with bipedal dinosaurs, hopping makes large animals increasingly susceptible to bone fractures and other impact or stress related injuries. Despite the higher efficiency of hopping, kangaroos are the largest extant animal to use saltatory locomotion.

6: Convergence

Of all the kangaroos, the largest was the *Procoptodon goliah* (Fig. 14.2), a more than two-metre-tall, quarter-tonne biped. It is one of more than twenty other extinct kangaroo species, with a few being man-sized, but most being in the range of two to two-and-a-half metres tall.

IN EXAMINING A list of marsupials and their convergent placental counterparts, it should be reasonably obvious that both mammal groups align or match with each other quite well. However, on the marsupial side, there is a blatantly obvious gap or omission. There are no marsupial analogues of the upper primates. Seemingly, there are no marsupial monkeys, apes or humans!

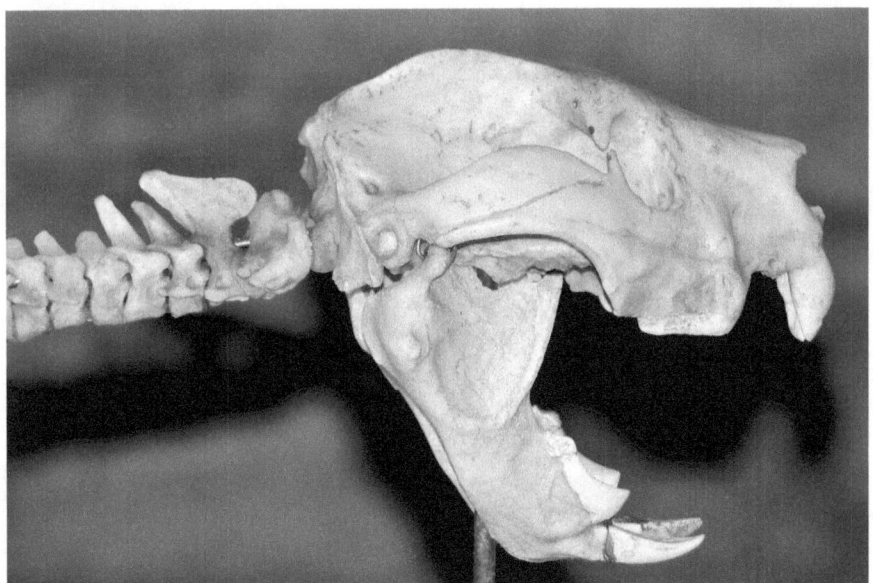

Fig. 6.3 Thylacoleo carnifex skull © Andras Deak

7: Octopus

ESTABLISHING A COMMUNITY network to investigate encounter reports is an important role for anyone who wishes to conduct this type of research. However, to be ethical the primary task should be to personally provide emotional and psychological support for the many confused and traumatised witnesses, by listening to their experiences and respectfully recording them as a demonstration of solidarity. This support also gives the opportunity to present alternative explanations and provide advice and assistance. This aspect is critical because unsupported witnesses typically have no one to turn to and would otherwise become self-censoring under the heavy barrage of scepticism and the social pressure to conform with the mob. There should also be respect for the cryptid's well-being, particularly concealing certain information such as current location that could be used by exploitative people. The network should also encourage participants to become involved at any organisational level, including research.

For anyone who is vaguely interested in this topic, or has watched television news reports on the subject, it should be blatantly obvious that this area of crypto study, like UFO research and many others, is taboo and an easy target for narrow-minded ridicule.[1] The stigma attached to these studies is real and a corrosive force that affects participants as well as the community, which seriously hinders any hope of uncensored scientific understanding and progress. Disappointingly, society and science have not advanced past these conformist medieval attitudes.

IF A COMMUNITY network is established sensitively and ethically it should, at a minimum, become self-sustaining because participants will see the value of their contributions and receive acknowledgment, know that the network is honest, unbiased and trustworthy and want to become a part of this reputable communal research group. The best way to achieve this is—old school![2]

For anyone who has had an encounter, the experience and memory of it are certainly challenging and usually traumatic as well. After many decades of research, it was obvious that there was no shortage of people who had witnessed an incident involving these 'marsupial hominoids'. Having been suddenly confronted by an unknown and threatening animal tends to forge a lasting impression. For these people, the overwhelming need is to tell someone but opening up is risky. Most people learn very quickly to keep their mouths shut. When this social rejection is allowed to persist, many witnesses choose to rationalise the incident as best as they can or try to repress the incident

---

[1] Ian Price often said that he never encountered personal ridicule because he had "credibility". The reality was that after people took a quick glance at this large, heavily-tattooed biker and perceived his 'smouldering intensity', they plied Ian with an appropriate amount of respect.

[2] Internet solutions tend to expose vulnerable witnesses to faceless trolls, sceptics, armchair experts and other controlling or editorialising people.

altogether.[3] Consequently, many encounter reports are lost forever. However, the typical desire of encounter witnesses is to obtain counselling, reassurance, an explanation and advice. These were our needs after we suddenly became Yowie-aware, after discovering that a large and powerful, seven-foot-tall biped was moving about in the bush a short distance from our front door every night. The common problems are: whom can you speak to, where can you get help from and what do you do about it? Having a sympathetic ear tends to make many of these problems and much of the anxiety, go away.

During our early period of research in 1993, we thought that having an encounter with these bipeds was highly unusual outside of our valley. Although I had had two previous experiences, one in rural outback NSW in 1981 and the other, also in the Blue Mountains in 1966, I didn't understand a tiny fraction of this phenomenon and hadn't formed an opinion regarding many things, particularly the frequency of encounters across the broad population and its settlements. However, a short time after becoming Yowie-aware it was becoming increasingly obvious that hairy-man encounters were much more common and broader in scope than what we had previously thought and other researchers had imagined. Although it was very difficult to quantify, the number of initial reports tended to suggest that encounters were common, perhaps as high as 1% of the Blue Mountains population. We initially thought that his estimate may have been inflated by the nature of our local community experiences.

---

[3] This is the opposite of the overused pareidolia argument used by sceptics to explain away the reality of a witness sighting. Instead, it is the tendency for a real audio and visual experience to have its meaningful interpretation downgraded, or stripped away, because of the need for psychological and social conformity.

The Blue Mountains is a section of an ancient mountain chain that runs along the eastern seaboard of the continent that overlooks Sydney. It is a geomorphologically diverse mountain system that was formed by the uplift of an archaic plateau that has been dissected to form many gorges and deep valleys. The sides of the main valleys are near vertical, brought about by the undercutting of soft layers beneath the upper, harder sandstone, which break off as large blocks, falling into the valley below. Over time these cliffs have retreated forming deep, wide and vertical valley systems. Flowing through the middle of the main Grose Valley is the Grose River, which flows into the Hawkesbury River system and then into the Pacific Ocean. Rain that falls on the western slope of the Blue Mountains travels through the Murray–Darling basin for more than 1200 km before draining into the Great Australian Bight.[4] On either side of the Grose Valley and River are a myriad of smaller, less steep valleys that interconnect through swamps and creeks that allow their runoff to flow into the main Grove River. However, south of the Great Western Highway,[5] the watershed redirects the runoff into the Warragamba Dam Catchment[6] area for use in Sydney's water supply network and also into a creek and river system that drains into the Nepean River.

---

[4] The Great Australian Bight is a semicircular bay that forms a major part of the southern Australian coastline. It marks the approximate boundary where the East Antarctica coast separated from Australia during the final breakup of Gondwana.

[5] The Great Western Highway #32, runs from Sydney to Bathurst in NSW, over the Blue Mountains, a total distance of 203 kms.

[6] Although much of the Blue Mountains is naturally inaccessible for many people, the Warragamba Dam Catchment, as well as many other areas associated with Sydney's water supply, are prohibited areas that are patrolled by Sydney Water.

7: Octopus

The settlement patterns of the Blue Mountains have been largely restrained by its ancient topography. Away from the small towns and villages, the secondary roads tend to branch out and follow the relatively level contour lines of the ridges. Houses tend to be built on the scarcely available flat land near the road edge, with the remainder of the allotted land usually running down into the valleys below. Consequently, most roads only have a single row of housing on each side, with the rear portion of properties directly interfacing with the native bushland and the interconnecting network of swamps, creeks and valleys.

Certainly, for the vast majority of densely populated suburban Sydney residents, the opportunities for an encounter are severely limited to those communities that directly interface with the bush on their outer perimeters, or where citizens are in the habit of taking camping trips in remote bushland for recreation or work. Interestingly, the percentage of Australia's population that was living in rural areas in 1911 was 43% (Australian Bureau of Statistics, 1999), which has now fallen to 14% in 2019 (World Bank, 2019).[7] Compared to the demographic situation one hundred years ago, these statistics highlight how the typical Australian today is potentially, far less bush savvy than what they would like to believe.

Apart from having a simple goal of obtaining encounter witness details, the main purpose of establishing a network should be to provide support for those affected by the incident. After we had the initial encounter with Fatfoot, we needed help and advice, but clearly, there was none readily available. Fortunately for us, Ian and Cheryl were excellent neighbours. Not only did Ian assist by engaging

---

[7] Some rubberiness exists with these statistics depending upon the source and time period referenced.

with the biped so that we could make better sense of what was going on but they both provided an opportunity for us to openly discuss the matter and try to develop a way to deal with it. Additionally, we had the support of the Parramatta Aboriginal Land Council who provided valuable advice on how to deal with the situation and what we should take precautions against.

Probably because we were providing assistance and advice to encounter witnesses and involving many of them in the research work, rather than just recording their stories, interest in the local network began to naturally flourish. Within a short time, the number of people who were part of the broader network was growing rapidly. After less than a year, the number of local residents from our valley system who were actively assisting with research was more than sixty people, which was in addition to those contributing from outside the local area. Most of these connections were made by word of mouth, with many current participants passing on new contacts that they had uncovered themselves through conversation, enquiry or by chance. To further facilitate this task, community members were encouraged to ask a few simple questions to anyone whom they thought might have had a higher than normal probability of having an encounter, for example, truck drivers, bushwalkers, forestry workers, people living in remote bushland, and others. On sensing success, a suggested starting point was, "Have you ever had an unusual experience in the bush?"

In order to demonstrate the continued efficacy of this line of fishing, in 2018 we purchased a dog from a breeder living on a large farm acreage at Dunedoo, in western New South Wales. After asking the 'standard question', we were not surprised to hear that the breeder's father had seen a monkey-like biped or Junjudee walk casually across the road in front of his ute during the middle of the day, turn its head and look towards him, before walking off into the bush. The dog breeder made the almost predictable comment, "Dad doesn't talk about that much." More recently, in 2021 our new neighbour Dan was employing a professional to fence his property. The fencer was in his late twenties. After asking him the 'standard question', he said that he had not seen anything recently, but when he was working for National Parks and Wildlife several years earlier, he had turned a corner on a track with another ranger and saw a large black figure, probably a Dooligahl, standing over the body of a kangaroo. The figure disappeared, but when they returned along the same track sometime later, the kangaroo was gone.

Shortly before writing this part of the chapter, I asked our patrolling Sydney Water ranger the same standard question. He knew of my reputation because I had taught both of his children. He whimsically replied, "Nope!"

We called this local network 'The Octopus' because its probing tentacles reached far and wide.

THE OCTOPUS RECEIVED many reports over its lengthy lifetime. Typically, these reports would be received by telephone with messages being left on our home answering machine, if the call was not taken in person. On other occasions, staff, students and occasionally parents would make personal contact at school. Otherwise, members of the local community or the many Octopus participants would find a suitable method to share their information. Ian and Cheryl Price also had their community feeds as did other local researchers.

At the height of its activity, from mid-1993 through to about 2015, the Octopus would typically receive up to three and a half encounter incidents per week at various peak periods. On average, the actual number would have been more like 1.7 incidents per week, spread over twenty-two years, making a total of about two thousand cases. Since the Blue Mountains population in 2020 was 79,195 residents, this lower figure would give an encounter rate of 2.5% across the 22 years that the Octopus was fully active. Clearly, this encounter rate is an underestimate, with many incidents remaining unreported outside of the Octopus for a variety of reasons. For whatever reason, the reported encounter rate has always seemed to be low. To be certain, the reported encounter rate was much greater than it would have been if the witnesses did not have access to a sympathetic and supportive local network.

A small number of local reports were historical accounts with dates exceeding one hundred years. Many accounts were very similar to each other and could be called basic or standard incidents, such as people being shadowed along a track or road and brief appearances in the bush.[8] Some reports identified clusters of activity, particularly when a rogue male Dooligahl or Quinkan was dominating a valley or broader area and generating multiple reports that could be identified and grouped according to their signature behaviours. A few cases provided unusual insights into police involvement, as happened in our valley. A few reports identified and highlighted certain characteristic behaviours, like looking through windows or intimidating bushwalkers and residents. Some were highly unusual or rare incidents, such as entering a house. Overall, the vast majority of cases reported a variety of current activity, usually having occurred within the previous day or two and frequently within a few hours of being observed.

---

[8] If you were the person who had experienced the encounter, there would be no way that you would categorise any aspect of the event as being 'basic' or 'standard'.

7: Octopus

Many of the encounter reports, though separate cases, were very similar in their nature and could be generally grouped. The most common type of encounter involved people who had been followed or pursued by a Dooligahl or Quinkan. The pattern of behaviour was typically very similar across Australia, with an individual or small group of people walking along a bush track or country road and being closely, but invisibly shadowed from behind, usually at an angle of about 30 to 45 degrees to the side. The mob was frequently reminded that as potential prey they can be easily run down, through occasional demonstrations of conspicuous bursts of high speed. This hunting behaviour typically concludes with some type of threat display, involving foot thumping,[9] rock smashing, tree shaking, branch breaking,[10] or rock throwing, with the targeted game mostly running away in fright in response to this intimidatory and predatory behaviour. The Dooligahl or Quinkan does not show itself but frequently alludes to its very powerful presence.

Below is a typical witness statement from 2003, of the encounter type described above. The statement was written by James, a film producer.

"We were staying out of town and were driven by taxi to Bermagui on the New South Wales South Coast to have

---

[9] A NSW South Coast Aboriginal term for 'foot thumping' is 'burminsticka' which means 'strikes the ground'.

[10] Branch breaking is one of my favourite intimidatory behaviours because it clearly demonstrates an understanding of the 'Theory of Mind'. Typically, after noisily breaking a very substantial branch from a tree, nothing further is heard. For a first time observer, something immediately seems very wrong with this scenario. The shock is perceived and the warning delivered sometime afterwards, with the realisation that no sound of the large branch hitting the ground was heard. Whatever broke the branch is still holding it! Another example is throwing a rock noisily in the opposite direction of travel, in order to deceive. Both are highly intelligent behaviours.

dinner and something to drink at the pub. We were having a good time, but we stayed at the pub too long and couldn't get a taxi to drive us back to our accommodation, so we started walking the 8 km back home. The problem started when the road entered a densely forested area around 12:30 A.M. We noticed that something heavy was walking behind us through the bush at an angle of about 30 degrees. Whatever it was, it seemed to be stalking us. We stopped and it made a loud stomping noise. We continued walking but stopped a further three times and on each occasion, it stomped back at us. It was not a kangaroo![11] It then approached closer to us and we became scared. We started to run very fast and continued until we were too tired, after running about 500 metres. We didn't see anything and didn't know what it could have been."

Another example of this encounter type happened to Damien, an audio technician in February 2011 at around midday. He had been riding his push bike along a bush track in Woodford, New South Wales, when he decided to stop for a rest at the lookout, 1.25 kilometres north of the very busy Great Western Highway. It was foggy, so he waited until the mist lifted so that he could see the view.

"I came down the road and then travelled along the bush track on my push bike towards the lookout around midday. The visibility was only about twenty metres, so I waited for it to clear. After the fog disappeared, I cooeed[12]

---

[11] Although James had no idea of what was stalking them, he was certain that it was not a kangaroo. The telltale indicator would have been that kangaroos don't stalk or herd people. They habitually tend to hop away from potential human predators.

[12] "Cooee" is a typical Australian bush call made into the distance to gain attention, particularly when lost because the sound travels a long way. The word comes from our local Dharug language and means, "Come here."

loudly into the valley. Almost immediately there was a reaction from the far side of the valley at the treeline[13] and I started to hear crashing noises coming up the hill towards me, but moving at an intercept angle towards the gate where I had just come from. It sounded a lot like a big boulder rolling through the bush, but moving uphill, like the scene out of *Raiders of the Lost Ark*. It also sounded like a four-wheel drive vehicle travelling uphill through the bush—not that I have actually heard one doing this! I took off as fast as I could on the bike towards the gate. When I got there, I jumped over the gate and sat down while I caught my breath. I started to hear movement along the ridge on the other side of the gate. Then a tree offloaded the water from its leaves, like when you heavily bump a wet tree with your hand. This happened several more times, one at a time, all in a row."[14]

Several aspects of this encounter highlight important behavioural and physical characteristics. It is interesting that the Dooligahl determined an intercept course in anticipation of Damien's retreat to the gate. During his return, Damien was cycling along the flat, wide ground, suitable for a vehicle, at an unknown top speed for a distance of 1250 metres, whereas the Dooligahl was moving uphill at an average slope of 5.6% and along a distance of 1335 metres. They both arrived at the locked gate at approximately the same time. Assuming that Damien was able to achieve a realistic speed of 40 km/hr, the Dooligahl was clearly moving faster. After an incident at our home where

---

[13] The distance from the lookout to the treeline on the far side of the valley is 470 metres. The altitude of the lookout is 600 metres and the treeline is 525 metres, a difference in altitude of 75 metres. The distance from the lookout to the locked gate on the highway is 1250 metres.

[14] See Mike Williams' YouTube channel for video interview. https://www.youtube.com/watch?v=SgkY3jTbW3c

Fatfoot was timed over a known distance when spooked by several 100-watt floodlights and two photo flash units, we determined a similar speed of 40 km/hr. In comparison, Usain Bolt has achieved a top speed of 44 km/hr during sprints.

A very large number of encounters had been reported to the Octopus from this area. A few incidents involved cyclists moving along the well-used bike trails of the Blue Mountains, where riders travelling at high speed would sometimes be pursued by a biped running alongside them at a matching pace through the thick scrub. The highly regarded 'Oaks Fire Trail' at Woodford would provide many similar pursuit encounters involving cyclists.[15]

According to traditional Aboriginal Elder knowledge reported by Andrew, a Northern Territory researcher, when entering an area, it is advised that visitors should not make any loud disturbing noises, because this could provoke a response from the spirits. Instead, "they should always treat the area as if they were entering a church, mosque, or synagogue." An Aboriginal Elder who worked as a teacher's aide at our school similarly mentioned that it was important to remain respectfully quiet when entering a suspected Hairy Man area and that in addition, he would rub the local soil over his body to help disguise his scent and avoid a confrontation.

DAMIEN WAS AN excellent encounter witness and Octopus contributor. Following his experience at the lookout, he actively participated in many research activities in the local area with other Octopus participants. As he gained

---

[15] A possible cause for the high rate of incidents could be the disturbing noise made by cyclists and their challenging speed.

experience, he teamed up with other researchers who were mainly active on the western side of the Blue Mountains, which further increased the reach of the Octopus. Like Damien, Adam and Emma were both extremely pleasant individuals and dedicated researchers.

As a result of his activity and the information gained through Adam and Emma, reports were received that pointed to a large 'marsupial hominoid' with a provocative personality, that was active in the Lithgow area. An early report from Adam and Emmas' friend Ben from the summer of 2010, was recounted by Damien.

"Adam's friend Ben was up in the bush near his house with a friend. It was dark and they were sitting on some rocks when something was heard 'walking' around above them at the top of the hill. He thought it was human at first, but then it charged down the hill towards him. He did the usual response and flew like a bullet back home. He asked Adam about Yowies and is now sure they exist."

In November 2011, some other friends of Adam and Emma were sitting at a secluded spot in Lithgow, near Hassans Walls lookout, which is the highest scenic lookout in the Blue Mountains at approximately 1,130 metres above sea level. Damien said:

"They were just sitting around at night gasbagging when a massive big black beast on two legs walked right passed them and by that, I mean just a metre or two away from them. They suffered a good look as it passed and froze stiff as they realised it was no man! It was massive, with black hair all over. The strange thing was that it just walked by them knowing they were there. It then increased its speed into the bush and vanished. The witnesses had heard of the rumours of Yowies, but never believed they were real. One of the blokes, a very shy, quiet, honest type, whom Emma knew from school, put it up on his Facebook

Fig. 7.1 One of a number of freshly damaged trees in the vicinity of the 'walk by' encounter at Lithgow, which has exposed the larval chamber. (Photo: Damien)

7: Octopus

and was hammered by the typical, redneck responses that Lithgow is famous for. He removed his post. He had no previous knowledge of Yowies and has never seen anything on the internet. He doesn't even know Emma and Adam are researchers. Adam, Emma and I went up to the area on Friday. Lo and behold, we found possible treebites and some very strange stick formations."

THE AREA SURROUNDING Damien's early lookout and bike encounter was well known by the Octopus as a source of frequent activity and I had worked the valley many times over the decades becoming familiar with the telltale clues. Numerous reports had been received by the Octopus that centred upon this location and I was certain that the culprit was Fatfoot, who had the newly discovered habit

Fig. 7.2 Location of the 'walk by' encounter, Lithgow. Many sandstone caves are located high up on the escarpment, which are believed to be the home of the Devil-devil. (Photo: Damien)

of moving her camp around.[16] It took some time to recognise this itinerant behaviour, which had previously led us to believe that Fatfoot may have died, been replaced, or moved permanently out of the area.[17]

Shortly after Damien's encounter I walked into the valley using the established bush tracks and was able to find the line of trees that we could see from the lookout. Looking up towards the lookout the vegetation was thick and confused. It would seem unimaginable that a person would be willing to similarly attempt such an ascent.[18] Moving further south along the track, it was becoming more 'untidy' and showing signs of increased traffic. Although it was a bush walking track, I had seen very few hikers along this particular section over the years, even though it eventually made its way up an incline towards the highway. Noticeably new to this section of path were a number of highly recognisable features that were characteristically typical of Fatfoot and Dooligahl behaviour in general. There was evidence of 'twig snapping' along the track edges, which we identified as 'track maintenance'. Foremost of these features were territorial markers which conspicuously demonstrated the strength of the resident individual, whilst strongly conveying to any intelligent

---

[16] There was a reasonable amount of uncertainty for nearly a decade leading up to Damien's encounter, where we considered that we might have been dealing with the 'Son of Fatfoot' instead; however, in short, our opinion reverted after repeated incidences of idiosyncratic behaviour that only Fatfoot would have persisted with. For example, continuing to foot thump in the swamp below our house during the early morning hours until I responded by turning the outside light on and off three times.

[17] There were other factors that may have influenced Fatfoot's behaviour, specifically the competitive challenges of Junjudee.

[18] In regard to moving through thick, thorny and generally nasty bush, Ian Price would say that, "The beastie either has thick protective fur, or it is indifferent to the pain."

7: Octopus

Fig. 7.3 Closeup of one of a number of trees that had been twisted, snapped and pushed behind other trees. See Mike Williams' 'Tree Breaks' YouTube video for a broader view of the damaged area. (Photo: Neil Frost)

visitor the warning to 'keep out'. For such damage to have been possible, the trees would need to be supple and therefore, alive at the time. The markers were from large diameter hardwood eucalypts, around 120 mm or greater, with heights of around fifteen metres, that had been very forcibly corkscrewed, snapped and generally abused on either side of the track and in such conspicuous numbers that they were more than readily noticeable as attention grabbers.[19]

Also in this area, about thirty metres below the track and very near to a water course, I could see what looked

---

[19] See Mike Williams' YouTube video, 'Tree Breaks', 13 March 2011: https://www.youtube.com/watch?v=IHHF83weTEk

potentially like a primitive bush shelter. My attention was initially drawn to this general area because of the territorial markers, however, on closer inspection, it was the overwhelming stench of urine and the generally disturbed nature of the nearby vegetation that highlighted an entrance in the wall of vegetation.

From a distance, the bush shelter was simply made from vines and other delicate, fine leafed plants that had been obviously dragged as a whole and pulled into position, leaving the surrounding area relatively deficient in vegetation. Some leaf litter was laying on top of the vines that suggested that the shelter had some age. As I moved from the main track down towards the creek, the canopy became thicker and the light was noticeably less. From experience, when the hairs on my arms started to stand up, it was

Fig. 7.4 The low entrance to a bush shelter below the main track. (Photo: Neil Frost)

always a good practice to take notice by being extra cautious. Even though I was deliberately over-announcing my presence, I was certain that if Fatfoot was in residence, I would have heard about it by now. Typically, at nighttime, if I approached Fatfoot too closely, I would hear growling and threatening noises. Regardless, I overplayed the sound of my approach to the shelter, just to be sure, before slowly inserting my head through the low opening. On looking inside a branch could be seen in the centre of the very dark enclosure that was clearly being used as a prop to maintain the ceiling of vines and other plant material. The internal size of the shelter was deceptively large, being more than six feet long and easily capable of accommodating several large men on either side of the supporting stick. Inside the shelter, the scent was more overpowering than on the outside. I didn't stay very long, because it wasn't much fun and I thought that I was probably being watched anyway.[20]

After proceeding a short distance further down the track, there was additional destruction on both sides of the path, mostly small broken branches on bushes along the edges of the track. As I passed through a partially cleared area where surface water from nearby rock seepage covered the ground and formed small pools, a number of bird feathers could be seen on the mud surface together with a partial wing tip and other small, unrecognisable bird parts. Obviously, something had been recently hunting there. After checking the view up to the lookout and looking for landmarks, this seemed to be the point where

---

[20] I can hear a commotion of people saying "Why didn't you look for any hair samples?" My reply is, "You can look for them!" We had not had success in having hair samples analysed in the past, for a variety of reasons that will be discussed later. Regardless, Ian was no longer living in the area and only a group of people were prepared to spend the necessary kamikaze time with me in the bush.

Fig. 7.5 A naïvely closed-off section of track by the Blue Mountains City Council. "This area is closed for erosion control order. Give me a break, I am trying to revegetate." (Photo: Neil Frost)

7: Octopus

Damien had observed Fatfoot start her ascent towards him at the lookout. However, directly ahead of me, a Blue Mountains Council sign blocked the way, which said, "This area is closed for erosion control—Give me a break I am trying to revegetate."

The wording on this sign was a source of great amusement. Clearly, the immediate damage and the further destruction that lay ahead were not the result of erosion or vandalism. The sign attempted to block the track as best it could when installed, but it was ineffectual as new paths had developed around it. Surprisingly, unlike the bush nearby, the sign itself had survived any supposed attempts at vandalism. If the intention of the sign was to plead with and modify the irresponsible behaviour of anarchistic hikers, then it had clearly failed. The sign was a waste of ratepayer taxes.

Though puzzled by this sign, I moved past it, but could soon appreciate the perceived need for it. The bush on either side of the path was extensively damaged with many old and new broken branches and twisted trees. It seemed like someone or something was making a statement or a warning that said, "Keep out." After walking a further hundred metres or so, the canopy started to close in and it was becoming dark and narrow. As was usual in situations like this, people tend to experience the classical feeling of 'dread'. After a few minutes, I began to hear subtle movement ahead, followed by the sound of branches braking, so I turned back. On thinking what Ian would do in this situation, I stopped and listened again. Ian would often say to me that you must not show any fear and if anything, it is important to be provocative. This was his advice learned after many years of experience trying to survive in maximum security prisons. So I turned around again and walked back towards the distant sound and stopped. I heard nothing more. After a short period of time, I began to slowly back out of the area and returned home. This strategy appears to show determination and courage and probably signals that you will not be intimidated, although I am sure that Fatfoot was cable of recognising me. One thing was certain, running away would have been futile and was not an option.

When I returned to the valley a week later the weather was overcast and cool. Researchers Rebecca Lang and Mike Williams, along with new Octopus participants Damien and Guy came to look at the damaged site and the associated artefacts. Mike concentrated on his own work, mainly by taking videos with commentary. The others inspected the unusually damaged vegetation and came to their own conclusions while Damien recounted various aspects of his recent encounter to the group.

A VERY COMMON type of encounter reported to the Octopus is centred around human settlement, whether permanent, like a house, or temporary, like a caravan, tent or swag.[21] All three 'marsupial hominoid' types are attracted to a human presence, where they spend lengthy periods of time observing the activity of the residents. During the day, their observations are initially made from afar using concealed positions and then at much closer distances that are more exposed as the light fades. While doing this, they are also alert for any feeding opportunities.

The causes of this attraction are related to food and entertainment. From Aboriginal tradition and knowledge, one of the basic food rules was to never cook after sundown. The reason for this seems obvious. Releasing cooking

---

[21] A swag is a traditional Australian lightweight bedroll consisting of a canvas outer sleeve with an internal mattress. They were designed to be carried by foot or on horseback and can be rolled up or out in seconds. Usually waterproof.

7: Octopus

aromas into the night air only serves as an irresistible attractant to any carnivore, hungry or not, that will then approach the settlement under the cover of dark, generating numerous and difficult to resolve problems or dangers. However, for most of us, sundown is only an advanced reminder to start thinking about the evening meal. Breaking this traditional rule was most probably a major factor that contributed towards our terrifying two-night encounter at the Blue Gum Forest when on a seven-day scout camp in 1966. Another food rule was to never leave meal leftovers out overnight. Like many campers, leaving the final vestiges of the lamb roast in the cast iron cooking pot[22] until morning or throwing the scraps into the fire does seem inconsequential for many Australians.[23]

Additionally, as these bipeds are very intelligent, they require mental stimulation. This is similar to sulphur-crested cockatoos that destroy property because they are bored or possibly because they enjoy the resulting chaos. As Ian often said, "What else are they going to do? They don't have satellite TV back in the cave. Even Toby[24] gets bored!" Usually, the animals will simply observe and not interact but sometimes they will become angry with the behaviour of the residents or will become provocative in order to elicit a reactive response.

OUR INITIAL CONTACT with Fatfoot began when we started building our house in late 1983. It was undoubtedly

---

[22] Bernie, Sandy's father, did this one night when on a fishing trip with his young son Michael. It resulted in a Dooligahl visiting the campsite and carrying off the cast iron pot and its contents. Also, fishing encourages unwanted inspections of the campsite, particularly when fishing stringers are used.

[23] Except for the possibility of food poisoning.

[24] Toby was Ian's German shepherd 'anti-personnel' dog.

our noisy and environmentally destructive activities that piqued her curiosity, resulting in a trail of footprints left across the soft soil after the foundations were excavated.[25] This was to be the only time that footprints were found on our land, however, many others were reported and found on adjacent properties, including Fatfoot's imprint that we cast. During the following decade, all of the typical indicators of Dooligahl activity around the house were present, though we were totally blind to them. Feeding dogs outside is a common attractant, where the pets are frequently intimidated or 'stood over', in order for the 'marsupial hominoid' to access the tasty food after the dog has been moved on. An interesting encounter incident received by the Octopus involved a woman observing a hairy hand in the act of retracting through a dog door with her pet's feed bowl, after hearing noises and turning on the downstairs light. Another early encounter instigator was our desire to educate and entertain our young children by feeding the local carnivorous birds in the late afternoon. As the many birds flew into our neighbourhood valley in the late afternoon to compete for their meat, all other observant carnivores would have become equally aware of the meat source. Also, in doing so, they would have become aware of our young children. This was a naïve activity by us that Aboriginal Elder Merve told us to stop immediately. More broadly, he told us not to feed any native animals.

Unlike the previous encounter type where the contact experiences of people were fairly similar, Wildman's visits to houses were much more varied, involving any combination of different activities and occasionally reporting

---

[25] The sudden rise in Junjudee activity in our valley from April 2017, was partly the result of noisy building construction. The sound of power tools and hammer blows carry a very long way into the distant bush valleys.

7: Octopus

incidents that had not been witnessed previously. Of the many incidents received by the Octopus, below is a list of commonly reported behaviours observed either individually or in combination, that were associated with houses.

- Red eyes observed in the bush at a consistently tall height
- Roaring, growling, grunting and other noises
- Walking and/or running in the bush nearby
- Objects thrown at the house
- Objects thrown at residents
- Tapping on windows
- Slapping of walls
- Eyes or faces seen through a window
- Breathing heard through windows, doors or external vents
- Movement heard underneath the house floor
- Floor joists slapped or knocked beneath the resident's location
- Outside movement shadowing internal activity
- Walking on verandah
- Climbing/walking on house roof (Junjudee)
- Objects stolen from outside
- External objects moved or damaged
- Dog/cat feeding bowls taken
- Bird seed scraped up from outside cage and eaten
- Foot/rock thumping heard in the bush surrounding the house
- Clothing taken from clothesline
- Attempting to, or opening of doors
- Entering homes
- External lighting damaged or interfered with
- Shiny objects or personal items taken
- Tracks leading to/surrounding house
- Jumping over fences

- Fences damaged
- Trees pushed over fences
- Climbing trees
- Daytime sightings
- Footprints found
- Trees damaged/ripped
- Hedges walked through
- Plants uprooted, gardens destroyed
- Alarm barking by dogs, hissing by cats
- Dogs fleeing their domestic territory or hiding under the house
- Cats/dogs disappearing
- Dog intimidation
- Flattened areas around the house
- Treebites
- Trees snapped/twisted
- Wood knocking—communication involving a 'mischief' of Junjudee
- 'Talking'—vocal communication involving any of the Wildmen
- Stick planting—symbolic communication between Junjudee and involving people
- Emitting strong, offensive odours—involving any of the Wildmen, but particularly Junjudee

Of all the listed behaviours experienced around homes, perhaps the most frequently reported incident to the Octopus, either singularly or as part of a cluster of experiences, involved residents seeing red eyes at night moving around their property in the bush. One night in 2003 while I was visiting Steve, an Aboriginal artist living in a neighbouring valley, he called his children outside onto the verandah to show them the red eyes that we had both just seen in the bush. He said that it was a good opportunity to educate his children regarding the detection and dangers of the Hairy Man. Later that night around midnight, we went for a walk through the valley with his dog and we exchanged what we knew on the topic.

Another very commonly witnessed behaviour was seeing red eyes or a large head looking through a window or door. This behaviour mostly involves children rather than adults, although this was my experience on the farm in 1981. Over the decades, the Octopus received many dozens of these reports with most being very similar to each other. Both of our children had experiences, but it was our son Drew who had the worst encounters because his bed was closest to the bedroom window and he was younger. This often required me to sleep with him after an incident in order to reduce his anxiety. As a consequence, we had to remove 'Big Ted' from his room because he was frightened by it, saying that it looked like the thing that he saw looking through the window. Another commonly reported behaviour is walking or running heard in the bush surrounding the house. A very high proportion of these incidences are rationalised away as being wallabies or kangaroos hopping because the noises can sound very similar to someone without much bush experience. These very common encounters only tend to be acknowledged on reflection after the witness has become Yowie-aware and can more easily differentiate between hopping and running. However, on occasions a naïve resident may valiantly decide to investigate the source of these sounds, only to be loudly roared at, as if by a lion. The least frequently reported behaviour involved a rural resident outside Mudgee, NSW, who heard and saw a *mischief*[26] of Junjudee climb

---

[26] We coined the term 'mischief' as the collective noun for Junjudee because the word summarises their overall demeanour. From our experience, Junjudee will destroy, sabotage, steal, relocate, burglarise caravans and attempt house entry, terrorise and more. Junjudee are typically

onto and over his farm shed roof after they were heard and seen looking through the window.

CAMPERS AND TO a lesser degree, free caravaners,[27] are a highly represented group of encounter witnesses. Prior to my last school, I was teaching in the Western Suburbs of Sydney from 1998 to 2002, where there weren't any extensive opportunities to tap into community experiences with these Hairy Men because it was a suburban environment and no one would have any inkling of what I was talking about. During this time, the Octopus maintained its bush connections in the mountains and continued to provide frequent encounter reports and people to talk to. However, it seemed that there might be hidden encounter stories going to waste within this suburban environment, as many of these people would spend part of their weekends or holidays in the bush.

Having been given a position at a Blue Mountains school in 2002, I spent my last day in teaching suburbia farewelling colleagues and friends. For five years, Helen had taught alongside me as a teacher's aide, assisting with the special needs students in my classes. Helen had also spent some time visiting my mother as they both shared a family connection with early Australian horse racing and enjoyed exchanging memories. On this last working day, for the first time, I told Helen about our Hairy Man research in the mountains. For the first time, Helen told

---

(cont.) encountered in tribal groups. More significantly, Aboriginal Elders would frequently describe Junjudee's behaviour as being 'mischievous', a very commonly used Aboriginal descriptor, from which the collective noun is respectfully derived. The word is also occasionally used to describe a group of rats.

[27] Staying at locations that do not charge a site cost or a camping fee for caravans.

7: Octopus

me about her Hairy Man encounters with her caravanning friends on the western side of the Blue Mountains!

Helen told me that they would meet with their free-camping caravan friends every year at the same camping ground west of the Blue Mountains. She said that it was a beautiful area located next to the Fish River, which was appropriately named because of the excellent trout and native fishing opportunities. She said that the location was very isolated, surrounded by farms and national parks.

Helen and her friends met at this camping site for many years with most people staying numerous nights. On most visits, they had no incidents to report, although, with hindsight, Helen said that they did see red eyes in the bush on a few occasions and wondered what was responsible for them. She said that one night, the year before their last stay, they were all sitting around the campfire when the people opposite her all started to stare off into the distance, followed by someone commenting about the red eyes that they could see moving about in the bush. Some people thought that it was strange, but others thought that they were simply watching possums moving about. However, the last time that they caravanned there, something scared them all. Helen said that one of the men took a shovel and walked off into the dark. He dug a hole and squatted over it for a little while when suddenly, something very big started to move next to him. According to Helen, the man said that it was "huge and walked on two legs." He also reported that "It walked so close to me that I could feel the heat coming off its body." The man hurriedly ran back to the campsite to report what had just happened. Regardless, they could all hear something very large walking noisily around in the bush, breaking vegetation and making various unfamiliar sounds. The campers all decided to hurriedly pack up and leave immediately. They never went back to that camping site.

Free-campers have a higher probability of experiencing an encounter compared to organised open campsites because of the lesser number of campers present and the more primitive conditions that provide greater opportunities for concealment. Further drawing upon the school situation above, I mentored a very promising senior student in his final examination year, across a number of related academic and non-related areas. Several years after we both had left the school, he looked me up in the telephone book to tell me that he and his girlfriend had an encounter at Euroka Clearing,[28] in the Glenbrook National Park at the base of the Blue Mountains. He said that a biped had walked heavily around their tent for most of the night. He had pitched the tent on the edge of a clearing.

After analysing many Octopus reports, a commonly perceived factor that is associated with camping encounters is pitching the tent on the boundary between bush and clearing. This is most probably because these animals prefer to remain concealed and consequently, tend to favour the bush and avoid cleared areas where they can. For some reason, people seem to prefer camping on the outer fringe rather than in the cleared centre. This is also what a colleague did when visiting friends on a 100-acre farm near Mudgee, NSW, in 2014. In wishing to provide privacy for all concerned, Gary drove his car through the paddocks and pitched his tent high up on the hill, a long distance from the farmhouse, at the edge of the clearing. After having had a large biped walk around his tent for

---

[28] Euroka Clearing has a long history of Hairy Man activity. A common theme recorded by the Octopus was a biped that regularly ran along the escarpment overlooking the Nepean River and campers being harassed at the clearing. Other stories centred on the river, where ducks were seen being chased along the banks, with some seemingly drowned by a Dooligahl.

many hours, I asked Gary if he had a look outside to see what the biped was like. He said, "Don't be silly! I didn't want my head ripped off!" I replied, "You were only in a nylon tent, Gary!" Gary also said, "I've listened to your stories for years, but I never expected to have an encounter myself."

AT SOME POINT during their nocturnal wanderings, these bipeds will encounter obstacles that hinder, restrict and redirect their movement. Many of these obstacles are man-made with typical barriers being roads, railways, bridges and fences. When circumventing these barriers, the most likely observers will be car and truck drivers during the late evening or early morning. This group of people form a substantial category of witnesses.

Perhaps the most remarkable encounter received by the Octopus was made by Susan, a young canteen assistant at our school in May 2002.[29] I usually arrived at school before everyone else did, an hour or more before starting time so that I could organise a relatively stress-free day. However, all too frequently, this early morning reliability was interpreted as a willingness to fill the hard to allocate, early morning or before-school playground duties. On this occasion playground duty was not mandated, so after arriving at school and leaving my bag in the staffroom, I left to do photocopying for my classes and make some changes to the room bookings. When I returned, Chris told me that Susan had come to the staffroom and urgently wanted to talk. When visiting the canteen shortly after, Susan was very excited, saying that she had "seen one" with her father!

---

[29] This date has been incorrectly given elsewhere as May 2000.

My first attempt to interview Susan was very short because she was still preparing meals for recess and lunch, although she gave a brief outline of the encounter and a description of the biped while she worked. Consequently, she agreed to have an in-depth interview after school and gave me permission to pass on her contact details to Paul Cropper, so that he could also record her interview to be used in their upcoming book, as well as share information for other researchers, as was our common practice.

Susan's father owned a farm at Cowra in central-western NSW and she was driving there with him at around 4 A.M. to complete some maintenance and tend to the animals when the encounter took place. They were driving west along the Great Western Highway and were approaching Park Road at Woodford[30] on their left, when Susan saw a large human-like figure coming down the footpath towards them on the opposite side of the road. At first they both thought that it was a person who was 'in trouble' because of the way that it was running, stopping, turning and looking around. There was no other traffic on the highway, so her father turned the ute towards the thing and stopped. In the high beam and with the spotlight and driving lights on as well, they could see that the biped was huge, probably about seven feet or more tall and it wasn't human. She said that it reminded her of a cow standing on its hind legs, because it was so big. The head and face looked like a normal human, except that the head might have been longer. The nose appeared normal but narrow. The eyes were set a long way back into the head, but they could not see anything more. The jaw was very

---

[30] Park Road crosses the Main Western Railway by overpass bridge, immediately before joining the highway. The next railway crossing point is an underpass 2.7 kms to the west. The railway corridor is fenced and maintained on both sides of the track.

7: Octopus

strong looking. I asked her if she could see any canines, but she said that they couldn't. It had no neck but incredibly wide shoulders. The main torso was much narrower than at the shoulders. The stationary arms were very long and extended down to the knees. The hands were large and the feet were massive. I asked if she could see any digits, but she said that it was too difficult. The body was covered all over with reddish-brown,[31] dirty hair except the face which was relatively hairless. No genitals or breasts[32] were visible.

Susan said that the biped had been dazzled by the bright lights, which reminded her of kangaroo hunting when they would go 'roo shooting during the night at Cowra. Like kangaroos, it just stood there until they turned the lights off, or moved away. She said that after turning the lights off, it turned and started to run down the middle of Woodbury Street. Her father mounted the central medium strip on the highway and followed after it. Susan said that it would run until we got close to it and turned the lights on, where it would stop and turn towards them. I asked Susan if either of them could see any red eye. She said that they didn't see any.[33] If they tried to get closer than about 20 metres from the biped, it would start to

---

[31] Red hair is an adaptation in humans and other primates that allows for the increased transmission of ultraviolet light through to the skin where vitamin D is produced. It is an adaption to low light environments, as with nocturnal animals, or living at high latitudes. The early hair samples, believed to be from Fatfoot and with excellent provenance, were red.

[32] Conspicuous breasts are a placental trait in females.

[33] Seeing the 'red eye' is not guaranteed. It requires a particularly accurate alignment of the light source, reflector (tapetum lucidum) and the observer to be seen, which in principle, is very similar to an astronomer looking through a telescope, or an optical technician evaluating a mirror using a Ronchi test. Any misalignment of these elements will result

back away and if they turned the lights off, it would make a break for it. Each time they turned the lights back on, the humanoid would stop and look very confused and disorientated. When it started running again, it tended to wander all over the road, as if it couldn't see properly. It also loped along as if it was having problems walking.³⁴ This sequence of action and reaction was repeated several times. When it could finally see the end of the road, it bolted off into the bush and began to push the trees about. They could see the trees moving about violently in the lights and could hear the dogs in the neighbourhood going crazy. When they returned home from Cowra, they stopped to look at the bush at the end of the road and saw that half a dozen trees had been snapped off about a metres from the ground.

Susan's detailed witness report was very interesting for multiple reasons and in having a duration of five minutes, it was the longest continuous encounter, under excellent viewing circumstances, that the Octopus had come across. Although we always maintained a bucket list of objectives to target and achieve should such a visual feast arise,

---

(cont.) in the star not being seen, or the figure of the mirror being impossible to determine. Similarly, photographing an 'Einstein–Chwolson Ring' using either the Hubble or Webb Space Telescopes, requires a perfect alignment of source, (gravity) lens, and observer.

Even with Fatfoot being present around our house on most occasions, we didn't see her eye shine every night and only managed to photograph it once. The success in photographing the eye shine was greatly increased by deliberately engaging with Fatfoot, 'mano a mano', with the conjoined flash (light source) and camera (observer) having no angular separation (in perfect alignment) and only requiring the camera/flash to be pointed at her head and Fatfoot looking at the camera (reflector), in order to complete the optimal geometric configuration. However, the important thing was to blind Fatfoot with the first flash from the camera in order to immobilise her.

³⁴ Ian and I had noticed that Fatfoot ran with an irregular gait, as if a leg or foot was injured or deformed.

Susan and her father did very well, nonetheless, to develop impromptu strategies to achieve what they did under stressful and naïve circumstances. Also, apart from knowing Susan personally, in confirming most of the known physical and behavioural characteristics of Dooligahl in a single report, there was no doubting her veracity.

Susan and her father both unmistakably realised that the biped was not human, with their description of the humanoid certainly confirming that it was not. However, having long arms is an arboreal trait and when further considered with the evidence of a grasping hallux from the foot cast of Fatfoot, this would confirm the Aboriginal knowledge provided by Merve and other Elders, that these bipeds are both tree and ground dwellers. Living within such a dual environment might also have resulted in adaptive compromises that could explain certain observed issues related to gait and locomotion, e.g. awkwardness and quadrupedalism. Furthermore, considering their length, the arms were stationary and did not swing, as long moving arms would be difficult to synchronise with stride and would be an energy inefficient activity as well. Having wide shoulders would be consistent with moving more easily through the trees and it is also a characteristic of all marsupials during their early stages of prenatal development, which requires overdeveloped forelimbs in order to climb from the birth canal to the marsupium.

Having a human-like face does not mean, by itself, that this biped is biologically related to *Homo*. Like us, it would be the result of many similar environmental factors that have shaped its convergent marsupial appearance. Although the Dooligahl's physical characteristics may seem to have emerged from out of a biological blender, it certainly parallels the antipodean pattern of life that has emerged from this continent, where the platypus is the classic example of this mixed type of Australian uniqueness.

However, the eyes of this 'marsupial hominoid' were different, being set "a long way back into the head". Even though they did not see any reflective glow, this night vision feature would have certainly been present because of how the biped reacted to the vehicle lights. Regardless, the presence of the red eye requires a specific optical alignment. Among the very few encounter witnesses that have seen the eyes, a former student also told me that the whole orb appeared to be black. This would suggest that Dooligahl have a dark sclera, instead of the white sclera that humans have. Many theories have been put forward to explain this difference, with dark sclera being difficult to see, particularly at night and being ideal for a predator. Alternatively, white sclera are believed to be useful for visual communication in social animals as it allows the visual gaze to be tracked.

The nose was "normal", but "narrow", which is a convergent adaptation to cold and dry environments, where inhaled cold air is heated and humidified through increased contact with warm blood vessels due to greater turbulence caused by the nasal cavity constriction. This was a pleasant surprise because no one had been able to describe the shape of the nose with such confidence and the photograph that we had taken nearly a decade earlier had suggested that Fatfoot also had a Greek nose.

The jaw was "strong looking" and even though they weren't able to detect any canines, its robustness would suggest a bite force capability that would be able to significantly damage hardwood trees. Having such a "strong looking" or robust mandible would suggest that a considerable amount of musculature must be attached to the jaw, which would result in the proportionate growth of dense bone. In regard to the head possibly being "longer" than in humans, this might be explained by the presence of a sagittal ridge, but it was something that Susan could not expand any further upon. When we observed a sagittal ridge on several occasions, we had either been using night vision binoculars or saw this feature when employing the indirect observing technique that utilises the additional rods used in peripheral vision.[35] Having a sagittal ridge provides additional tendon attachments with longer and more sustainably powerful muscles which, combined with the shorter muscles of the lower jaw, greatly increase the bite force of the mandible.

Although the biped had a "human-like face", Susan and her father were both convinced that it was not human.

ANOTHER IMPORTANT ASPECT of this previous encounter was the discovery of an unrevealed arrival path and the projected departure route made by the humanoid. When the Dooligahl was first detected, it was attempting to navigate an easterly path along the main arterial highway that links Sydney with the western rural districts of New South Wales. It was trying to do this in the early hours of the morning, during a time of least human activity. After being discovered, the consequences of its initial anxiety were its frantic and unsuccessful attempts to quickly find an exit path through the roadblock formed by the fenced housing along the highway edge. As this was undoubtedly

---

[35] Though only marginally more effective than direct observation, this observing technique is mainly used in astronomy to see faint galaxies and nebulae beyond the human eye's normal sensitivity limit. When using this alternative method, the observer looks directly away from the area of interest, whilst indirectly concentrating their conscious attention on the subject using the rod-rich part (more light-sensitive areas) of their peripheral vision instead. In addition to having pupils capable of dilating more widely, it is reasonable to assume that these primarily nocturnal predators would have greater densities of rods in their retina compared to humans. Since colour vision is rarely found in nocturnal animals cones would not be required which would further increase rod densities.

a transitory phase in its overall journey, from somewhere to somewhere else, the questions are, where had it come from and where was it going?

Susan first saw the Dooligahl opposite Park Road in Woodford. This was the same location where the school bus would briefly stop and disembark its passengers every school day afternoon. The students would then make their rapid and perilous attempts to cross the highway on foot and en masse, like a gaggle of garrulous geese. The solitary Dooligahl had made the same journey, but far less perilously during the less busy hours of the morning.

The southern side of the Great Western Highway is flanked by the Main Western Railway, which was a major engineering project during the latter half of the 1800s, that originally required the construction of a 'Little Zig Zag' at Lapstone and a 'Great Zig Zag' at Lithgow, so that steam engines could more easily climb the steep 1:33 (3%) gradients across the Blue Mountains. Presently, a dual line track transports passengers and freight across the mountains, ultimately reaching Perth in Western Australia on the *Indian-Pacific,* after a four-day journey of 4,352 kilometres. Both sides of the railroad corridor are fenced to a height of six feet and well-maintained to prevent illegal trespass.

The intersection of Park Road and the Great Western Highway has a bridge forming part of the road junction, which passes over the railway. After travelling over the bridge, the road splits three ways, allowing access for the residents on the southern side. Like most residential areas in the mountains, properties back directly onto the thick bush, with views extending into the distant, remote valleys. The sealed section of Park Road continues south for a further 3.7 kilometres into the wilderness but after having lost most of its residential settlement halfway along its length. It then reverts to an unsealed track for a further half kilometre, before terminating on a ridge. The only settlement south of this ridge is Mittagong, 78.5 kilometres away, across the forested, untouched wilderness. Like the local residents, it would seem reasonable to assume that this Dooligahl was gaining access to the southern part of the mountains via this isolated bridge.

The next available access point for humans and 'marsupial hominoids' is the railway underpass 2.7 kilometres to the west at Oaklands Road in Hazelbrook. It was directly across from this underpass, also on the southern side, that several encounter reports were received by the Octopus. One particularly interesting encounter report was made by Jenny and Jonno that was recorded by Mike Williams and is available on his YouTube channel: 'Yowies Talking' [https://www.youtube.com/watch?v=eFSHLaON6Yw] The interconnecting and frequent reports made to the Octopus serve to highlight the importance of this type of community network and its ability to uncover many hidden relationships and patterns of behaviour.

At the opposite end of the encounter, where Susan and her father saw the Dooligahl disappear into the bush at the end of Woodbury Street, is a track that descends into the valley. Further along this track are a number of damaged trees. Almost nine years later, a council sign banning entry into the area would be erected because of environmental damage done to the track and from this location, an angry Dooligahl would ascend towards Damien in response to his 'cooee' call.

JOHN APPLETON WAS a work colleague and friend. We had worked together for several decades and over the years we had discussed many things and in particular from 1993, the latest encounter reports from the Octopus. John was a very good researcher because he was familiar with many aspects of the research and he would follow up any lead that he came across. He would also ask the 'standard question' to anyone whom he suspected should know something.

In April 2005, John heard a man at the back of a tour bus that they were travelling in, discuss an encounter that he recently had. John walked to the back of the bus and spoke to the man who said that he was returning home in his car along the Great Western Highway near Valley

Fig. 7.6 Outline sketch of a Dooligahl seen by two boys at Warrimoo in a cave below the Great Western Highway, approximately one kilometre from Green Parade, in 1995. (Source: Neil Frost)

7: Octopus

Heights at 2 A.M. Craig had just completed a job interview for a security officer vacancy at the Oriental Hotel in Springwood, NSW. Craig was driving east and had slowed his car down as he came into a section of road that is locally known as a fog-prone area. As he entered the dip in the road, he dimmed his headlights to low beam to reduce the glare and was monitoring his speed as he approached the fixed speed camera ahead. On Craig's driver's or right-hand side was Green Parade which has a railway underpass that provides access to the southern settlements, similar to the Park Road overpass further to the west. One hundred metres ahead on the left was the speed camera and immediately below it was a large section of vacant bushland that slopes away into the rural settlement of Sun Valley. The bush in this area has many gullies that feed into Sun Valley Creek and to the north is Yellow Rock, a broad area that heavily reported encounters to the Octopus from its very beginning. The next road is 320 metres past the speed camera on the left.

As Craig was driving through the fog towards the speed camera, "someone" stepped out in front of his car from the left, forcing him to heavily apply the brakes and resulting in the car skidding.[36] By the time the car had closed the gap, the individual had taken a long stride into the next road lane. Craig said that the incident seemed to occur in slow motion as the individual went past his driver's

---

[36] The Octopus received a historical report of a similar incident that occurred along Mulgoa Road, Penrith—east of the Nepean River during the early 1980s. As the car swerved to miss the large biped crossing the road, the 'marsupial hominoid' extended its long arms out onto the car bonnet to prevent it from being hit. The vehicle left the road and crashed into a ditch. When the police arrived at the scene, the driver anticipated being ridiculed so he told them that he had attempted to avoid a cow crossing the road.

window. It was then that Craig realised that it was not a person. He said that he could mainly see its back and it was huge, seven-and-a-half feet tall, with shaggy dark brown hair, long legs, a straight body and arms. Craig didn't remember seeing anything else. The car continued to skid on until it hit the guard railing. He got out of the car and in spite of what he had just seen, he called out asking for help across the highway. Craig heard rustling noises coming through the fog on the far side of the road. He got back into the car and continued to drive home. John joked with Craig, saying that it was a pity that he wasn't speeding at the time, otherwise, he might have gotten a photograph of the animal from Roads and Maritime Services NSW.

As with many other road encounters, it is common to have a cluster of associated reports with long histories.[37] These 'marsupial hominoids' are crossing the road for a particular reason, usually because that location provides one or more of the best available options, including secrecy and cover, a shortcut, the easiest route, the only access point for many kilometres and other considerations. These habitual crossing locations create a pinch point where collateral encounters within the surrounding neighbourhood can also be discovered.

Craig's sighting occurred at a location where the combined highway and fenced railway barriers could be successfully crossed during the early hours of the morning,

---

[37] The sketch of a Dooligahl (Fig. 7.6) was made by one of two boys who saw a large human-like animal standing side-on in a cave below the Warrimoo Bush Fire Brigade building, while they were catching snakes to sell. The cave had vines growing across the entrance. The boys had seen eye shine on one occasion, coming from out of the dark cave.

7: Octopus

usually with minimal risk and effort. However, when attempting to cross a road during less favourable circumstances, kangaroos, cats and many other animals with excellent night vision can be blinded by the oncoming vehicle headlights. This usually results in a brief period of indecision, followed by a growing sense of panic that forces a reaction to the rapidly closing opportunity. Most nighttime drivers would have experienced this furtive animal behaviour. It would seem that this near miss with the car by the 'marsupial hominoid' was typical.

Like the experience of Susan and her father, Craig witnessed a road crossing where the 'marsupial hominoid' was attempting to get to the territory on the other side of the railway. From our experience gained from many clustered Octopus reports, it was always worthwhile to further investigate any encounter involving a road crossing because it usually meant that it was a preferred location, for one reason or another and there would probably be other related incidents awaiting to be discovered. The underpass at Green Parade allows vehicles, people and other bipeds to move under the fenced railway to the settlements and wilderness on the southern side. As tends to happen around these restrictive chutes, the Octopus initially received two associated encounter reports from the small linear settlement in the immediate vicinity of the Green Parade underpass.

Of several encounter reports from the area, the more detailed ones were received from a couple that I got to know very well. They lived close to the railway underpass and like many other residents, they had experiences with a Dooligahl for many years and were concerned about its regular activities.

Greg's latest reported encounter, amongst many others, occurred on Wednesday, 6 November 2013, at 10:30 P.M., with the sighting lasting for about four seconds. After hearing loud barking from a usually quiet dog from across their road, Greg turned on the outdoor light and opened the front door. He saw a figure standing near low-density bush on the front, easterly edge of their property at a distance of sixteen metres and looking towards his house. It turned and walked silently away.

Greg had been having regular sightings around their house, as had their neighbours for many years. I would have spoken to Greg regarding their encounters about a dozen times or more, with most sightings having involved a tall figure walking, running at high speed, breaking branches off trees, running through mounds of rubbish and jumping over fences. There have also been visits to their bedroom windows and similar incidents involving his neighbours. I spoke to a distant neighbour who had been independently reported to the Octopus, who told me of his similar experiences, also over many years.

Greg is a highly intelligent and very reliable professional witness. Using his artistic and scientific abilities, he was able to very accurately describe and measure what he saw by using me as a comparative model. In confirmation of what he had seen, he used Google Images to identify three drawings that most closely resembled what he had seen, ultimately deciding upon the same drawing[38] that was independently chosen by a family of witnesses five and a half kilometres away, along the same valley system.

The description was:

> HEIGHT: About seven feet or slightly taller
> SHOULDER WIDTH: about 27" (686 mm)

---

[38] A drawing by Katrina Tucker after a sighting made at Acacia Hills, Northern Territory, in 2010.

7: Octopus

ABDOMEN THICKNESS THROUGH THE UPPER CHEST: about 18" (457 mm)
UPPER ARM THICKNESS: about 8" (203 mm)—muscular, but in proportion to the body
HEAD: rounded but slightly pointed towards the top and merged into the shoulders, forming a triangular-like profile
BODY SHAPE: slightly stooped; muscular—in proportion; profile looked like it was wearing body-fitting clothing
MOVEMENT: silently walking on two legs, turned from front to side as it walked away
ILLUMINATION: mostly backlit by distant streetlights

Greg did not speculate on weight, other than to say that he thought that some of the "French Rugby Team or Arnold Schwarzenegger would have come closest in size, except in height". At some stage, Arnold stood 6'2" (1.87 m) and weighed 118 kg (261 lbs).

AROUND THE SAME time, the Octopus received encounter contact details from Pam, a colleague from school. Many teaching colleagues referred incidents to me after speaking to students, their parents or their neighbours which would have amounted to more than a third of the total teaching and support staff over a decade. This encounter was not associated with a pinch point like a bridge. It was a typical, even common encounter that must occur on a daily basis along the suburban/bush interface in the Blue Mountains. The family of witnesses live 5.4 kilometres from Craig's road-crossing incident in 2005, as well as Greg's and other's encounters surrounding Green Parade. Similarly, the location is less than five kilometres from the long-term Pendlebury encounters. More significantly, the main incident experienced by Peter occurred on Saturday, 19 October 2013, just eighteen days before Greg's encounter at the front of his house.

My interview with the family took place at their home. They took me for a guided tour of their property and the associated bushland and we inspected a number of damaged trees and bushes that were consistent with our accumulated experiences and knowledge. Peter, aged 13 years, was an intelligent, respectful and very well-mannered teenager. He was good enough to provide an encounter statement and some sketches of his encounters, while the others spoke separately of their involvement.

"I was walking the dog along the fire trail at the back of our house. After a few minutes I arrived at my comfortable rock, where I usually sit. I was about to let my dog

Fig. 7.7 Peter's family pointing to the location of the encounter behind their house in 2013. (Photo: Neil Frost)

off the leash so that he could run around, when suddenly I saw this big, black, furry thing. It was about thirty metres away, at the creek at the bottom of the gully. It ran on all fours extremely fast and disappeared into the very tall ferns at the bottom of the slope. My dog, Christmas, looked at it in fright, then looked at me, then we both bolted back home. This was the second time that I had seen this animal.

"The first time that I saw it was a few months earlier in the June 2013 school holidays. My two cousins, Chris and Harry were staying for a couple of days. Chris, Harry, four of my brothers (not Michael, he had stayed in the house) and myself were all playing Dr. Who down in the bush. We had gone right down to the bottom of the gully, over the creek and into the next gully. It was a foggy day with bits of fog between the trees and we were playing a game about the fog ghost. We were all in a group together and kept hearing loud rustling sounds and the snapping of twigs. We all thought that it was a kangaroo or something because we had seen kangaroos there before. Chris said, 'Let's split up and find the fog ghost.' So we split up into pairs. My little brother Joseph was on his own. I was with Brendan and after a couple of minutes, when we had finished looking around, we came back. Chris and Dominic were already back. Joseph came back next and said that there was something following him.[39] We thought that Joseph was just pretending that the fog ghost was following him. Anthony and Harry came back next. When we finished playing we heard the faint voice of my mother calling out, 'Lunchtime.' We all came back to the house and the others were a little ahead at the trampoline at the

---

[39] This association with small children was something that Merve had specifically warned Sandy and I about, back in 1993 after a yellow baby's '000' jumpsuit was found in the swamp.

bottom of our garden. I was still on the fire trail about ten metres behind them when I again heard branches snapping and leaves rustling. I turned around and through the saplings and branches caught a glimpse of a large, black, furry animal walking through the trees. It had its side to me and was about fifteen metres away. I thought that it must have been some big emu or kangaroo[40] or something because I had not seen it clearly through the trees and the undergrowth. I started to follow along with it. Then it came into a small section that was not so thickly wooded. It stopped there and I saw it properly for the first time. I was completely freaked out. It was not an emu or kangaroo. It had two legs with very long arms hanging down its side with huge, thick claws.[41] It was covered in thick brown fur and was about seven feet tall. It had a hunched back. I couldn't see its head, that part was behind a tree. That's when my heart started beating violently and I turned around and ran past the trampoline, up into the house and onto the deck. From the deck, I saw it moving away through the trees on all fours. Later on I asked Joseph if he was just pretending that something was following him to make the game more exciting. He said, 'No, something really was following me.'"

This encounter statement is very important for a number of reasons. Firstly, Peter saw the 'marsupial hominoid' "move on all fours", run "on all fours extremely fast" and "moving away through the trees on all fours", on three separate daytime occasions. From our experience with

---

[40] This was a daytime encounter. With most nighttime encounters, kangaroos are the default explanation, unless the encounter is extreme enough to exceed the limits of a person's life experiences.

[41] All marsupials have well developed front legs with claws, that are used by the embryo to pull and crawl from the birth canal to the marsupium.

7: Octopus

Fatfoot, quadrupedal locomotion was used to rapidly exit an area whilst maintaining a very low profile. Ian and I had suspected that this was the behaviour during many nighttime pursuits because Fatfoot's height should have towered above most of the vegetation if she was running bipedally. Also, quadrupedal locomotion is faster than bipedal, which her estimated speed of slightly more than 40 km/hr, seemed to suggest. Additionally, the sound of locomotion or gait was different when moving fast. It wasn't until Steve Ruston bought us Russian night vision binoculars that we could finally see this behaviour, though only barely. On the first occasion, I had been monitoring the relatively deep and cleared area on the edge of the swamp from a location that was about thirty metres distant. As I traced the sound of Fatfoot moving to the south, this was followed by the sound of a rock or other object landing noisily ahead of the apparent direction of travel. As I scanned back in the opposite direction through the cleared area with the binoculars, I caught the 'impression' of Fatfoot bent heavily forward, running to the north. The image was not very clear and only gave a 'feeling' of what was actually occurring.[42] The first comment that Ian made was that the beastie is intelligent. It uses deception. He went on to say that what I had seen, seemed to confirm our

---

[42] A night-vision device (NVD) amplifies the available light to create a brighter image and is dependent upon the viewed objects having different reflectivity in order to create contrast. Unfortunately, an NVD only allows vision up to the point of the first obstruction, e.g., a leaf or mist. Anything beyond that remains invisible, unlike forward-looking infrared (FLIR) devices (particularly medium wave IR) which can usually see beyond or through these translucent barriers. Additionally, a FLIR camera is a detection device because it can discriminate between hot, living animals and most other colder objects.

belief that Fatfoot was both quadrupedal and bipedal.[43] Clearly, by utilising both methods of locomotion, Fatfoot could take advantage of the speed of quadrupedalism on the ground and its increased dexterity in the treetops, or alternatively, the energy efficiency of bipedal walking.

The second very influential observation made by Peter was the "very long arms hanging down its side with huge, thick claws." Having "huge, thick claws", which other witnesses have further described as being "black", definitively identifies this animal as not being a primate. Primates, like humans, have flat nails or plates on the tips of their fingers and toes. Human nails tend to be pink. Claws are, however, a definitive characteristic of all marsupials, which have well-developed front legs or arms with dark or black claws, that are used by the embryo to crawl from the birth canal to the marsupium.

When we all went for a walk in the valley below Peter's house, I could identify a number of markers in the vegetation that suggested that this Dooligahl was making regular visits to the valley. Specifically, young acacias had been split down the middle to reveal the contents of their insect and larval chambers, which clearly would have required two hands to pull apart. There were also the archetypal snapped-off tree trunks, non-avian treebites and snapped branch tips.

THE CRITICAL ADVANTAGE of the Octopus was its high reporting resolution, particularly involving local incidents. In comparison with one-off historical records, the increased quantity, frequency and detail of reports received did not miss many things and allowed a greater understanding of

---

[43] The 'marsupial hominoids' can be further labelled as such because they are not limited to bipedalism as their only method of locomotion.

interconnecting incidents, which would be invisible otherwise.

Not all crossing points were bridges, overpasses or underpasses. Like Peter's encounter, some were simply a part of a habitual route, as was the situation with the crossing observed by Brad Crofts, where the Quinkan was making its nightly pilgrimage to the O'Connor residence.

A large number of observed road crossings are shortcuts that connect with well-established trails. They conserve energy by avoiding having to travel the long way around. Occasionally witnesses mention seeing a humanoid figure standing by the side of the road. Some add to their description by mentioning that they saw "red eyes" or "long arms," that the dark figure appeared to be in trouble, or that the biped looked like it was wearing bulky clothing, such as a "greatcoat." A few commented on seeing in the rear vision mirror, a figure running across the road after the vehicle had passed by.[44] Sometimes in our valley, I will get out of bed during the early hours after hearing the dogs barking and wait for something to happen locally. Occasionally, for example, a series of motion-sensing lights will activate somewhere across the valley, or a conga line of neighbourhood dogs will join together in communal barking, which on probability, means that Fatfoot is taking a shortcut through a neighbour's property.

---

[44] Our daughter, Avril, saw a large dark figure standing by the side of the road near a pedestrian railroad level crossing while travelling along a local back road at about 2 A.M. After she passed by, Avril looked into her rear vision mirror and saw the large figure move quickly into the middle of the road, balk and then run back, before making a final attempt to cross the road. Similarly, Bev, a school assistant, was driving past her girlfriend's house as she was about to exit her driveway early in the morning when her attention was ripped away from her friend's car by the presence of a huge, tall, human-like figure standing close by in their neighbour's driveway and only separated by a border shrub.

In May 2017, Michael and his family moved to the Blue Mountains. As a courier driver, he needed to leave very early in the morning so that he could travel to his pickup depot and load, while trying to avoid the morning traffic rush and hopefully, have enough remaining time to have a quick nap. Winter was always the worst time because it was darker for longer and Michael would usually return home after sunset.

Recounting the story of his encounter, Michael wrote:

"Our family moved from Western Sydney to the Blue Mountains in May 2017. We bought a double-storey house on a small acreage because we wanted to have a more peaceful life. I operate a very successful courier business, however, I needed to drive to Sydney, a distance of sixty

Fig. 7.8 Fatfoot crossing the road in front of Michael, a courier driver sitting in his van while preparing it, before leaving for work at 4:30 A.M. It took very long steps as it walked across the road—Winter, 2018. (Source: Michael)

7: Octopus

kilometres each day. I typically get out of bed at 4 A.M. and get ready for work. On the morning of the encounter, it was about 4:30 A.M. during the winter of 2018. The weather was clear and cold. The visibility was very good because there was a street light in the distance. I walked in the dark to the van parked on the street and started to get the vehicle prepared for the day. I normally turn on the van camera before I drive off and was just about to do this when something black and huge walked out of our neighbour's driveway and onto the roadside. Our neighbours weren't home because they were travelling in their caravan and their property was dark. This very large human-like figure walked across the ten-metre-wide road in front of me, only taking five, long steps. It then disappeared into light bush and along a track that leads to the next valley. The animal was about seven to eight feet tall and very large. I then turned on the collision camera and drove to work."

Michael bought a house in our valley where the previous inhabitants had had very little experience with these 'marsupial hominoids', except for their teenage son, Evan. Also living nearby were three similarly aged teenagers, Steven, Tom and Ben. The four boys would frequently play Spotlight, which is similar to Paintball, except that no guns or paintballs are used. Instead, the boys would hunt each other in the bush using torches, which inevitably led to contact with Fatfoot and her family. One evening in 1990, Steve ran home to tell his stepfather that there was something big in the bush with red eyes that would growl at them. Steven's stepfather told him that they were probably interacting with wallabies and that he should "grow some gonads". Steven's stepfather was Ian Price.

Three years after this incident, Evan showed interest in our local research, as did several of the other young men. Evan would often sit in the bush below his house and listen but always from behind their rear wire fence. When I questioned Evan one day on what he had discovered, he told me that, "They walk with the wind." This was a very perceptive comment that we had also observed. We had particularly noticed that on windy nights, Fatfoot would patiently stand still and wait for the next wind gust to come through the valley, before briefly resuming her walk. If someone was not aware of this behaviour it was likely that they would find it difficult to separate the two sources of noise and Fatfoot could continue to move about with impunity. From time to time, Evan, Steven, Tom and Ben would report incidents that they came across when talking to their peers.

Although Michael's encounter was very brief, it confirmed several things that we had suspected. With Michael's estimated height of Fatfoot[45] being between seven to eight feet, we were more comfortable with his lower estimate, but with his encounter being so close, he did draw attention to her very large size. Much more significantly, Michael said that the biped took five very long steps[46] to cross the ten-metre wide sealed road, giving her a step length of two metres, or about 2.4 times the step length of a typical male human. His estimate is consistent with many other observer statements, particularly Craig, who described the passage of the biped during their encounter, saying that by the time the car had closed the gap, the individual had taken a long stride into the next road lane.

---

[45] For various reasons, in 2018 we were confident that we were only dealing with Fatfoot, after many years of uncertainty.

[46] On his drawing, Michael used the colloquial word 'strides', but after seeking clarification, he actually meant steps. A 'stride' is equal to two steps.

7: Octopus

Having a long stride is a key physical characteristic of extant kangaroos. When a long stride is combined with a low to medium stride frequency, greater energy efficiency of locomotion is achieved. At higher stride frequencies, faster speeds are obtained, but only at a slightly greater energy cost. From a standing start, kangaroos can easily jump two metres and when at full speed can achieve bounding distances of about nine metres.

Michael and his two sons have also become involved on many occasions with our local research since the initial encounter. One evening at about 9 P.M. in April 2021, Michael called me on his mobile phone to say that they were following the sound of walking on their land that was heading away from them towards the swamp. His son Josh had a new LED torch that he had been testing, which had a beam that was currently incinerating the bush. On moving outside, I immediately picked up on the sound of heavy walking as it travelled parallel to me and a nearby wire fence. Just inside our door, I had a fully charged FLIR[47] that would have enabled me to clearly see the biped, but I didn't attempt to get it, despite it being so close, as it would have taken more time than I had available to prepare it. Consequently, I delved into my bucket list of alternative research tasks and immediately began to count steps and started the process of visualising a start and stop point along the fence line, as I listened to the walking while witnessing the oncoming visual assault from Josh's light sabre.

---

[47] Sceptics typically bewail and bemoan witnesses for not being able to photograph or video a cryptid in this age of smartphones and digital cameras. Most encounters happen spontaneously and terrify the witness into relative inaction. Smartphones need to be logged into, applications found, selected, opened and options chosen. Then you might be able to take a video, but wait, no, it is gone.

I only counted ten steps until the sound of walking was no longer heard. Considering its direction of travel away from the light, the biped would have intersected with the wire fence. Dooligahl normally step over fences with ease, however, this fence would have been much easier, because it was very poorly constructed and very low, being only 800 mm in height. As it had repeatedly demonstrated, it was not even capable of containing goats effectively. Having stepped over the wire, the sound of walking would have been absorbed by the softer ground on that side, making any continued aural tracking impossible and an easy escape to the swamp for the 'marsupial hominoid'.

8: Dooligahl

EVEN THOUGH I had had encounters in 1966 and 1981, we remained ignorant of these Wildmen until Sandy and I became Yowie-aware in February 1993. During the intervening twelve years, there were other incidents that involved the experiences of family, friends and students, that we would later become aware of.

In early 1986, Sandy received an urgent phone call from her sister. The night before, her niece Zoe had run away from home. We immediately packed the van and drove to Carol's house to offer comfort and support.

After interrogating Zoe's friends, Sandy obtained information on her likely location and soon found her. During discussions with Zoe and her parents, it was decided that she would live with us and enrol at our local high school. This was especially advantageous because she could recommence life in a new environment and as I taught at the school, my colleagues and I could keep her under close surveillance during the day.

The following months were not easy and there were many confrontations with Zoe. One particular disagreement involved her smoking in her upstairs bedroom, which we were strongly opposed to. Unfortunately, this altercation occurred at the beginning of a long weekend holiday, which resulted in Zoe defiantly staying in her room for three days and destroying any hope that we had of having a holiday. Consequently, fearing that Zoe would run away again, Sandy and I took turns standing guard. We called this incident 'The Siege'.

After 'The Siege' was broken, Zoe agreed unconditionally to our terms. As the cost to us had been very high, we insisted that she must not smoke inside the house and threw in some additional conditions regarding other matters that had been concerning us for some time! True to her word, Zoe never smoked in the house again. Instead, she would climb out of the second-storey window and have a cigarette while sitting on the verandah roof.

Like most mountain residents, we were aware of the eccentric 'Y' word that had gained some notoriety during the mid to late 1970s, mainly because of the prolific number of newspaper articles written by Blue Mountains researcher, Rex Gilroy,[1] who had propagated the use of this proprietary term and provided various accounts of the phenomenon. Although his writing certainly raised non-indigenous awareness, unfortunately, in my opinion, it did much more to increase public scepticism. Prior to the books and newspaper articles written by Rex, these 'marsupial hominoids' were more typically referred to using Australian colonial terminology such as 'Hairy Men' or 'Bush Apes' and formed a large part of 'Australian bush mythology.' As such, any attempt to make sense of the phenomenon was typically met with inane comments and criticisms that suggested that encounters were largely exaggerated and distorted stories told by bush raconteurs. Consequently, studies of the 'Yowie' were shared and grouped with fantasied

---

[1] Rex Gilroy was a pioneer Yowie researcher who was brave enough to write about this taboo topic. Unfortunately, Rex did not help his situation or research generally because of his controversial methods and opinions.

8: Dooligahl

stories of 'Drop Bears'[2] and other misrepresented animals, that were used to highlight and attack the supposed gullibility and stupidity of those who dared to mention them. Even though most 'myths' are said to contain a kernel of truth, encounters with these Wildmen failed to gain any traction, even when the veracity of the witnesses seemed to be impeccable. Any potentially worthwhile attempts at discussion were typically cut short. The use of the word would instantly initiate ridicule and become synonymous with a fool.

However, from an Aboriginal perspective, there was no problem. Every tribe held trusted and detailed traditional knowledge of these spiritual animals and has at least one generic name for them or more when other anthropomorphic species have been separately identified. As a result of the collective investigations of many researchers, it seems that 'Yowie' is the non-indigenous corruption of the Aboriginal word 'Yowrie' or any number of other similar-sounding possibilities. Consequently, the name 'Yowie' has too many negative connotations and should be rejected in favour of more legitimate Aboriginal or Colonial Bush names.

While Zoe was at my school in 1986, an incident occurred that was very memorable for many people.[3] On arriving at school for early morning playground duty, I

---

[2] Typically used to terrify tourists, the 'drop bear' has a high probability of being based on fact. It may have been a large, ambush, predatory marsupial, like *Thylacoleo carnifex* or marsupial lion. The origins of this story are impossible to trace, but may be related to Aboriginal cultural memories dating back 30 kya, when they became extinct. Alternatively or jointly, the three marsupial hominoids are also tree dwelling and are ambush predators.

[3] In 2021, I spoke to Peta, a former school captain from a few years after this incident. She commented on how brave Troy had been and how badly he had been bullied for speaking up about his encounter.

found a large group of students encircling Troy, who were unmercifully interrogating him about his recent Yowie encounter that had been reported in the local newspaper. He was receiving plenty of ridicule when I saw him, which he clearly wasn't enjoying.

As Troy was having a difficult time with the other students, I broke up the large group and told Troy to keep close to me until the bell rang for the start of school. On arriving home, Zoe was able to give me a student's perspective of the morning event. It seemed that Troy had given an interview to the *Blue Mountains Gazette* and everyone had read the newspaper. Wherever Troy was seen at school, or in the wider community, other students would loudly call 'Yowie'[4] which greatly affected him.

A short time after becoming Yowie-aware, I contacted Troy. He was so happy to hear from me that he immediately left his work and drove to our home to talk. He was extremely glad to discuss what had happened to him with someone who understood, as he had learnt to keep quiet over the past seven years. This is typical of the experiences of most witnesses, even after they attempt to broach the topic under safe circumstances with a few close friends and relatives.[5] In Troy's case, he had an extremely supportive father. Ray was fully prepared to back his son and so he spent many nights at the same contact location until he also had an encounter, which he similarly reported to the *Blue Mountains Gazette*.

---

[4] Cowardly students would similarly shout this to me from a distance but only briefly after 1993, until the research started to gain respectability, mostly because of the wide community involvement with the Octopus.

[5] As similarly reported by Mary after mentioning her encounter to friends: "I told one or two people about it, but you know the kind of reception you get. They think: 'She'll be talking about UFOs next!' So I didn't elaborate any more . . ." Healy and Cropper, *op. cit.*, p. 114.

8: Dooligahl

On the night of Troy's sighting, he had caught the last passenger train up the mountain, sometime after midnight. After notifying the guard to stop the train at the unmanned station, he exited the last carriage at Linden, crossed the highway and began walking east on a side road parallel and close to the old main highway. There was a small strip of bush separating the two roads at that time with extensive areas of National Park on either side that extended far beyond the horizon. Hearing a noise from behind, he turned to see a very large, dark humanoid figure faintly illuminated by the street light, crossing the road and turning to look at him. Fearing what he had seen and that the figure might pursue him, he ran the remaining distance home.

The unmanned Linden Station was highly overrepresented amongst railway stations on the Main Western Railway Line in reports to the Octopus. The most common reports mentioned by waiting passengers were growling sounds coming from the bush surrounding Burke Road on the northern side of the Sydney platform at nighttime. A variation of this included a few reports where vegetation was heard being thrashed about while tall eucalypts swayed in the background, most likely because the idle Dooligahl was becoming increasingly impatient to cross the tracks. It seems that Troy's encounter site was an earlier crossing-over point for at least one Dooligahl, most probably involving movement through the Linden Station pedestrian cutting that finishes at the edge of the highway. A historical and corroborative report obtained by the Octopus was told by Pam and Brian who had a large biped repeatedly pass through their property during the early hours, seemingly taking a shortcut from the highway into another valley. However, major highway reconstruction was carried out between the years 1991 and 1993 which would have hindered the viability of this route as new railway fencing and masonry work was initially installed and further upgraded by 2004. Interestingly, a several-metre-tall 'community painting' of a Dooligahl's face was displayed above the eastern or Sydney lane of the Great Western Highway, on the Station Street overpass bridge at Woodford for many years until it was overpainted by a humourless local council.

However, not all students are easily intimidated by confronting incidents and some clearly thrive on the adrenalin. Andrew was one of my senior Industrial Technology students in 2010. Andrew was a delightful larrikin with tattoos and an intimidatory presence, who owned the respect of the entire student population. In many ways, Andrew reminded me of a much younger Ian, which could have explained why I liked him.

For many weeks we heard late-night explosions in the next valley, which sent substantial shockwaves bouncing through the valley system that woke the sleeping communities. The detonations became louder and more powerful over time as the perpetrator perfected his shenanigans. Then, early one morning, Andrew excitedly came to see me in the staffroom before school. He said that he had an encounter with a Yowie. During the early hours of that morning, he had ignited the fuse on his bomb and was running down a remote bush track to safety when he was forced to stop. Andrew said that a huge beast was standing across the path blocking his way and it didn't look happy! In its two hands, it was holding a thick but short tree branch that was being held horizontally across its chest. Since he wasn't going to pass it, he turned around and ran back the way he had come, past the activated bomb and along another track that headed in the right direction. Unfortunately, Andrew was not able to provide a detailed description of the beast but his interest in the topic was certainly piqued. A short time after this, Andrew came to see me while I was on playground duty. He said, "You know the Year 10 boys who have a smoke behind the water tank? I heard them putting s**t on you about the Yowie. You don't need to worry about them Sir—they are believers now!"

It was probably not surprising that Andrew was confronted by Fatfoot late one evening, after repeatedly detonating bombs on a local bush track but it was surprising that the police had not arrested him. Dooligahl, Quinkan and sleeping residents do not like loud noises. Andrew would often overtake my ute on the way home after school, repeatedly beeping his car horn and frenetically waving his hand from out of his driver's window.

At three in the morning in 2021, I pulled into a garage to fill the ute up with distillate and the service attendant recognised me as his former teacher from more than a decade earlier. Jake told me that, in addition to getting married later in the year to another student that I knew, he had an encounter recently when taking an early morning shortcut through the scrub on the way to work at the service station. It was the same track where Andrew had had an encounter with the large biped. Jake said that the very large animal was blocking his way, so he slowly turned around and returned the way that he had come.

Also a short time after becoming Yowie-aware, during a family visit in 1993, we told the adult Zoe of our recent encounters. To our complete surprise, Zoe spoke of several experiences while she was living in our home. She mentioned her frequent evening cigarette that was smoked on the verandah roof, something that we were not aware of. She then expanded on her revelation by saying that she had often heard walking in the bush below the verandah. On a few occasions she had also seen 'fluorescent' red eyes looking up at her and occasionally, she could make out a dark figure sitting in the bush on Ian and Cheryls' land.

Of greater concern was an incident that occurred towards the end of Zoe's stay with us, around December 1986, during the late afternoon. We had been very strict in our supervision of Zoe but we had allowed her boyfriend from school to visit a few times during the day, but not stay in her room. Having snuck away from the house, they had walked to the bush at the edge of the swamp. Hearing movement as they arrived, they saw a large, dark figure squatting among the reeds. James abandoned Zoe by immediately fleeing the swamp but, Zoe maintained her presence long enough to get a brief glimpse of the animal. She said that the eyes were "dark and quite human" but, seemed to be "too close together in relation to the size of the head." She commented that the appearance of the face was very aggressive, seeming to say, "Don't mess with me!"

Zoe further recounted a much earlier incident that she, her brother Gabriel and cousins Natalie and Sharlene had experienced as young teenagers when living on her isolated farm in Nerrigundah,[6] New South Wales. This was another encounter that we were not aware of. It was summer, around Christmas 1982, after lunch. The children had been playing on a bush track when they heard, what they thought was the sound of people approaching them. Thinking that children from a nearby farm were in the area, they pointed sticks in that direction and shouted that they would shoot if they didn't stop. Coming out of the valley, diagonally towards them from their left, at a

---

[6] Zoe's mother Carol told me that the local farmers around the Nerrigundah area (NSW South Coast) would regularly report the loss of cattle and calves over a period of more than twenty years, from about 1970. Nerrigundah farmers believed that a professional butcher was slaughtering the cattle for their prime cuts and leaving the carcass remains in the fields. This imaginary scenario was very similar to the background information provided to Ray, a retired NSW South Coast Police Senior Detective, who investigated the loss of lambs in the Goulburn district in 1990.

8: Dooligahl

distance of about fifty metres, was a very large, tall, dark, hairy, manlike animal that crossed in front of them a short distance away. Zoe said that "the hominoid was about eight feet tall with a human-looking face, though it clearly wasn't a person." It had "disproportionately long legs that took slow but, very long steps, covering the distance very quickly." She said that the animal didn't seem to be in a hurry and only casually glanced at them as they walked past. When it reentered the bush they could hear it loudly breaking branches as it moved away into the next valley. At the time they had been very frightened and had spoken loudly about the "guns that they were carrying," even though they weren't because "it seemed like they should try to protect themselves somehow!"

A few minutes after this, their adult neighbours appeared on the scene enquiring about the commotion and the talk about guns. Zoe said that they explained what had happened but, the adults were not particularly concerned. She now thinks that either the neighbours did not believe their story, or they had some prior knowledge and possibly had some previous experience.

Speaking to Gabriel about their encounter, he also said that the animal was "very large and powerfully built, about eight feet tall." He commented that it seemed to "glide along" and gave the children little attention except for a quick glance when they shouted at it. Gabriel also remembered that the animal made a lot of noise as it walked, with twigs and branches snapping loudly under its considerable weight, although it left no obvious sign of where it had been. Adding to the description, Gabriel clearly remembered that, "The hair covering the body was ginger-brown because it was the same colour as our dog."[7]

---

[7] Their dog's name was 'Red'. Red hair is an adaptation for low light environments which assists with the production of vitamin D.

EVERY ABORIGINAL TRIBE has a least one name for these Australian humanoids. Most tribes differentiate between the various types and their sizes, by providing two and sometimes three names. Dooligahl is a tribal name that was adopted by us after speaking to Merve, an Aboriginal Elder from the South Coast of New South Wales. After we had become Yowie-aware in late February 1993, we sort assistance and reassurance from the Aboriginal community and Merve was the first person to provide us with answers and a culturally appropriate name that matched the specific characteristics of our local Wildman.

I spoke to Merve over the phone on two occasions. Although we knew a few things about our humanoid, Merve placed this information into context and introduced us to many more details, particularly what we should be aware of. It was also very apparent that Merve was trying to help us in any way that he could.

Merve began by describing our land without having seen it. He asked questions about our house and its location. With hindsight, it is now possible to see what he was concerned about across a broad number of areas. He was able to determine that our house was isolated and therefore, it was prone to visits by inquisitive Dooligahl that could approach on all sides, thereby allowing close access to every external part of the house. We were also surrounded by thick bush which would have assisted a Dooligahl with a stealthy approach and departure. Consequently, Merve wanted to know about any tracks that existed on our land and particularly, if we had built the house over any of them. I told him that we had several tracks that had passed through our house site and that we had found a series of large footprints in the soft soil, shortly after we had excavated the foundations in 1983. Merve said that his tribe regard Dooligahl as being the "Protectors of the Environment,"[8] because they react violently to any clearing of the

8: Dooligahl

land, burning off and more recently, fencing and other farming activities. He said that they have habitual movement patterns and do not like having to alter their routes.

I described the appearance of the footprints to him as best I could, considering that ten years had passed and we were not very interested in them at the time. We thought that we had determined their origin to be a combination of dog and horse tracks. Our tracking analysis was faulty but it had been based on the knowledge that a horse, which was stalled further down the valley, would very frequently shy and bolt from its stable and fenced paddock during the night[9] and walk throughout the valley until it was haltered the next morning. With hindsight, the response of the horse is understandable, as was the behaviour of our collie dog, Patch, who went missing early one evening in the late 1980s and was found many hours later after a search, exhausted and no longer able to bark, as he laid in the corner of the empty horse stable, in a hole that he had dug himself. I had to carry Patch home through the bush and over several barbed fences because he couldn't walk and was clearly in shock.

Like other members of the Parramatta Aboriginal Land Council, Merve told us that the use of light affected their behaviour. He told us that light will attract Dooligahl from far away and that it will also repel them at close range. He told us to stop feeding the local native wildlife, especially the carnivorous birds, as this will attract the attention of Dooligahl with the promise of an easy meal. By extension,

---

[8] Having spoken to a few North American First Nations people, they similarly say that their Sasquatch are also regarded as the 'Protectors of the Environment'. They also said that Sasquatch bite maple trees to get the sweet sap.

[9] Local historical records from a century ago mention that horses would frequently balk, or refuse to proceed down many local roads at night.

we should have stopped feeding our dog outside on the path leading to our front door but I don't think that we would have tolerated the mess inside the house.

Knowingly, Merve asked how far from the swamp our house was. He told me that his tribe are wary of Dooligahl when near or passing through a swamp. He said that Dooligahl would sometimes reach out and touch the young girls and women on the legs as they moved along the thickly vegetated swamp tracks. For some reason, this was considered 'good luck'.

Of all the information that Merve had passed on to us, the most disturbing was related to children. Over the years there were many encounters that confirmed this malevolent intent,[10] with Merve's own story of a child being taken from around their campsite, during the time of the 'Stolen Generations', being the most distressing account. However, in the past anthropologists would frequently interpret such losses as infanticide, which is what I was told when at university. I could never understand why Aboriginal mothers should be any different but it was most probably the distorted accounts from religious fanatics like Mrs Charles Meredith[11] during the 1840s, that portrayed Aboriginal women as 'wretched and unnatural mothers' and at the same time, relegating the real existence of Yahoo and Devil-devil to the hidden, unreal and taboo world of evil spirits and superstition. After all, these animals don't exist, so the Aboriginal mothers must have killed their own children!

---

[10] Examples include: Peter's encounter report where his young brother Joseph was closely followed; the '000 yellow jumpsuit incident' and the home invasion at Wilhelmina's house; and children being watched through their bedroom windows. Many other accounts demonstrate the close association between children and Dooligahl.

[11] Mrs. Charles Meredith. 1973. *Notes and Sketches of New South Wales: During a Residence in that Colony from 1839 to 1844.* Sydney: Ure Smith.

8: Dooligahl

THE EVENTS THAT followed us becoming Yowie-aware were traumatic and paradigm-shifting. Sandy and I were extremely fortunate to have very supportive people like Merve and neighbours Ian and Cheryl Price. Without their contributions, the course of events would certainly have been much different.

When Ian accompanied me down to our house after asking him for help, we weren't expecting him to immediately attempt to ameliorate the situation. We would have been happy if Ian was simply able to confirm the reality of our predicament. Instead, Ian engaged with the Dooligahl, an action that clearly surprised me and the Dooligahl more so. As was to become a regular conclusion to the night's engagement, we would return to our verandah to have a smoke and a cup of coffee and discuss what had happened, while the biped would return to the house and watch the proceedings from the edge of the bush.

During this inaugural post-chase evaluation, we discussed the nature of the animal and we thinly drew upon our knowledge of the Yowie. From Ian's immediate experience, he said that the beastie sounded like it was extremely large, the equivalent of several front row footballers, yet, it was also extremely fast, being able to quickly outrun his pursuit and clear fences in the process, that he had to climb over. He also commented on its gait, saying "It was not quite right." Ian was also highly surprised, even disappointed, that his old antipersonnel dog Toby was frightened by the confrontation and refused to follow his commands.

The next night we both walked down to the swamp. It was here that we heard our first foot thump at very close range. Ian thought at the time that the beast was 'thumping its chest'. I thought that it might have been a large rock or tree stump being forcefully flung to the ground or the biped was thumping its foot. To be certain, the purpose of the thump, in this case, was intimidation.[12] As soon as Ian turned on his quartz-iodine spotlight, the 'marsupial hominoid' took off to the north at high speed.

Over the following nights, Ian and I spent time engaging with the biped, frequently up to and beyond midnight. Consequently, Ian decided to limit his involvement to Friday nights because he was finding it difficult to function at work. However, Ian would arrive on our verandah most afternoons when Cheryl was not home for a cup of coffee and a chance to discuss the weekly events and plan our strategies for the upcoming Friday night.

From these early interventions, we had confirmed that the biped was repelled by bright light and we could, therefore, use light as a weapon to protect ourselves. This was good news for me because I had been carrying a tomahawk down the back of my pants during our nighttime pursuits as personal protection and at the best of times it was very uncomfortable. In order to reassure me in his typical style, Ian made several comforting comments about me carrying the tomahawk. In particular, I remember him saying "It's probably going to hurt but not for very long" and "Using the tomahawk on the beastie is only going to make it angry and that you would be better off using the axe on yourself because it would be far quicker and less painful."

I was a stay-at-home Dad, looking after our children and educating them during the day before daytime childcare had become popular and both parents needed to be fully employed to afford the preschool fees. Around this time, Avril was attending preschool a few days in the week so that she could socialise and I was looking after Drew during the day. When Drew had his midday sleep, I would

---

[12] As a consequence of this characteristic behaviour, we coined 'intimidation' as the collective noun for Dooligahl.

8: Dooligahl

search the area around our house, noting changes to the vegetation and mapping the constant evolution of the tracks. On a few occasions, I thought that I felt the presence of the Dooligahl in the swamp even though it was during the day. This was something that Merve had told us to be aware of.

As each Friday approached, we would make plans for the night's excursion. Sometimes preparations needed to be made, depending upon what had been decided during the week. As the weeks went by, we started to notice some changes. Perhaps the most noticeable was that the biped was not trying as hard to escape our pursuits. When we first started, the biped would flee at breakneck speed and could be heard disappearing into the distance. We were progressively finding that pursuits were becoming quicker and shorter for us and the Dooligahl. As a consequence, we started to strongly suspect that we were over-chasing our quarry and that it was hiding in the bush as we passed by. One evening, Sandy and Cheryl were sitting on the steps along the path to the driveway, listening to our chase. When we returned home, they asked us if we knew that the biped had been following close behind us along the track. They said that they could hear us walking slowly on the track and also the synchronised walking of the biped, a short distance behind. Ian and I weren't aware of this. Now knowing that this 'marsupial hominoid' was extremely cunning we devised another plan. Ian and I went back to the track and immediately ran along it, which forced the biped to run away. After we abandoned the chase, we both started walking back to the house, but this time we agreed to stop on the count of three. Having stopped, we could hear the biped's continued step a short distance behind and so, we turned and resumed the chase. After we returned to the house for the final time, we all sat on the verandah and discussed what had happened. We realised that, just as we had learnt from the animal, it had learnt from us. Ian's attempt to 'crash tackle' the biped and my attempt to follow him and witness his brutal dismemberment was no longer a possibility. Instead, the pursuits had become a game and its name was 'hide and seek'.

Meanwhile, during the weekdays, I would continue with my work. A major part was mapping and determining the active paths so that we could better understand what the biped was doing and where it was going. As we didn't have anything that could be used for this purpose, I thought about a low-technology solution that was also very passive. I stole a reel of thin cotton thread from Sandy's textile collection to use as a tripwire to monitor movement around the swamp. After tying some lengths to trees on either side of a track, it became obvious when 'test walking' that anything moving along the path would feel its presence and become immediately suspicious. The thread could not be strong, like linen thread, because it would cut into whatever was pushing against it and become detectable. Thin, inferior-grade cotton, like Chinese thread, was best because it was easily broken if any force was applied to it. A modification was made to the thread's installation where the cotton thread was tied to bushes rather than solid objects like trees but this resulted in the attached and remote vegetation moving about in an unnatural and suspicious manner in a direct response to whatever was passing through the trap. After several days of experimentation, the optimal arrangement was anchoring one end of the thread to a stout tree and then freely draping an excessive amount of the cotton across the intervening vegetation, leaving the other end unattached. When test walking through the thread, it didn't seem overly suspicious, feeling very similar to a fine tree vine or an insect web. It was important to have one end fixed because otherwise, the thread would continue with whatever had snagged it and not be found. Having a fixed end caused the cotton to slip around the animal's body but still maintained the location of one of the end reference points. On finding the line, it first indicated that something had interfered with it. Secondly, depending upon the installation, the thread indicated minimum animal height and thirdly, its trailing pattern gave an accurate indication of the direction of movement along the track. Over time further modifications were made, mainly adjustments in the height of single lines, using multiple threads at the same location to more accurately determine the height and using different coloured cotton to identify the date of thread installation when threads weren't immediately affected.

The main problem with doing this, or any other installation work, was that it took a lot of time in the swamp to complete and additional time checking and following up on the activity. Merve had warned us that these 'marsupial hominoids' are mostly nocturnal, but are also diurnal, so it was very important to remain constantly aware, especially with anything involving our children. Consequently, during these long periods of work in the swamp, there were occasions when it seemed like the biped was watching my activities. Confirmation of this happening occurred for the first time during the autumn of 1993, when I suspected movement to my north at around 4:30 P.M. After paying closer attention to the bush in his area, I noticed a band of brown 'jenny wrens', or mostly female wrens known as Superb Fairy-wren *(Malurus cyaneus)*, in the dense shrubs, that were making an angry chatter, which is the typical behaviour that we had observed when a Carpet Python *(Morelia spilota)* or Green Tree Snake *(Dendrelaphis punctulatus)* is present. Listening further, I could hear movement which suggested that, like legless lizards, this was no snake! As I moved pointedly in the direction of the fairy-wrens, the band immediately in front of me erupted into flight, followed by the sound of 'running' as the next band of wrens also took flight. As I ran along the track on the edge of the swamp, ahead I could see other bands of wrens take to the air, which gave me an accurate indication of which fork along the many paths to take. As I turned a bend in the track that headed towards the swamp and having only run fifty metres or so, I could see a hole in the vegetation still in the process of closing in. I couldn't hear the sound of running anymore, but it was a fair guess that the Dooligahl had taken refuge amongst an island of thick vegetation ahead. Waiting outside the newly formed bush entrance, I thought about what I should do. Being a time when mobile phones weren't in use I couldn't tell anyone where I was, particularly Sandy who was at home and similarly with Ian. I guessed that if I left the area to get backup, then the animal would simply slip away.

Standing on the transitional edge of this forested wetland was a middle-sized Scribbly Gum *(Eucalyptus haemastoma)* which had some branches that grew over this swampy island. Having smooth bark made it more comfortable to climb and so I slid out over the vegetation and peered down, hoping to see the biped. Instead, I heard a long growl. After climbing down, I walked around to the far side of the hide where the growling continued, but there was nothing further to see from there. Finally, I returned to the front entrance and contemplated my options. After a few minutes, I slowly began to approach the entrance and was growled at another time. I waited until sunset then returned home.

For a few months after becoming Yowie-aware, we approached many colleagues, friends and neighbours to see if any of them knew something about the phenomenon. With them, there was generally no need to be coy or to use the 'standard question', because they knew us. This marked the beginning of the Octopus.

John and I were colleagues from school and good friends. In 1991, I helped John get established with his

8: Dooligahl

Blue Mountain curry house business, Rijsttafel, from the Dutch word meaning 'rice table', by doing the desktop publishing for his business and working as his sous chef. After talking to John in 1993, he told me that his son David had had an encounter one evening about seven years earlier and had been heavily ridiculed when he spoke about the incident. He said that David, together with a group of his school friends had had an encounter at the top of our land, where a very large bipedal animal charged them at high speed from behind, through the bush on the side of the road, before sweeping back down into the valley, grunting as it ran. Although this was a common type of encounter, the significant thing about David's experience was that they heard the biped grunt in time or phase with its gait. This suggested that the diaphragm, which is responsible for breathing, was involuntarily instigating the sound. This is a trait that some people mention after a Dooligahl has made a high-speed run. This synchronised breathing is also a characteristic of kangaroos and wallabies, that must time their inhalation and exhalation with their hopping gait, in order to achieve efficient respiration.

John also told me to see Dave and Peter, both local residents. Dave lived towards the end of our valley and he told me that they frequently heard someone outside their bedroom at night. When this happened, his wife would slide their bedroom door open, while Dave would run out swinging his baseball bat. Dave said, "I can never catch the bugger, he is too fast!"

The other local resident lived on another street at the head of our valley, not very far from Dave. He said that someone regularly walked through their vegetable garden at night, ripping up the plants and leaving very large footprints. Thirty years later, another couple, Michael and Sandra, now own this property.

"Sandra and I bought the land in October 2001. The property was just land and a shed at the time, which Peter, the previous owner had been living rough in. I was told that his vegetable garden had been regularly trashed and they had found unusual footprints, as well as having something moving around in their backyard at night.

"Within the first six months of being there, I found the wallaby carcass. What appeared strange to me at the time was that it looked as though it had been cleaved in half, or one leg ripped off, and it was no small wallaby. I was stumped because I knew no fox could have done that. We had previously lived close by for ten years or so and had heard experiences from neighbours, and my brother living next door, but a possible Yowie attack didn't come immediately to mind with the wallaby.

"The other thing I found strange was a small and well-worn track through our boundary with neighbours. It was close to where the swamp snakes pass through the gully, but I couldn't see a logical reason why a person would walk here. We had plenty of experience with residents of the lower part of the road walking down the driveway next door and cutting through our place to walk to the shops or station, but this was on a completely different part of the property.

"Over the years we have had plenty of experiences of our dogs joining the chorus of barking from dogs down the end of the valley, working their way up to the top as something moves through the bushes. We have also had some strange stripping of bark on some of the trees,[13] but not sure whether to put this down to birds, or others."

During this formative period of early research, John was frequently present during the afternoon. He would help with setting up the cotton, looking for trailing thread and doing some test walking. It was also beneficial to have

---

[13] This bark stripping occurred around the same time that Junjudee were active in our valley.

8: Dooligahl

someone with whom I could openly talk about these fringe topics and who would lend his support when the Police were carrying out visits and investigations.

Like Troy's father, John had been affected by the encounter that his son David had experienced seven years earlier which resulted in him being heavily ridiculed. John was keen to learn about these animals, although he particularly wanted to have an experience himself. By late June, John had participated in many activities around the valley. He had heard and witnessed many things related to the 'marsupial hominoid' and wanted to have a better experience. On July 1st, John decided to sit on the edge of the bush to the northeast of our house for about an hour. He heard many of the usual sounds, including footsteps and twigs breaking. He came back inside and told us what he had experienced, so I went outside with him to support any further experience that he might have. We heard much of the same except that we could hear the loud sound of bush being parted. It sounded very deliberate to me like it was another invitation to play.

The next evening, John arrived at the house at sundown. After it became dark, I joined John who was sitting beside our car which was parked at the end of the path to our driveway. We heard movement at various distances for the next half hour before we decided to go back inside. As I was in the process of opening the front door for John, we both heard very loud movement that was meant to gain our full and immediate attention. John was very excited by what he had heard. He said, "I now totally believe that these things exist."[14] He also told Sandy and me "If the Aborigines are right and this thing takes one of your children, no one is going to believe you! Like Lindy

---

[14] Like Troy's father, I think that John now believed enough to be able to fully support his son David, if the need arose.

Chamberlain, they are going to lock you up and throw away the key!" This was the most frightening thing that anyone has said to Sandy and me.

APART FROM USING cotton thread, another way to monitor movement was to leave an active voice-operated tape recorder (VOR) in the bush and listen for walking during playback. The VOR that I originally used was lent to me by Robert, the husband of Kerry. Robert thought that I might be able to use the recorder after we had shared two encounters on June 5th and 18th. On the second occasion, the Dooligahl had roared in my face after we had tracked her down.[15] Having had several encounters with Robert during that time, I first got to use his VOR on July 12, when I simply hung the recorder by its wrist strap on a broken branch of a scribbly gum, next to a track on the edge of the swamp. When I played the recording back, multiple sounds of footsteps could be heard, which were each interrupted by breaks in the voice-operated recording. After several of these steps were played, there was a continuous recording of something noisily rushing the position of the recorder, followed by the sound of the recorder being very forcefully handled. It sounded like the recorder was being crushed by a forceful hand for several more seconds, after which the recording ceased. After experimenting with the buttons on the recorder, depressing the stop button, or most of the other buttons randomly, but simultaneously, both had the effect of cancelling the

---

[15] We did not know at this time that there was more than one 'marsupial hominoid' in the area, even though Merve had told us that our Dooligahl was a female because of her interest in children's clothing. Another very interesting observation a few minutes earlier was the sound of panting. Kangaroos and wallabies pant through their nose with their mouth closed when hot from exertion or when stressed.

8: Dooligahl

recording. I think that the recorder was misidentified by the 'marsupial hominoid' as prey, with the misleading factors being the wrist strap (tail), VOR case (body), the red LED recording light (reflective eye) and the sound made by this mechanical drive (movement). She never made this same mistake again![16]

As a consequence of this incident and because winter was approaching, several important modifications were made over a short period. Fresh batteries were always used on every recording occasion because battery life is significantly shortened during cold weather. The recorder was wrapped in cloth for thermal insulation, then in plastic wrap for waterproofing and finally, buried in a metal container for soundproofing[17] and possibly to counter my scent.[18] An external microphone was attached and placed at an appropriate height. This new setup worked very well for many years until the 'Bucket Experiment'[19] in 2000, where the recorder was heard by the O'Connors' Quinkan because it had been placed under a half log, rather than buried.[20]

---

[16] You only get one chance with these animals. Ian and I learnt this very early on with our research. It is essential to keep them naïve for as long as possible and only use passive methods of observation to avoid detection and minimise learning by them.

[17] Sound proofing became superfluous after digital VORs became available.

[18] An Aboriginal Elder told me to rub soil all over myself if I wanted to avoid detection. Also, Gary and his football mates, saw a Dooligahl with its back towards them, raise its head and sniff out their presence in the very remote wilderness, at the confluence of the Cox's and Kowmung Rivers in 1989.

[19] See YouTube: Jerry and Neil: Bucket Experiment, https://www.youtube.com/watch?v=8McBSdlIILY

[20] As should be expected, Dooligahl and Quinkan have very sensitive hearing.

Cassette tape voice-operated recorders were used during this period of early research until digital VORs became more readily available about a decade later. For us, they were relatively expensive, so only Robert's recorder was used until it was returned and we purchased a replacement. They were originally used cautiously as tracking devices and only if I felt like following the correct procedures to maintain their stealth. Consequently, they tended to remain supplemental to the passive cotton thread detection method for some time.

However, the VOR was sometimes repurposed for other uses. During the final days leading up to the police stakeout, we were given a request from the police to obtain fingerprints of the 'intruder'. This waste of effort was required because the police had not ruled out the possibility that we were dealing with a criminal, child molester, or hermit and I think that the sergeant wanted to have evidence that they had taken all reasonable steps to eliminate human suspects, just in case someone got shot during the stakeout. I also think that Graham didn't want to appear to be a 'Yowie convert' to his overwhelmingly cynical, work colleagues.[21] As we wanted to remain compliant, we did what we were asked. As requested, I hung a wine glass in a tree fork and an empty four-litre wine cask bladder securely taped to a bush alongside a swamp track.[22] To complete the police request, I placed the VOR near the swamp edge and hung a bright chromium-plated

---

[21] Each police participant had cartoon drawings and notes left on their desks during the time of the stakeout.

[22] These items remained untouched until after the police stakeout, with the wine bladder being bitten through the middle, leaving two canine holes, sixty millimetres apart, on September 22, after I had done something to annoy Fatfoot the previous night, by attempting to take two sneaky flash photographs.

microphone at head height on Wednesday, July 21. The next morning I retrieved the cassette which strongly suggested that Fatfoot had handled the chrome-plated microphone. I took the bagged microphone and the cassette to the police station. After playing the recording, the police decided that they did not need fingerprints.[23]

Several weeks after the stakeout and the earlier successful recording using the VOR, I repositioned the recorder in the middle of a swamp track on September 8 and 10. During the latter recording, Fatfoot touched the microphone before a twin-engine commuter flight to Bathurst flew overhead and again handled the microphone after it had flown into the distance a minute and a half later. This scheduled flight flew overhead at 7:30 P.M. each weekday. This was very useful as the VOR could be used to determine a specific time for an activity, or in combination with other audio markers, a window or range of time. Some of the local audio events that we used as time stamps included scheduled aeroplane flights, like this 7:30 P.M. flyover, timetabled passenger train services and regular early morning coal trains. All of these activities could be used to establish a time frame for VOR tape recordings. Random events were also used, for example, the sound of an emergency vehicle siren and the explosion of an electrical power transformer. Usually, these audio-time references were scribbled as a note on paper next to my bed.

Another amazing aspect of this recording was Fatfoot noisily handling the microphone and then pausing her movement while a twin-propeller aircraft flew overhead. The sound of the microphone being handled resumed after a ninety-second pause, as the aeroplane faded into the distance. We speculated on how Fatfoot was interpreting

---

[23] I didn't record this incident in the log.

the flyover during this one-and-a-half-minute period of contemplation. However, it reminded me of a visit that I made to the New Guinea Highlands in 1975 where I saw tribesmen who were wearing a mix of replica and authentic Australian and U.S. military uniforms and carrying air force paraphernalia, in the sympathetic belief that free goods would again fall from the sky as they did during World War II resupply drops. Like most beliefs held by the locals, this 'Cargo Cult' was taken very seriously and was expressed through many rituals. Although I didn't get to see it myself, I was told that detailed replica aircraft runways and simulated infrastructure had been cut from the forests, in an attempt to induce a resupply landing. Cargo cult members could be easily identified as they typically wore a U.S. Air Force officer's cap as they walked around in general society. We wondered how Fatfoot would have interpreted these flying objects as with the other strange objects that she regularly encountered during her wanderings.

We also used the early VOR to troubleshoot issues that we were having with our automatic still camera. For a while, the passive infrared sensor used to trigger the camera was taking photographs of nothing in the targeted field of view. When we finally installed the VOR alongside the camera, we could hear the camera mechanisms at work, which were preceded by the sound of movement. It seemed apparent that the Dooligahl was, once again, playing games with us!

During this early research work, we were very lucky to have a sponsor. As it would have been useful to remotely listen to live movement in the swamp from the house, so that we could effectively plan any imminent action, Steve Ruston bought us a hundred-metre high-quality microphone extension cable that allowed us to both record and passively monitor live activity. This was a much simpler

8: Dooligahl

and cheaper alternative for us when compared to radio technology, although we did make one simple test attempt using a walkie-talkie that had its transmit button taped down. The problem with radio technology was the high cost of batteries.

Perhaps the best use of VOR technology was still many years away. When digital recorders became more widely available, compared to analogue tape recorders, they were cheaper, smaller, had greater audio storage, allowed a range of file formats, provided better protection from water ingress, were more energy efficient and most importantly, they were silent. Being passive meant that they could achieve much more without the need for complicated setup precautions. Being cheaper and easier to deploy meant that more than one was likely to be used at a time. As a consequence, some excellent recordings would be made, particularly those where communication between individuals were recorded.

FROM THE VERY beginning of our research, cameras and standard photograph equipment were used in order to obtain further information about our Dooligahl. Initially, I carried my old, but reliable Olympus OM2n f/1.4 camera with me, but very soon realised the futility of the action, preferring instead to carry a tomahawk down the back of my pants for protection. Ian, on the other hand, preferred to not carry anything, however, we started to realise that we were never going to be in a position to catch the Dooligahl, let alone be able to take a photograph whilst on the move.

After one of our afternoon planning sessions, Ian decided that the best way to get a photograph was for me to shepherd the biped towards him, as he hid in the undergrowth along one of the many tracks. The plan then called for Ian to spring into action as the humanoid thundered towards him and take a photograph as it passed by. It sounded like a simple plan and we gave it a try, but steering Fatfoot along our desired path was never going to happen. It was like herding cats. I think that Fatfoot could easily see, hear and smell our presence, as well as read our intent.

From early on, Ian decided to limit his active involvement to mostly Friday nights, because he was not coping that well at work with the late nights. During the week days, I prepared the cotton threads and monitored the evolution of the tracks and flattened areas. On weeknights I carried on with the interactions as best I could, mainly by moving around the bush boundaries of our property. During these early encounters I would carry my Olympus camera and T20 flash unit with me, in the hope that a golden opportunity might eventuate. In order to be ready for an immediate response, I left the camera turned on with the automatic setting selected and the flash unit charged and on.

After many nights spent circling around the house, I eventually came upon an opportunity on the western side of the house, half way towards the swamp. To my right, I could hear walking along the boundary and moving away from me. The movement wasn't that far into the dense bush that the vegetation might block a photograph. After listening carefully to determine the location of the walking, I fired off a shot, with the flash unit becoming almost fully discharged in its attempt to achieve the correct exposure. Almost immediately came a reply from the bush as a medium-sized root ball came rapidly spinning through the air towards me, landing with a thud on the ground close to my left. With the range of the trajectory being accurate, I hoped that the near miss was intentional rather than a miscalculated throw. As Ian would say many times during the months ahead, "If it wanted to kill us, it would have

8: Dooligahl

Fig. 8.1 Root ball with only half the original stem remaining attached. (Photo: Neil Frost)

done it by now!" I turned and unashamedly ran back to the house.

In the morning I inspected the camera with the intention of getting the film roll developed quickly, regardless of the number of remaining exposures. Ironically, having taken a risk trying to obtain the photograph, I found that the film chamber was empty. However, I returned to the grass slope and immediately found an isolated tree ball root.[24] Still attached was the bush stem that was snapped half way along its length, which I thought was consistent with a break caused by a substantial inertial force being equally applied across the weak point, midway between the root ball and the end of the throwing handle. Only the bottom half of the throwing handle remained fully attached to the root ball.

Several things were immediately obvious as a result of this very early encounter. Firstly, the responsible animal had a grasping hand. Kangaroos and wallabies also have grasping hands that can skilfully manipulate objects but they do not throw them. Also, they are not intelligent animals. The difference in ability between this animal and these other macropods is that the root ball was thrown, something that these particular marsupials, with their short forearms, are mostly incapable of achieving. Whether the root ball was lobbed or tossed probably doesn't matter very much but considering the tall surrounding scrub, it was most likely lobbed high into the air with an arching motion to order to provide some clearance of the vegetation bordering the clearing. Such a throwing action would lie outside of the kinematic parameters of kangaroos and wallabies.

---

[24] The root ball appeared to be from a *Banksia oblongifolia*, a typically three-metre tall shrub.

Secondly, we were dealing with a biped with a manipulative ability. Being bipedal and having dextrous hands are regarded as prerequisites for intelligent behaviour.[25] By throwing the root ball it also proved its spatial reasoning ability, by determining the distance and the required direction of throw. More significantly, the biped demonstrated an understanding of 'Theory of Mind'—I throw the root ball and the little man runs away!

As novice researchers, we had no experience with wildlife photography, nor did we have the specialist equipment required to achieve a worthwhile result. Night photography added another thick layer of complexity to the problem. Additionally, we were dealing with a highly intelligent biped that does not want to be seen.[26]

Photography was always a problem for me because my first camera was an East German Praktica LLC that I bought for my first trip to Papua New Guinea and Melanesia in early 1975. As a manual camera, it required the use of a light meter and various calculations to be made to determine the exposure settings for the camera. Small errors tended to become big problems resulting in the wrong choice of f-stop or exposure time. Unlike digital photography, any exposure problems wouldn't be discovered until after the film roll had been developed and printed, sometime after the important event had concluded.

Apart from checking to see if the camera had a roll of film loaded into its chamber, another lesson of nighttime wildlife photography was trying to get as close to the subject as possible before it moved away, to gain the maximum benefit from the flash. For me, this prospect

---

[25] A dextrous beak and foot is also a prerequisite for intelligent birds, e.g., crows, ravens, kea and cockatoos generally.

[26] Daytime photography of snow leopards in Nepal is professionally regarded as being 'notoriously difficult'.

was somewhat daunting because I would typically attempt this during weeknights when Ian was not available. It was also very important not to overuse the flash during futile attempts to get a photograph because this only resulted in increased camera shyness. Regardless, as a consequence of using the camera and attached flash, or simply having the gear in my possession, it was becoming obvious that Fatfoot was learning to recognise the equipment in advance and anticipating the possibility of becoming blinded by it.

There were two likely characteristics that were responsible for the camera's early detection. From the very beginning, we had introduced Fatfoot to Ian's powerful quartz-iodine spotlight during our early encounter pursuits. When the torch was turned off, it could be easily recognised in the dark by its very large reflector and lens. In structure, the torch was very similar to a floodlight and a flash unit. The camera, however, was only partially similar because it had a lens with a reflective outer surface but lacked an internal reflector. Later on, Fatfoot was able to detect passive infrared detectors (PIR) which we used to activate external lighting and the homemade 'game camera', as well as above-ground VORs. Additionally, the other characteristic that betrayed the presence of the camera was the sound generated by the flash unit. By its nature, the internal capacitors of the flash would emit an audible buzz as they charged and maintained their power.

The multiple obstacles standing in the way of our photographic progress brought about many rethinks on how we should be going about the work. It was very obvious that attempting to photograph Fatfoot while we were in pursuit or lying in ambush along a track, was futile. Also, we were slowly realising that despite her seven-foot stature, we were not seeing her upper body along the tracks, except under very rare circumstances. Other clues regarding her unusual gait on certain occasions, suggested

8: Dooligahl

that she might be a facultative quadruped, which is an animal that is adapted to walking on two legs but on occasion, is also able to move on four. As an immediate fix to part of the problem, we started to hold the camera at arm's length overhead when taking photographs. This is what I did when taking the photograph of the red eyes and it was having greater success.

There were also other participants who were trying to obtain photographs and video during this early period. After having had several shared encounters, Robert and his friend Iain spent some time running power into the bush and setting up floodlights so that they could video any nighttime activity. They had further encounters, particularly Iain who "heard four footsteps moving west towards swamp and saw briefly a reflective disk of an eye for a fraction of a second as if the head turned towards him" [see Log for June 26]. The lights that Robert and Iain had set up were pushed off their tree attachments by Fatfoot, using a stick on July 14 [see Log and photographs].

Perhaps the most successful change in photographic strategy occurred on June 16. For some time prior to this, I had contemplated ambushing Fatfoot by stealth and blinding her with a flash from the camera, which we thought would immobilise her.[27] Ian and I had found that Fatfoot's weakness was light because she avoided it, even at a distance, and could easily recognise anything that had previously emitted it—anything with a reflector seemed to be the principal giveaway. When blinded, she seemed to remain in the immediate area.

Wednesday, June 16th, 1993, was a busy day. It was the first time that I started making detailed entries in a

---

[27] This is what happened when Sue and her father blinded Fatfoot with their ute's high beam and driving lights, causing her to stand still until the lights were removed.

log that the local police asked me to keep. In the morning our daughter needed to be driven to preschool and our son accompanied us on the journey. After that, there were the domestic tasks that needed attending to and Drew needed playtime, feeding and sleep.

Several ongoing research projects required maintenance. Tracking the whereabouts and movement patterns of Fatfoot was something that I did during the week, so that Ian and I had something to work with on Friday night, when we would do the majority of our joint pursuits.

Newly established tracks were obvious (see Fig. 4.7 and Fig. 4.9). Also, the frequent use of any track would show signs of wear and widening, mainly because Fatfoot would undertake track maintenance by breaking off the tips of branches, which we speculated would make movement easier and quieter. Tracks that were no longer used were not immediately obvious, as vegetation would take time to slowly reclaim the path. If unbroken, cotton threads across the path helped solve this puzzle. Determining the direction of travel was difficult but not impossible, as occasionally grass or branches would be swept forward with the movement. Taking photographs of the tracks helped in establishing their history but, this was still the age of film when results weren't immediate and economic decisions had to be made regarding the worthy expenditure of a frame from a roll of twenty-four, so exposures were sparingly taken.

After Sandy got home from school, I could indulge in a few more tasks. That afternoon a colleague arrived at our home at around 3:45 P.M. to help with the effort. John was very interested in what was happening because his son David had an encounter in front of Roy's property, on the street adjacent to our house, with a group of his friends, one afternoon during the early 1980s. As they walked along the road, they could hear something

8: Dooligahl

approaching out of the valley, from behind and to their right. David said that the animal was running at a very high speed, grunting in time with its gait. It ran close by as it overtook them, though they could not see anything in the daylight because of the dense bush.[28] David said that it seemed to be attacking them until it veered off to the right and sped down into the valley below, continuing to grunt as it ran. John said that the encounter had affected his son, who had been heavily ridiculed at school.

John and I went for a walk along the edge of the swamp. John said that "some of the tracks seemed to be enlarged further", compared to the previous Sunday. He also noted that "the banana was missing". This was a reference to a ripe banana that I had strategically suspended by cotton thread from a tree branch high above a track in the swamp, that only Fatfoot could access. I had similarly noticed that the banana was missing about two hours earlier. Several days after this, I found the banana skin with attached cotton thread under the railing at the front steps of the house.

As we continued our walk, John found a purple cotton thread that I had suspended across the clearing next to the swamp. He found the free end, twenty metres down a track on Ian's land. By tracing the thread along its length we found the fixed end that marked the starting point, at a river red gum. Attaching it at one end only stops the thread from being completely carried away but still allows the direction of movement and path taken to be easily traced because of the cotton slippage across the animal's body. It was always important to loosely suspend the cotton at a uniform height by using intermediate

---

[28] I strongly suspect that the Dooligahl was running as a quadruped at this point in its approach in order to reduce its height profile.

vegetation and trying to prevent potential snagging. Being purple would have told me the layout details of the cotton when I installed it because I recorded these details and the results daily. The details included the thread colour, time and date, starting or anchorage point, average thread height and any other relevant information. Thread height was important because it acted as a filter, where all animals would trail a low-height thread but only a Dooligahl would snag a high one. Thread tension had to be slack. Early attempts at using the cotton thread had both ends fixed and under tension. No threads were broken and testing the layout by walking into it, instantly revealed the problem. The thread could be unnaturally felt as the tension built prior to breaking. Also, the surrounding vegetation would suspiciously move about.

After looking around, I showed John areas where vegetation had been recently flattened and he said that he would return tomorrow to make further comparisons. By now it was becoming dark and being mid-winter, quite cold as well.

By 6:30 P.M., it had been dark for over an hour and the dogs had already started their barking relay up the valley. A short time later, Gordon's dog started barking, followed by the Good Dog—a highly reliable watchdog, very soon after. By 6:40 P.M. the foot thumping had started on the northern side of the house. Our dog Bess started barking and then ran away. The foot thumping continued.

For sometime prior to this, I had contemplated ambushing Fatfoot by stealth and blinding her with a flash from the camera, to immobilise her. Ian and I had found that Fatfoot's weakness was light because she avoided it at a distance and could easily recognise anything that had previously emitted it—anything with a reflector seemed to be the main give-away. When confidently blinded, she seemed to remain in that area, as if she could not see.

8: Dooligahl

On the evening of June 16, after the dogs had finished barking, Fatfoot had commenced foot thumping in the bush to the north of our house at about 6:40 P.M. I moved to our bedroom, on the opposite side of the house and dressed in full black clothing, including a balaclava. I wore black socks and soft slippers that allowed me to feel sticks and anything else under foot, that might make a noise and give away my position. I checked the camera for film, because a few months earlier I had taken a risk and after having had a root ball thrown at me, discovered that there was no film in the camera. I mainly used Ilford XP2 400, a B&W film, because of its high speed and fine grain. Less fortunately, on this occasion, I was using Kodak Gold 200, which was a colour film with fine grain and high sharpness. I then attached my fastest lens, at 50mm and f/1.4. I pre-charged the flash and then installed fresh, high-quality batteries, to maximise the life of the batteries and minimise the recharge time. The flash was going to be a major concern because of the long recharge time and similarly, the two bright LEDs and the buzzing noise that it emitted whilst charged and charging the capacitors. I covered the power and charge LEDs with several layers of black electrical tape. The best solution for the noise was to place the whole camera under my jumper and coat, in order to muffle the sound. I placed a torch in my pocket, to be used if the situation became difficult.

Listening from the bedroom, I could hear the foot thumping continuing on the far side of the house. I slid open the sliding door and walked to the southeastern corner of the house and briefly waited. The next stopping point was the verandah post, two metres in front of me. I walked very slowly to the post. Having thought about this problem in advance, I believed that the faint ambient light from houses to the south might show me in silhouette with Fatfoot's night vision and that the best option would be to move very slowly so the Fatfoot's peripheral vision, which should be more sensitive to light and motion, wouldn't detect me. Ahead was the bush house, the next stopping point, about ten metres away. After a cautiously slow journey, I reached the bush house and was relieved to hear a thump soon after. The next objective was the driveway and to reach it I had to circumnavigate a thin vail of bush vegetation alongside the road, taking time with every step to feel the ground for sticks through my thin slippers, prior to committing my full weight. Moving slowly, I arrived at the edge of the driveway and was further relieved to hear another foot thump. Stopping there I took brief solace before excitement overtook me, having realised that I had successfully stalked the master of stealth.

I hadn't planned the next phase because I never thought that I would get this far! Standing on the edge of the cleared driveway, I gave the prospect some thought. I knew that the game would be up as soon as I stepped onto the road. I also knew that stealth was no longer necessary and the best option was to have the finger on the trigger and walk quickly towards Fatfoot and see what happened?

As soon as I stepped onto the road, I heard walking commence. This movement allowed me to fine-tune Fatfoot's position and range, which was straight ahead and at the width of our driveway plus a few metres. I fired the flash for the first time, with the camera held high over my head, hoping to clear any vegetation. I was also hoping that she was looking my way without obstructions. Fatfoot's movement continued, but not at the usual high speed, so I felt reasonably confident that she had been partially blinded. Walking slowly towards Fatfoot I had to wait for the flash to recharge. The camera had been set at automatic with the aperture held fully open at f/1.4, so the flash discharge had been very deep and took forever to completely recover. During this time I felt totally defenceless. It was an extremely scary event!

8: Dooligahl

I was now walking very slowly along the bush track from our driveway, that led into the thick bush on our neighbour's land. This track had an intensive history of activity and would continue to provide more. As soon as the flash had recovered, I fired off the second shot. Once again, I felt naked as the flash recharged and wondered if I should get the torch out of my pocket, in case something happened! While I was waiting I could hear slow walking ahead of me that was starting to rotate clockwise. Again I knew where Fatfoot was because the northerly track that I was on, joined with an east-west trail at a T-intersection. I fired off the third and final exposure, with the camera held overhead.

By now I was starting to sense that Fatfoot was becoming a bit agitated! With the flash recharged and holding the camera at arm's length in front of me, like a light sabre, I continued walking and was soon standing at the tee intersection where Fatfoot had been a very short time before when I took the third photograph. Continuing on, Fatfoot had rotated fully clockwise and was now positioned between me and the house. It was starting to become serious and I could not accurately locate her position. It reminded me of the time that Fatfoot foot thumped directly behind me without warning. Drawing upon my war gaming experience when I was at university, I applied the old maxim, "Attack is the best form of defence." Though not completely convinced, I walked directly toward Fatfoot's anticipated location. The only thing that I could hear was the loud and fast beating of my heart. I also attempted to reassure myself by thinking, "Ian does this crazy sort of thing all the time and he is still alive!"

As I started to move towards Fatfoot, she began to move anti-clockwise, back the way she had come. The distance between us diminished to about twenty feet, so I made a concessional deviation to the right and back onto the main northern track. Soon, I was enjoying the relative safety provided by the driveway. I could hear Fatfoot slowly walking into the thicker bush and then on towards the north. The time was about 7:00 P.M. I went inside the house and told Sandy what had just occurred. She had not realised that I was outside and told me never to do that again! By 7:45 P.M. we heard the dogs barking in the valley.

Early the following morning I went outside with a ready camera after hearing the dogs barking and suspicious noises in the bush to the east of the house. I did find a new area of flattened grass on the southern side of the driveway. Later that morning I decided to have the film developed at Kmart, even though the twenty-four frame roll contained only three exposures.

Arriving at Kmart, I gave the roll of film to the young female assistant. To prepare her, I told her that there were only three exposures on the roll and that there was no need to print the other twenty-one frames. On receiving the extremely thin photo wallet, which contained three nighttime photographs of bush images, she said nothing whilst giving me a puzzled look.

Sitting in the Kmart carpark, I took a long look at each of the exposures. The first two prints were nighttime photographs of an uninteresting bush setting. The third print was of a similarly boring, nighttime bush setting, except that two red eyes could be seen in the lower left corner. The finer details of this photograph, including calculations and analysis, can be read in Chapter 13.

On the short drive home, I stopped at the local police station and spoke to Constable Gilpin. He advised me that, in addition to keeping a log of events, "I should also telephone the police every time that there was a disturbance, in order to keep a record."

At 12:45 P.M., the police arrived to inspect the area around our house. Having just completed a number of calculations, I showed the police sergeant the photograph

8: Dooligahl

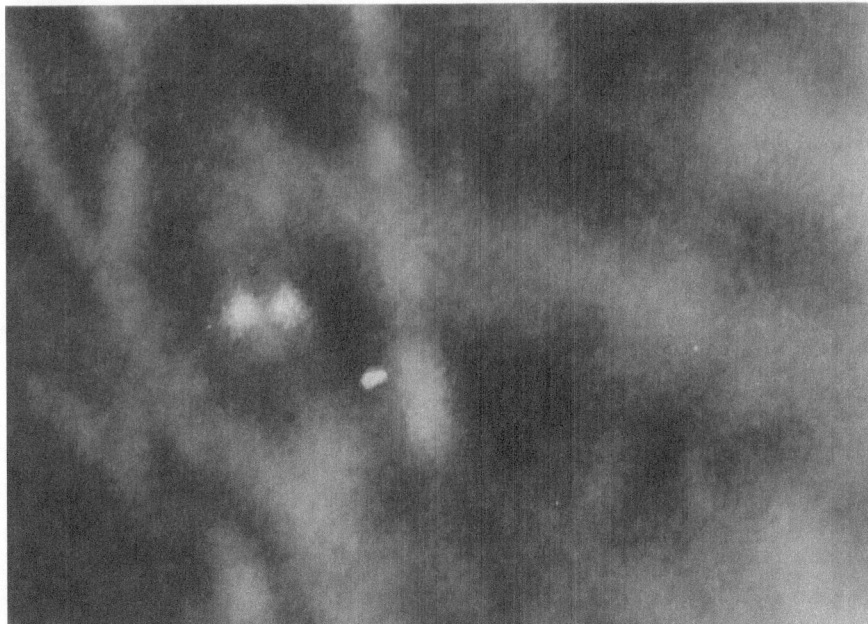

Figs. 8.2 a and *b* Third photograph showing Fatfoot's red eyes at the track T-intersection. Camera held overhead at arms length, tilted slightly up. Distance to eyes = 13500 mm. Eye height from ground = 1850 mm. (Photo: Neil Frost; view full color at http://tinyurl.com/red-eye-fatfoot)

of the eyes. Looking at the photograph, the sergeant said, "It's a fox!" Rather than argue, I told the sergeant to take the camera, that he might like to use the chair because I was holding the camera overhead at the time and that he should use his constable as a model. After about fifteen minutes of colourful discussion and careful movements, the sergeant approached me and said, "This thing is six foot high!" After a brief discussion, the police left.

Later that afternoon, John returned to have a look at the newly flattened areas. By 5:30 P.M., the dogs in the valley had started to bark. Shortly after, John and I heard vague noises in the bush to the north. We then moved down to the swamp where we heard walking travelling south. The dogs continued to bark but, much more intensively. The Good Dog was barking continuously and loudly, resulting in his owner coming outside several times to quieten him. We then heard a long, deep and obviously disapproving growl from the swamp that fluctuated up and down in tone. The movement continued south, the dogs barked and we went back to the house.

Over the following days several other people inspected the photographs. Most had no opinion to offer, however, Basim, who was a marine biologist, colleague and nearby resident, said that the eyes looked like those of deep-sea animals with reflective retinas. He also said that the eyes were forward facing which suggested that it was a predator because of the need for stereoscopic depth perception. Peter, another local resident and electrician, came to our house to discuss the electrical requirements for a film game camera that I was designing and building. He thought that the eyes could have been from a swamp wallaby, though he changed his mind after I outlined the circumstances leading up to the photograph being taken.

I also started to think about the two other photographs that I had taken, which didn't show any eyes. I thought that the first photograph should have blinded her because it was taken at close range and after Fatfoot became aware of my presence. She should have been looking at or towards the camera. Her immediate behaviour after the flash did seem to suggest that, at least, she was blinded by the indirect flash. With the second photograph, it would be reasonable to assume that Fatfoot was moving away from the camera, making it unlikely that the eyes would be captured on film. The third photograph showed two red eyes. In this case Fatfoot was already in position, having turned fully about and standing at the end of the northern track, near to the T-intersection and looking directly at me. It would be certain that she was blinded by the third flash. In humans, it takes about ten minutes for rhodopsin to significantly recover from light exposure and the resultant bleaching. This arrangement of elements reminded me of testing mirrors for telescopes, where the flash is the light source, Fatfoot's tapetum lucidum is the mirror and the film camera is the observer.

Two days after photographing the red eyes, there was another major incident. The incident occurred on Friday, 18 June 1993, when I was accompanied by Robert as we investigated a disturbance outside. Particularly after taking the red eye photograph a few nights before, I interpreted this incident as a warning.

Robert and his partner Kerry regularly visited our home and were partially aware of our predicament. Prior to this encounter, they had both socially visited our house on the 5th and 11th of June and we had told them some of our updated community stories. Robert and Kerry were regular diners at our home. We did this because it was too difficult to dine with friends at a restaurant with two very young children, who also needed to go to bed early in the night. As an owner of a Mazda MX-3 sports car, Robert would not drive down our long, unsealed road because of the vehicle's low ground clearance and his fear of damaging the undercarriage.

About an hour before their arrival, the dogs had been barking but noticeably fewer and much quieter than usual. On this visit, they brought their teenage daughter, Kate. When they arrived at the front door, it was obvious that something was wrong. Robert hesitantly approached me and said that someone had shadowed them in the bush as they walked down the driveway. Although Robert and Kerry were aware of our unusual circumstances to some degree, Robert seemed particularly concerned by the event. I told them that we would explain the situation in more detail over dinner.

As we were eating a Thai deep-fried snapper in green curry, with coconut milk, kaffir lime and extra bird's-eye chillies, alongside basmati rice plus a few side dishes, Robert, Kerry and Kate commented on the sound of someone "splitting timber for firewood" on the land to our north. This was a fair description, especially because it was mid-winter and the sound was very similar. We had also likened the thumping to the noise of timber being split and also to the sound of a Holden or Falcon car door being forcefully slammed. The percussion had the unusual characteristics of being simultaneously transmitted through the air and the ground, as there was an apparent lag between the arrival of the two sounds, even over a short distance. We asked our diners a few simple questions. What is directly across from our driveway and who would be splitting timber, on our or the neighbour's land, in the dark and at this time of the evening? On asking them to identify the source of the sound on the next occasion, they correctly indicated that the location was nearby, to the north of the house. On further interrogation, they agreed that it could not be a person splitting timber in the dark

8: Dooligahl

on the vacant land next to our house. Consequently, for the remainder of the meal, we paused after every thump and speculated on how the sound was being generated and what the intent was. By combining the wood-splitting inquiry with the question of who was following them down the driveway, we commenced the lengthy process of explaining our difficult and highly unusual circumstances in slightly greater detail. By the time that we had made a satisfactory attempt at reconciling the two encounter experiences, the meal was finished.

After we opened the lounge room door, we could all hear the very deliberate sound of tree branches being noisily broken in the scrub across from our driveway. We had experienced this attention-seeking behaviour before but this attempt was more persistent and provocative. As Robert was interested in these highly unusual circumstances and since the front door would be under close observation from the northeast, I told Robert that we would exit the house via the main bedroom sliding door on the other side of the house instead. I had used this exit two nights before when starting my stealthy approach to photograph Fatfoot, as apparently, this door wasn't under surveillance at this early hour of the evening. Robert and I carefully climbed off the southwestern verandah corner and very slowly walked a short distance on the soft grass. As with two nights before, I thought that Fatfoot's peripheral vision might be more sensitive to light and motion and by taking our journey slowly we might avoid being easily detected. As this had been part of a successful strategy to prevent detection and photograph Fatfoot, it seemed wise to use these exit tactics for a second time, although Fatfoot may have become aware of this alternative approach.

We took a position on the edge of the clearing, just above the absorption trench and behind a line of trees. Sitting there we heard movement to our north and south.

Although hearing movement from both directions was puzzling, if true, the implications were obvious. Ian and I had suspected that there was more than one 'marsupial hominoid' in our valley because we thought that there was subtle activity further away in the background during pursuits. Sometimes Ian would ask me if I had heard other movement while Fatfoot was clearly in front of us. However, we hadn't realised how special this find was because we had not considered that these bipeds were marsupial and that these other individuals might be Fatfoot's joeys. Confirmation of two and most likely three Dooligahl was achieved during a pursuit on the 6th of August, with Phil observing yellow eyes near his house while seeing Ian and me following another individual into the swamp.

As Robert and I listened further to the sound of walking nearby and various other disturbing noises, Robert leant across to my ear and said "WTF do you have walking around in your backyard?" Replying into Robert's ear, I suggested that he should wait where he was seated while I returned to the house and retrieved the camera. Having a touch of déjà vu to the situation, Robert informally protested against my recommendation for leave and in accepting the logic and overwhelming persuasiveness of his verbal argument, we decided to proceed without the camera.

While the individual to the south was noisy and less subtle in its movements, perhaps even clumsy, the other biped to the north had become less threatening and had quietly moved down the northern track to take up a new position across from us, about thirty metres away. The northern biped might have been monitoring us. For this reason, we later thought that we had been deliberately played early on, possibly in an attempt to divert our attention away from the southern individual.

Shortly after this, Robert and I heard a loud crashing noise moving through the bush at a distance of about ten metres from the back of our southern neighbour's house. The loud sound was repeated a minute later followed by a separate but quieter movement coming from the northern track. After speaking to the neighbours a few weeks later, we discovered that they had several fruit trees and a vegetable garden in this rear area.

As this was happening, we could see our neighbour washing up at his kitchen sink through the window. We started to hear very slow walking that was moving from behind the neighbour's house towards the two of us for several minutes. This was followed by the sound of more branches being broken. Then we heard the misplaced call of a currawong coming from a nearby tree which was very strange to hear during the evening. Hearing a bird call at night was a very rare event that normally occurs when a possum or predatory owl is very close by. There was the sound of a tree being variably and violently shaken with distressing calls coming from the currawong. Finally, we heard flapping, as the bird desperately attempted to fly after losing its perch and the accompanying sound of its wings striking the leaves and branches of the tree as it fell towards the ground and was not heard anymore.

Suddenly, the silence was broken by the sharp sound of our neighbour loudly striking the sink with a frypan that he was cleaning and had dropped. This caused the startled biped to rapidly move diagonally towards us for several seconds before gradually slowing down and then stopping a short distance away. This incident caused Robert to more clearly reiterate his previous exclamation into my ear.

Over the next few minutes, Robert and I could audibly track the very slow movement of the biped walking quietly but steadily away from the house and towards the swamp. It was obvious that the southern animal, unlike the other, had not detected our presence. Knowing that this was an excellent and rare opportunity to follow and study the biped's behaviour at a close distance, I again suggested that I should get extra equipment from the house and phone Ian for support. Like before, Robert's reply in my ear was very succinct. However, he did say that he was prepared to accompany me into the swamp from a safe distance behind.[29]

This decision necessitated ongoing changes to the action plan. Without the camera and Ian's support, I had to totally rely on myself and try to look after Robert who had no experience. Listening to the steady movement, it was clear that Fatfoot or, most likely, another Dooligahl was returning to the safety of the swamp where any further pursuit would become unlikely. After a minute spent accurately determining the acoustic bearing, I was confident of the Dooligahl's position near our border with Ian's land and at the edge of the swamp. Abandoning Robert to his own initiative, he followed somewhere behind me and I started walking on a straight intercept line towards the last movement heard. Knowing that we would be instantly detected by both bipeds, particularly with an inexperienced person in tow, meant that stealth was not a major concern and that the best that we could hope to achieve would be a physical or some behavioural observation. Consequently, the movement ahead of us stopped. However, we continued walking for a further fifty metres through the lower clearing and then, I briefly turned on the torch pointing it confidently at the last heard position. I expected that the Dooligahl would immediately run away. Instead, we heard growling and panting[30] coming from a different position

---

[29] What is a safe distance behind? It didn't seem that Robert was aware of how fast I could run because the First Law of Yowie-hunting is "Always go Yowie-hunting with someone who is a slower runner than you."

[30] The panting was loud and I would compare it to a very large and exhausted dog. Kangaroos sweat when moving and pant when stationary to avoid overheating.

ahead of us, at an additional distance of about twenty metres to the northwest. After walking directly towards the new location, I turned to Robert and warned him that I was about to turn on the torch and to expect a strong reaction. Raising the torch at arm's length over my head so that the beam would clear the taller vegetation in front of me, I turned on the quartz-halogen spotlight.

Immediately there was a very loud roar. It sounded remarkably like a lion's or perhaps a bear's roar. The volume was extremely loud and was coming from a big set of lungs. As it roared, the Dooligahl rose up in front of me from behind a bush. It had been crouching, only six feet away. As it did, the Dooligahl was standing at a slight angle, with its right side closest to me. It then leant forward over the spiky dagger hakea and swamp grasses and after matching my height it roared into my face. The main beam from the torch continued to shine above the animal's large and dark head.

Describing my reaction to this highly intimidatory threat is difficult. Unlike the encounter a few nights before, there was no apprehension leading up to this incident because it was a spontaneous event and as it occurred so rapidly, I didn't have much time to think about it or react defensively. Also, considering our previous pursuit history where many close encounters had not resulted in any interaction, I had not anticipated a physical engagement on this occasion.

Ian and I had previously discussed the possibility of such a close physical encounter and the potential consequences. For Ian, the danger was real but probably because of his broader life experiences, it was not an issue that he was accustomed to worrying about. For me, lacking his alternative grounding meant that the thought of being maimed or worse was usually at the forefront of my consciousness. However, as Ian would reassuringly say "If it wanted to kill us, it has had plenty of opportunities to do it!"

After this, the Dooligahl turned and ran away towards the swamp, roaring for a few more seconds as it did. The dogs along the entire eight-hundred-metre valley started to fire up, something that I had never heard before. I pushed around the insubstantial hakea that had previously separated us and followed behind the 'marsupial hominoid' as it cut a wide swathe through the dense swamp vegetation. It was like following a compact Bobcat track loader being driven by a demented sceptic into the dense vegetation. Some of the larger vegetation was flung to the side, while others became airborne. From what I can best recall, this ill-advised pursuit lasted for a few seconds or maybe as long as ten seconds. After speaking severely to myself and coming to my senses during the process, I turned and hurriedly exited the swamp the way I had entered.

I remember running past Robert, who was standing in the lower clearing. I attempted to shine the spotlight in the direction of the grunting coming from the far side of the swamp. By now, all of the dogs that could, were barking. Some of the nearby dogs were barking in a continuous and rapid manner, running up to or along their boundary fences, while a few could be heard running until their chain or other restraint reached its limit and throttled their response.

Aiming the spotlight at the head of the biped as it ploughed through the thick vegetation, we could clearly see the red eyes each time that it snapped its head around to look at us. It reminded me of 'roo shooting with the grazier's sons in Northern Victoria when I was a boy. Since the trajectory of the torchlight was very flat on the lower clearing and because bush obstructions were becoming difficult to avoid, I kept the beam on target as I backed up the cleared area to higher ground. As a consequence, I was able to keep the light fixed on the Dooligahl's head for the majority of its difficult escape, as far as the fence line on

8: Dooligahl

the other side of the valley or a distance of approximately fifty metres.

By now, a large number of dogs in the immediate vicinity were barking loudly and incessantly. The dogs further away had become much quieter, reducing the diameter of frenzied activity to a few hundred metres. The active dogs were also aiming their barking directly at the Dooligahl, as they also tended to do when we were pursuing Fatfoot. This resulted in a few of the dog owners coming outside in an attempt to quieten their pets, totally oblivious to what we were observing on the other side of their fences. Some had chosen to chasten their dogs at the precise moment when the Dooligahl had run behind their wooden paling fence a few metres away, grunting and noisily destroying bush while a spotlight from the opposite side of the valley was following the pursuit by illuminating the biped's head.

Although Ian and I had heard growling and grunting on many occasions, panting was something new and for a brief period, it sounded like we were listening to an extremely large and exhausted dog. When we heard the panting, it gave the impression that the biped had a very large lung capacity. Basim, a colleague and marine biologist, made a similar comment after hearing a recording of Fatfoot, when he said, "Got a big set of lungs. Has a respiratory problem. Sounds like a gorilla. Where did you get this?" Dr. John, a highly trained neurosurgeon and resident of a neighbouring area, gave his professional opinion on the lung capacity of their local Dooligahl after listening to it breathing from outside their bedroom window on several occasions. He said, "Compared to people, this thing has at least, double the lung capacity."

Another interesting occurrence was the grunting, which was synchronised with the running of the biped. With every high-speed stride, there was an audible intake and outtake of air that produced a rhythmic, two-part grunt, with each part being at a slightly different sound frequency. This grunting was a dominant component of the various sounds made by the Dooligahl as it fled the lower clearing towards the swamp and ran along the western fence line. We never heard these sounds from Fatfoot, even during intensive pursuits. Additionally, using the torch on this latest pursuit resulted in a very aggressive physical response, accompanied by a roar. As this hadn't happened a few nights before, I thought that this behaviour was unlike Fatfoot.[31]

However, this grunting noise was heard on another occasion by David, the son of John, together with a group of his friends along our road sometime in 1986. When I spoke to David, his description of the grunting seemed very similar. Other out-of-area and interstate witnesses have occasionally commented on this association between the grunting and running.

It seemed obvious that the grunting was an auditory indicator of breathing during strenuous running and an involuntary response. Perhaps it is the result of the diaphragm being forced up and down by the inertial mass of the abdomen during rapid movement or is due to muscular contractions caused by running. Interestingly, the breathing of kangaroos is synchronised with their hopping, where the air is expelled from the lungs on landing and drawn in on the rebound.

THE NEW MOBILE photographic strategy that we had applied to obtain a photograph of Fatfoot's red eyes had

---

[31] During this early period, we had not come to the final realisation that these 'hominoids' were marsupials. However, from the observed aggressive behaviours, we thought that we were possibly pursuing Fatfoot's male partner or male offspring. We didn't know.

8: Dooligahl

worked exceedingly well. However, as we had learnt from the experience, so Fatfoot had as well! From this point onwards, Fatfoot would not hang around if there was the slightest hint that we were staging a flash or light ambush. If suspected, she would move off into the bush and cautiously observe from this safer distance. Consequently, this experience was another component in the formative process that was being developed by all of the participants. For us, we frequently spoke the mantra, "You only get one chance."

Having discussed the many aspects of the problem with Sandy and Ian for several weeks, the agreed solution was to make our own automated camera. Something that we could camouflage and set up in advance of an incident. At this time, there were no automated film cameras available in Australia or possibly elsewhere that we knew of. Having made a simple sketch of the required components, I asked Ian if he could obtain a 12-volt solenoid from a car starter motor assembly. I also asked Peter, our local electrician, if he could give me a second-hand 240-volt relay from his spare parts. We already had the remaining components.

It was hoped that having an automated camera might be more successful because it could be easily set up in a variety of locations and configurations, for long periods of time and without anyone needing to be present and taking a risk. For the system to work, it needed a passive infrared (PIR) detector, like those typically used with outdoor motion-sensing lights. I already had an old PIR light with two 100-watt sockets. A 240-volt relay that our local electrician had donated to support the cause, was then connected to the light circuit, with the output being used to close the 12-volt circuit, connecting the battery to the 12-volt starter solenoid. When the solenoid was charged, it depressed the connected remote cable release that fired the shutter and flash, taking a photograph. Unfortunately, the system only took one photograph per cycle, because the negative needed to be mechanically advanced on the roll to the next exposure frame. To achieve this, the system needed to have a motor drive attached as a future upgrade.

The system had failed to take a photograph on the first night of operation. However, the first successful use of the automatic camera occurred sometime during the evening of June 27-28 when a photograph of a human-looking face was captured. It was disappointing that the 'red eye' was not captured during the exposure but this was most probably because of the imperfect alignment of the light

Fig. 8.3 Homemade 35mm film game camera. From left to right: car starter motor casing with attached 12-volt solenoid; shutter remote cable release; Olympus OM2n camera with 50 mm f/1.4 lens; 12-volt car battery; 240-volt relay; PIR sensor; 240-volt remote sensor and 2 x light holders; 240-volt 100-watt floodlight (only one light installed); various power cables. (Photo: Neil Frost)

source, subject and the observer (camera). Additionally, the system was electromechanical, which created a brief half-second internal delay between the floodlight coming on and the camera taking the photograph, which would have allowed enough lag for the animal's eyes to close in reaction to the light.

On identifying and inspecting the site where the photograph of the face was taken, the sense of smell was immediately engaged. The area had the strong fungal odour of freshly disturbed humus. Looking more closely, a rotten log could be seen that seemed to have been recently trodden on and partially crushed, leaving several cream-coloured compression marks on its upper surface and side, where the outer layer of decayed timber had been scuffed off. It seemed to Sandy and me that something heavy had

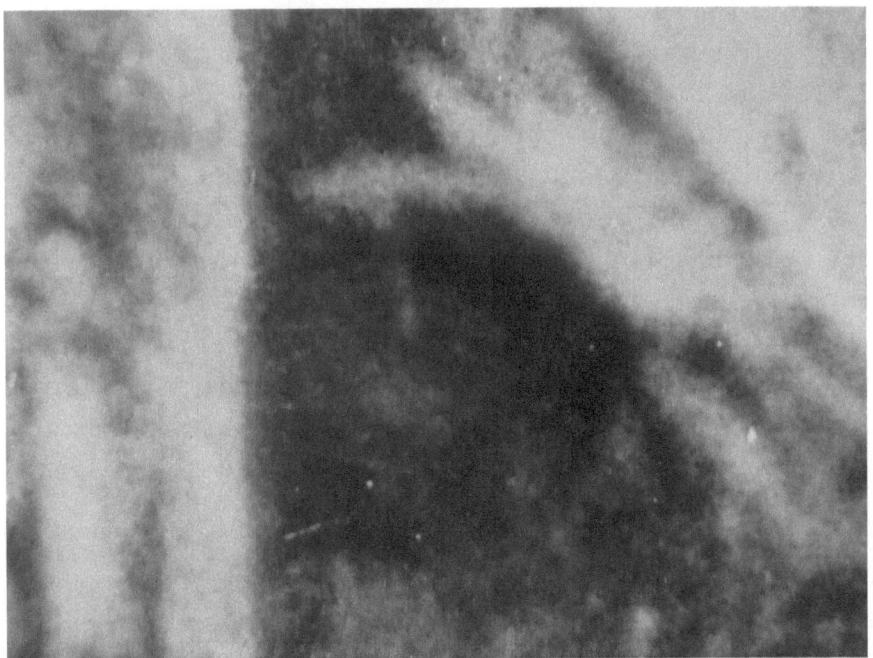

Fig. 8.4 Fatfoot's face, taken using 35mm film on a homemade game camera. (Photo: Neil Frost)

been standing on the log. There was also the typical compression and disturbance of the native grass around the base of the tree. As we typically did, we checked to see if there was any strategic significance to this particular location and so we positioned ourselves next to the log. With the undergrowth being dense enough to restrict most attempts to clearly see through to the house, we noted that there was a small window of opportunity conveniently located at this location. The clandestine observation of our family members by Fatfoot was a typical behaviour that we had noted at many locations around our house, which were readily identifiable by areas of flattened native grass at the base of trees that opposed the residence. Sometimes the surrounding trees had treebite damage at a typical height of about 1.8 metres which suggested to us that Fatfoot was having a snack while waiting.

After the success of the automatic film camera, I sought the professional assistance of Ralph, a local photographer. I drove to his house and explained our unique situation. Fortunately, Ralph was intelligent and open minded. After I had explained matters to him, I showed him the two significant photographs. With the photograph of the red eyes, he spent a lot of time reprinting the image while experimenting with the negative exposure settings to gain the best result. He then produced a number of prints of various sizes, including some carefully crafted enlargements. By chance, Ralph also used Olympus cameras and equipment which made his suggestions and help relevant and more easily understood.

I also showed Ralph the photograph of the face. As he had done with the other negative, Ralph spent much time trying to obtain the best print image possible. He later concluded that the range of my old Olympus T20 flash was not as expected because the capacitors had significantly deteriorated. As he thought that I would have a greater

8: Dooligahl

chance of photographic success, he lent me his Olympus T45 flash, which could work with the T20 attached as well. Unfortunately, the T45 flash used four 'C' cell alkaline batteries which were very expensive to replace on a short and regular basis. Since I had been experimenting with making my own remote sensing camera, he also gave me an Olympus E2 motor drive so that more than one frame could be automatically taken each night. It also allowed continuous shooting with speeds of 2.5 fps. As a replacement for the mechanical shutter release cable, he lent me an electrical remote cable. As a consequence of these potential upgrades, the newly created automatic camera needed further modification. The starter motor solenoid, 12-volt battery and mechanical shutter release were no longer required. Instead, the electrical remote was wired directly into the output side of the 240-volt relay. With

Fig. 8.5 Flattened vegetation and rotten log with scuffed fungal surface found next to the tree seen in the photograph. (Photo: Neil Frost)

the Winder 2 motor drive connected together with the T20 and T45 flash units, the automatic camera was much more formidable with a greatly increased range, reduced cycling time and the ability to take multiple frames.

With all of these substantial improvements, the increased chance of success seemed assured! However, it was apparent that the combined light from the two floodlights and two flash units was driving Fatfoot further back into the bush, probably because the extra hardware was more detectable in making more noise from the motor drive and the reflectors of the two flash units being more conspicuous at greater distances.

Ironically, the greatly improved automated system was decreasing our photographic opportunities. As usual, the solution seemed to be camouflage and stealth. After some discussion, it was decided that the main advantages of the new system were the ability to photograph larger areas, recycle the flash units more quickly and the ability to take many photographs in a sequence.

An ideal location for the new automated camera seemed to be on the mound to the east of the house that overlooked a sweeping section of our driveway. The driveway, like any road, would require Fatfoot to cross an uncovered section which would leave her exposed to the view of a camera for many metres. The mound was a raised mass of earth that we had placed along the driveway to control water runoff, that we had also planted out with many native plants. To help with the repeated installation of the new bulky camera setup and allow for the rapid removal of the equipment when rain threatened, I welded up a steel square section mount that could permanently accommodate many of the components, which could then be easily and securely fixed to a post or tree. Having fixed the steel mount to a medium-sized scribbly gum of about 200 mm diameter, with two 100 mm coach screws, the majority of the automated camera was securely in place. To set up

8: Dooligahl

the equipment, I positioned the PIR sensor so that it was facing the midpoint of the driveway, at about 22 degrees NNE. Similarly, one of the 100-watt floodlights, the camera lens and the two flash units were also positioned to cover this position. The other 100-watt floodlight was pointed to 270 degrees W, towards the front door of our house, where it was meant to act as a warning light at the activation of the PIR sensor. The 240-volt power supply was hidden along the garden. With this in place, the remaining task was to camouflage the components as best as possible with repositioned and cut foliage. True to our mantra, the Winder 2 motor drive was set to continuous firing at 2.5 fps with a 24-exposure roll of 400 ASA colour film, as we only expected this new location to be activated one time.

As night fell, we became alerted to the sound of foot thumping to the northeast of the house. Soon afterwards, the curtains behind the front door lit up, indicating that the camera had been activated. As I immediately run out of the front door, I was met by the glare of the floodlight, but I could still see that the bush thirty metres in front of me was variously illuminated by the other floodlight and the intermittent firing of the two flash units. At the same time, the sound of the motor drive could be heard as it was working its way through the film roll. Very shortly after this, the sound of extremely heavy running could be heard moving south towards our 1200-mm-high gate and fence, followed by the sound of the top wire strand connecting with a foot as the biped leapt across onto the other side. The violence of the wire being struck resonated loudly for a few seconds afterwards, across a number of star pickets.[32] As I turned left off the verandah

---

[32] Steel fence posts with a three-pointed star cross-section, used for the affordable construction of wire fences.

Fig. 8.6 Film frame #1 of 24. (Photo: Neil Frost)
The short video can be viewed at http://tinyurl.com/camerabump

and began to run down the northern side of the house, Fatfoot had already made her way down the southern side of our house and had probably travelled a distance of about thirty-five metres from the mound. By the time that I reached the western corner of the house, Fatfoot could be heard entering the swamp below, before stopping silently. At the time, I estimated the distance travelled by Fatfoot to be slightly more than one hundred metres, including the distance from the mound to the fence. The total time taken for the journey, including clearing the fence, was somewhere around ten seconds. This time estimate was very difficult to determine, considering the circumstances, because there seemed to be a delay before Fatfoot commenced her run. From the photographic evidence, the initial lag before the running commenced was

Fig. 8.7 Film frame #10 of 24 (four seconds after the motor drive commenced the film sequence @2.5fps). Note the displacement of the tree on the left of frame. (Photo: Neil Frost)

at least four seconds. This gives an average speed of about forty kilometres per hour,[33] which is close to the top speed obtained by Usain Bolt during a hundred-metre race. The more amazing thing about this achievement is that the speed was accomplished when moving through thick and spiky bushes rather than on a level running track.

The two film frames, Figs. 8.6 and 8.7, show a night-time scene of a dirt road in the foreground with the residential lights from neighbours seventy metres away in the background. The camera and its accessories were mounted on a steel chassis which was bolted to a 200-mm-diameter scribbly gum located atop of a one metre high soil mound.

---

[33] This is the same speed that was estimated when Damien was being pursued on his bicycle.

The gate and wire fence were situated behind and to the right of the camera field of view (FoV).

Looking at the two film frames above, the second 100-watt floodlight was mounted on the left-hand side of the camera, facing forward towards the central vanishing point of the frame. The T20 flash was centrally located and pointed along this major axis. The T45 flash was located on the right-hand side of the camera, facing forward towards the central vanishing point of the frame. When the first photograph was taken, a light bloom appeared on the left-hand side of the frame, which suggested that an object outside of the field of view had intercepted the light from the floodlight, reflecting it back at the camera lens and causing the aberration. By frame #10, four seconds later, the aberration has migrated across the sequence of frames, producing a much larger bloom that is most likely caused by the interaction of a nearby object with the light from

Fig. 8.8 Testing the cotton net detection system with a stick. The brick wall seen in the photograph was built as a physical barrier to keep Fatfoot away from the house. (Photo: Neil Frost)

the T20 flash unit, reflecting it back at the camera lens. Also, during the filming of the sequence, the location of the camera has been shifted to the right by a considerable amount of force, resulting in the relative position of the small scribbly gum seen in frame #1, now appearing closer to the left margin of the photograph in frame #10. This physical shift of the camera is compatible with a heavy impact upon the main assembly unit, pushing it towards the gate and fence to the right of the frame and consistent with the observed direction of movement. There is also a considerable amount of other lesser movement visible across a number of other frames. With the camera and its chassis bolted about 600 mm above the top of the mound, it seems likely that the combined 1600 mm rise above ground level was enough clearance to allow Fatfoot to escape beneath the light, probably by using quadrupedal locomotion, which it tends to adopt when attempting fast, stealthy escapes.

Clearly, the new camera setup had worked very well. More significantly, it demonstrated how good camouflage was perhaps, the most critical part of the entire system. Over the following weeks we would set up the equipment in new locations because, from experience, Fatfoot would not fall for the same trick, at the same place. In most situations, it was obvious that she would simply avoid the area all together. Having chosen a new location near to our bedroom, the camera was set up and camouflaged as best we could.

Over the following days the camera had not been triggered, even though the area around our bedroom was usually very active. After the previous setup, the motor drive was reset to take single exposures, mainly because the combined cost of the high-speed film, developing and printing was too expensive for us to maintain at that rate. Also, my excessive visits to the Kmart Photo Centre were visibly raising the curiosity of the attendants, as I collected another slim wallet of obscure nighttime photographs of bushland.[34]

As we were eager to discover why the camera was not actively taking photographs, I placed one of the tape VOR devices next to the camera motor drive. As the camera had not taken a photograph the following night, there was no corresponding audio recording from the Winder, but to our surprise, we discovered that the recorder was capturing the sound of movement in the bush around the camera. After borrowing Cheryl's VOR tape recorder and placing both in the bush behind the camera, it seemed obvious that Fatfoot knew which side of the automatic system emitted the light and which side was innocuous. After some adjustments and many more attempts, the camera started to take photographs again. However, there was nothing of any significance captured on of the frames!

We interpreted these 'empty photographs' as Fatfoot's attempt to test the limits of the PIR detection network.[35] In an attempt to try a more precise detection method, I made a cotton net that was attached at three of its corners with tension springs and the other with an electrical earth switch, which was connected to the electrical remote cable for the motor drive. The cotton net which sat slightly above the ground, less than fully matched the distorted two dimensional area covering the field of view of the

---

[34] In order to minimise the repeated awkwardness involved with frequent photo-processing, I rotated my business around a number of local film processors.

[35] From here on, we started to understand that 'testing the limits' was a typical behaviour used by Fatfoot with PIR motion-sensing lights. As we found out about a decade later, Junjudee also do this, but are much more destructive and prone to acts of sabotage. The Dooligahl that Dr. John was dealing with, had amazingly learnt how to move within a PIR sensor area without triggering the lights.

Olympus lens. The aim of this complex arrangement was to guarantee that anything that tripped the earth switch would be photographed, well within the FoV.[36]

At the time this alternative sensor network seemed like a good idea, however, apart from its greater complexity and higher maintenance requirements, the system took too long to set up and was over sensitive to new triggering agents that would fire the camera. Being a ground-based detection method, the camera was now taking photographs of terrestrial animals that, until now, had not been a problem. These included dogs, cats, foxes, rats and many other representatives of the local wildlife menagerie.

After abandoning the cotton net detection system, a significant event occurred on Friday, 16 July 1993—ten days before the police stakeout. For a considerable time, Sandy and I had been very well tuned into the activities and whereabouts of Fatfoot during the evening. The main indicators had been the relay barking of the dogs throughout the valley, particularly as their bark changed from a warning to a threat. Also, Fatfoot liked to draw attention to her presence by foot thumping and making various other attention seeking sounds.

Friday, July 16
"Basim heard dogs early in the evening. Heard Bess and noise to the east, investigated—8:00 p.m. and found police coming quietly down driveway. Talked to police—Matt and Graham. Showed them the swamp, branch up tree, photos, tapes of walking, etc. They went at 10:25 p.m. I walked them up to the car, up on road. Graham saw light in tree tops across the road. I investigated, thought it was Ian—wasn't, called to Sandy. Sandy thought that I

---

[36] Later attempts to contain the sensor pattern to within the FoV used cardboard 'blinkers' attached to the sides of the PIR detector.

was nearer. Sandy called out that something was walking across driveway, above east of house. Ran up to police, they drove down driveway very fast with lights on, jumped out and we investigated bush around—found/heard nothing. Police left at about 11:00 p.m. Dogs barked to south house. 11:05 p.m. dogs barked across valley—west. I heard a 'growl—woof' noise—shortly after 11:05 p.m. Dogs barked at 11:15 p.m., 11:25 p.m. Light rain falling at 11:30 p.m. Heard nothing all night."

Entry by Sandy: "Footsteps were extremely loud. I thought that Neil had heavy shoes on. They began and ended with the same volume on the driveway. Stopped when I shone the torch in direction of steps."

ON THIS OCCASION, our collie Bess had been barking at the front of the house and we could hear other sounds coming from the area. We were both certain that something was in the bush to the east of the house. As I usually did, I exited a side door with the camera and flash and slowly began to move around the circumference of our front lawn, stopping frequently to listen. It soon became obvious that something was moving slowly down the brick pathway leading to the house. As I continued to rotate around the outside of the grass, I was coming very close to the brick path at the edge of the bush where I could hear walking. I raise the camera in front of me and started to walk directly towards the sound of movement. Suddenly I was brightly illuminated by the light from two police Maglite torches. A police officer whom I didn't know, was extremely surprised. He said, "How did you know that we were here? I replied, "Because we are very used to reading what's happening outside!" Although it was dark, I think that he was impressed with how serious I was.

I mainly spent the next two and a half hours speaking to Graham, who was a local Detective Sergeant. As I walked him around the land, he asked many questions. After they spoke to Sandy, I walked Graham and Matt up the driveway and along the road a short distance to their 4WD, a total distance of about 150 metres from our house. As they were preparing to leave, Graham noticed a light in the treetops on the far side of the road, opposite our driveway. I walked down the road to our driveway and could see torchlight. Thinking that it was Ian, I called out, "Is that you Ian?" and Sandy replied, "Is that you Neil?" I replied, "Yes." Sandy said, "If you're up there, then what is down here with me?"

By now the police were starting to drive up our road, so I ran after them and managed to slap the back of their vehicle with my hand. I said to Graham, "He's down there now!" They spun the 4WD around, turned on all of their driving and spot lights and drove at high speed down our driveway. As they slid to a stop, Matt and Graham jumped from the vehicle. Matt ran straight into the bush. Graham stopped and yelled out to Sandy, saying "What are you doing? Don't you have two young children? Get inside and look after them!" After Graham had searched around the house, he spoke to Sandy again. He asked her to recount what had happened. Sandy added to the audio description by saying that, "They were really heavy footsteps. I thought that it was Neil returning and he was disappointed because they were slow and heavy. The closest sound to what I heard would be the noise made by a car tyre moving across a gravel road. I had walked up the path towards the noise because I wanted to light up the driveway for Neil."

Perhaps the most interesting sequence of still images was made four days before the police stakeout, during the evening of Thursday, 22 July 1993. Having moved the location of the automatic camera many times prior to this, I decided to try a new area near to the swamp that overlooked one of the main tracks. Contained within the camera FoV were the two items that the police had asked me to display, in the futile hope of obtaining human fingerprints—a wine glass and wine cask bladder.[37] Supporting this new camera position required the 240-volt power supply to be relocated about hundred metres from the house and recovered with leaf litter. Fortunately, Ian had arrived home from work and helped with the task.[38] The electrical cable was then closely secured to the top edge of the scribbly gum branches, in a further attempt to minimise detection. The power was then connected to the PIR sensor. Clearly, as camouflage was considered to be the main precondition of success, I used an extension ladder to position the simplified camera about 4.8 metres up into the tree canopy,[39] in the hope that the extra height might further conceal its presence. Previous photographic attempts had always been much lower down. The motorised camera and T45 flash were tilted down at an angle towards the targeted area in an attempt to reduce reflection from their various optical surfaces. The Winder was set to take a single photograph with each detection. Some small cut branches were then inserted into any camera or flash cavity that would accommodate them. The only concern that I had at the time of installation was the late start, because there was always a reasonable risk, or certainty, that

---

[37] On September 22, the previously inflated wine cask bladder was found on the ground, with two puncture holes, 60 mm apart and bitten through the middle, leaving an impression of a jaw. This occurred about two months after the stakeout.

[38] I didn't have enough electrical extension leads to complete the run, so Ian lent me one of his. He also placed electrical tape around the plug and socket to make the connections more waterproof and to help keep them together.

[39] This was the same location that the producer of the documentary *Animal X* placed the Sony thermal imaging camera in 1997. At that early time, the PIR camera cost $AUD80,000 to purchase.

Fatfoot was watching from somewhere nearby. Finally, as the Sun had set about thirty minutes earlier and I was running short of time, I simply placed a tape VOR device about five metres east of the centre of the FoV in a nearby bush, with an attached external microphone at a height of about five feet above the ground.[40] With the valley rapidly becoming dark, it was time to leave. Following is the log entry for this event.

Thursday, July 22
"Heard dogs at 3:15 p.m. Set up camera in new position. Started at about 3:30 p.m, finished basic setup by 4:30 p.m. Placed wine cask bladder in bush to right of centre of camera frame. Set tape recorder in bush to the NE of camera about 5 m distant at 5:45 p.m. Lights came on at 6:50 p.m.—4 times over about 10 minutes. Investigated with Ian, heard and saw nothing, except for twig breaking to the north—from camera position. Went up to area near driveway, heard twig(?) Went inside to listen to tape retrieved from near camera. Lights came on several more times—10 X. Listened to tape, heard suspicious noises. No lights from about 7:50 p.m. to 9:45 p.m. Dogs barked at 9:30 p.m. to 9:45 p.m., dogs 9:50 p.m., 10:00 p.m. I went outside at 9:59 p.m. to check on sound heard on tape—paint can.[41] Good Dog 10:10 p.m. Bess barked at 10:12 p.m., dogs 10:33 p.m.—Gordon's. Placed two tape recorders, Robert's on SE corner and Cheryl's on SW

---

[40] The original analogue tape recordings for this and some other events, still remain. They were digitised and converted to mp3 recordings by Damien, who is a professional sound engineer.

[41] The noise heard sounded like an empty paint can being placed on the ground and the wire handle being allowed to fall down against the side of the metal container where it rapidly, but briefly rebounded. The sound was unmistakably from a paint can, which makes me think that Fatfoot was carrying one and then placing it down.

corner of verandah at 10:30 P.M. Went to bed at 10:45 P.M. Woke at 3:05 A.M. to loud thumps (?) listened, heard 2 more but gentle thumps at 3:47 A.M.—SE. Heard various noises. At 5:10 A.M. heard movement on SE corner of verandah. Heard 2 loud thumps at 5:18 A.M. to the SE. Heard various noises until about 5:30 A.M."

SHOWN IN FIGS. 8.9 and 8.10 are two automatic camera photographs. The wine cask bladder, with its prominent dark shadow can be seen in the bush slightly right of the photographic centre. The clear wine glass is located on a short twig found at the base of the small branch on the left scribbly gum. If the smaller scribbly gum tree on the right hand side of each photograph is compared to the other, it may be noticed that the tape VOR device can be seen on

Fig. 8.9 Automatic film camera photograph #3 of 4. (Photo: Neil Frost)

the ground at the base of the tree in photograph #4. The recorder had originally been hung in the bush immediately above where it had landed, because of a lack of available time to install it properly.

At the time that these photographs were printed, the above-mentioned items were the limit of the known detail seen, which was very unfortunate considering the upcoming police stakeout that we were facing. This remained the case for many years until I applied the 'blink comparator' technique that I use to examine astronomical photographs, when looking for subtle changes or movement amongst the stars.

If each of these photographs Fig. 8.9 and Fig. 8.10 are separately scanned at a high resolution and accurately cropped to the *same frame size,* then opened together in

Fig. 8.10 Automatic film camera photograph #4 of 4. (Photo: Neil Frost)

the same graphical application, the images can be easily and rapidly compared by toggling between them. This is like using an astronomical 'blink comparator' used to show the movement of planets, comets and asteroids against the stellar background. When this has been done, a very large, black, prostrate figure can be seen moving beneath the bushes, which is otherwise, invisible. This is another reason why it is difficult to photograph these 'marsupial hominoids'! (The reader can view this pair of images, toggled, in a short video at http://tinyurl.com/fatfootblink)

Either by measuring directly off the screen or by counting the number of pixels by dragging the mouse across the screen in 'Preview', an estimate of size can be made. From my estimates, the left to right 'length' of the figure is about 600 pixels and the top to bottom 'width' is about 400 pixels. In comparison, the trunk width of the left side scribbly gum tree is about 100 pixels. Thirty years later, this scrubby gum is twice this apparent size and after a hazard reduction burn, it is the only tree remaining in the original FoV, making a baseline measurement between the two trees impossible to determine. Therefore, the best estimate for the diameter of the tree at that time is a maximum of 400 mm. Since the 'length' of the figure is six times the tree diameter, this gives a physical length of 2.4 metres and similarly, the 'width' is four times the tree diameter or a physical width of 1.6 metres.

Using an alternate method, the tape VOR recorder fell conveniently to the ground along its length. The exposed end of the recorder was measured at 90 mm and from the photograph, it was 24 pixels wide. Dividing 600 pixels in length by 24 pixels gives 25 times by 90 mm, which equals 2.25 metres 'long'. Similarly, dividing 400 pixels in width by 24 pixels gives 16.7 times 90 mm which equals 1.5 metres 'wide'. Both methods are reasonably consistent across their dual estimates. Whatever the actual measurements are, it seems to be very obvious that this animal is quite big! I think that the 'width' is exaggerated because it appears to include an arm in the measurement.

From the accompanying log entry it was noted that the lights from the camera were seen four times over a ten minute period, from 6:50 P.M. Using the blink comparator technique, the first two photographs showed nothing of interest. From the log entry which mentioned the dogs barking, it would seem that Fatfoot had been watching us for fifteen minutes before we started work and was aware of the camera's presence from the beginning. Consequently, as she had done elsewhere, she was simply testing the limits of the PIR sensor with the first two photographs, by approaching the automatic camera from behind, a behaviour that was earlier proven through the use of two VOR tape devices. I strongly suspect that the third photograph was a miscalculated error, where the overlying vegetation cover was insufficient to fully hide her presence and the speed of her movement was probably too fast, causing the PIR detector to fire. I know from my own attempts to defeat a PIR detector, that using intermediate bush cover makes an undetected advance much easier, as does adopting a suitably slow speed of approach and maintaining a small profile. It seems to me that the fourth photograph was also a mistake, brought about by Fatfoot's overwhelming desire to obtain the shiny microphone from the overhead bush and indirectly, the tape recorder next to it, which subsequently fell to the ground! I would imagine that after the fourth flash photograph, her adventurous will to proceed was overcome by her need to preserve her crucial night vision. The only feasible exit from this situation therefore, would have been to back out the way that she had entered. Undoubtedly, the remaining ten camera activations were nothing but deliberate attempts to regain the attention of Ian and myself, so that we would return to the swamp and possibly engage with her in play. Another behaviour that supports this opinion is the additional "attention seeking behaviour", where deliberate attempts were repeatedly made to give away her location in the bush, by loudly causing twigs to snap and by foot thumping.[42]

OUR EARLY PERIOD of research was extremely busy and varied, with our attention randomly shifting towards new concerns as Octopus reports came in that deviated from our expectations. The learning curve, for all of us, was intense, demanding and confusing at times. Although we were making good progress, some things remained puzzling, for example, the different foot morphology descriptions that we were receiving from across the expanding Octopus network. Containing all of these incoming inconsistencies within the singular Dooligahl framework was becoming difficult.

Sandy and I would frequently reflect upon our earlier work and wished that we could start again because the outcomes would have been a lot different and much better. As conventional wildlife photography had reached an obvious stalemate, much of this work was put on hold. Even though we had tried to use passive methods of investigation with everything, until forward-looking infrared cameras became more available and much cheaper,[43]

---

[42] As a special note to sceptics, I would hope that it has finally become obvious why photographing these hominoids is not a simple case of 'point-and-click' with your iPhone, that is, after you have entered your passcode, opened the camera application, selected the mode and then, if it is still around, take the photograph. Can I suggest that sceptics get out of their armchairs and walk the 'Six Foot Track' in the Blue Mountains, staying the night at Alum Creek Camping Ground?

[43] The first time that we used a forward-looking infrared camera was in July 1997.

8: Dooligahl

nothing was going to make much of a difference in the meantime. More significantly, the Octopus was starting to increase its network traffic and gain local respectability. With the increasing volume of reports, more time was required in following up on incidents and providing general counselling and advice to the witnesses.

With the police stakeout completed, we could all relax again! Around the house there was less urgency and much less attention was being paid to what was typically going on outside. From about here onwards, the main focus of our attention was shifting towards the Octopus.

The foot thumping and other attention-seeking behaviours remained, but some things did shift. One of the early changes happened after I had gone outside to investigate multiple noises between our house and the neighbour's property. After listening for a minute or so, I realised that Cheryl was removing washing from her clothesline after coming home from her night shift at the Queen Victoria Hospital. Having spoken out loud to Cheryl so that she was immediately aware of my presence, she stepped over the fence and walked over to stand with me on our driveway. We had a talk about our respective days for a few minutes until we both noticed movement a short distance into the bush from where we were standing. The call "Mook, mook, mook" was then heard. Amazed at what we had both heard, we thought that Fatfoot must be attempting to communicate with us. I told Cheryl that I was going to repeat the call back to her, which I did. Immediately Fatfoot returned the call but at a greatly increased volume. Again I told Cheryl that I was going to do the same but with a raised volume. Immediately Fatfoot repeated the call back, but this time at an impossibly high volume. This was not the first time that I had heard this call, having heard it at 2 A.M. on Friday, July 9, coming from the bush to the WSW of the house. I said to Cheryl that I wasn't sure what we were saying and that the increasing volume suggested that she was becoming angry with my response. Cheryl said, "I think that it likes you!" Cheryl then asked me to walk her back home.

The area centred around the clothesline became a very active site over the following months and years. This was not surprising because it was located midway between our houses. It was also an area that Sandy had reforested over the previous decade with native trees and plants after the previous owners of Ian and Cheryls' property had unilaterally applied an agricultural policy of collective land ownership and used 'slash and burn' techniques that I was familiar with from my studies in New Guinea and Melanesia. Consequently, the mature-sized trees from this area now provided plenty of cover for a biped that was intent on closely observing the occupants of both houses.

Fatfoot was particularly fond of Ian. His kamikaze style of interactive research was undoubtedly interpreted as a willingness to engage in enthusiastic play. As an example, one Friday evening Ian and I had pursued Fatfoot into the swamp and had noticed that she had caught the attention of Gordon, a neighbour on the diagonal, far side of the swamp. After Gordon had shone his torch in the treetops, I called out to him asking what he was looking for. Already knowing the answer, he yelled back to say, "Something bloody big just ran past me!" Having obtained his address from across the swamp, Ian and I got into the ute and drove around to Gordon's house, picking up John along the way. We then spoke to Gordon for a few minutes, determining Fatfoot's last location and predicting her soon-to-be displayed behaviour to Gordon and his assembled family. Ian then motioned to me to follow him. We climbed over the rear fence and then got further instructions from Gordon who suggested that the animal was last heard from behind a large tree to our left. Ian then climbed over an adjacent fence and immediately there was a reaction from behind the large tree. Within seconds, Ian had disappeared into the thick bush although the sound of Fatfoot's gait quickly became the dominant noise heard as they both moved further into the swamp. As my presence was superfluous, I climbed back through the fence and took a seat alongside the rest of the assembling audience, including John and members of Gordon's family, that were comfortably seated on the large decked area and enjoying the live entertainment. The pursuit continued for lengthy periods of time, only to be punctuated by brief pauses in the running, which Ian later insisted were the result of his attempts to determine Fatfoot's location. The chase didn't have any obvious objective, with both participants repeatedly running around in wide circles. Occasionally the two bipeds would sweep high up along the edge of the swamp, past the upstanding and cheering audience. With the pointlessness of the exercise having been well exceeded, the chase came to a slow conclusion on the final circumnavigation of the swamp, after almost half an hour. It seemed obvious to the audience that Fatfoot still had plenty of energy, whereas Ian was utterly farnarkeled.[44]

Consequently, Fatfoot would frequently stand behind one of Sandy's mature eucalypts opposite Ian and Cheryls' elevated deck and back door, in an attempt to catch a glimpse inside. When Ian hadn't noticed Fatfoot's presence after a while, Ian and Cheryl would often hear growling, grunting and various other noises coming from behind the Tallowwood Eucalypt *(Eucalyptus microcorys)*. During an evening visit to their house, I also witnessed this behaviour. However, the most disturbing example of

---

[44] It was patently obvious that Fatfoot was in command of the situation. It reminded me of early settler accounts where English hunting dogs would be unable to run down kangaroos over long distances.

this attention-seeking activity was having one of the potted plants pulled off the deck which fell onto the concrete path below. This incident was also witnessed by Debbie, who was Ian's cousin. When Ian, Cheryl and Debbie went outside onto the deck to investigate, they heard various other noises. After turning on the outside light, they heard growling coming from below them at the base of the tallowwood tree, followed by the sound of running away to the northwest. A short time later, the dogs in the valley began barking, which I investigated without any success. It was interesting to note that Debbie lived on a farm in the central-west of New South Wales. She mentioned that she had had similar experiences with strange activities occurring around her farm. For example, Debbie said that her working dogs would sometimes show extreme fear at night and hide under her house. She highlighted this comment by saying that this was an extremely uncharacteristic behaviour, because her working dogs were a very tough pack that normally showed no fear of anything![45] This pot-destroying behaviour was repeated on another occasion when a large terra cotta pot was smashed after falling off the deck.

By spending time near to the rear kitchen door, it was not surprising that Toby's food bowl would be targeted by Fatfoot. After several dog bowls went missing, Ian attempted

---

[45] This uncharacteristic fear displayed by working and hunting dogs was often mentioned by people who should know. In particular, having interviewed several pig hunters who had encounters with something other than what they were familiar with, said that they were surprised to see that their pig dogs were afraid and unwilling to engage. Two pig hunters visited after an incident where they shot their rifle and hit a Dooligahl in the buttocks which they initially thought was a pig because it was moving about quadrupedally. The dogs had previously fled the scene. The hunters also became terrified after the animal stood up on two legs and roared at them. The Dooligahl would have certainly died from its injuries. Ian's anti-personnel dog Toby is another example.

8: Dooligahl

to remedy the problem by chaining a two handled cooking pot to the verandah post. This strategy was partially successful, because only the contents were subsequently taken, with the pot typically being found hanging by its chain over the fence. Similarly, our dog Bess had her food bowls regularly taken and not recovered, although Ian and I were not initially aware of our shared problem. For us, proof that the suspected culprit was responsible was tested one evening after I had left a chicken carcass outside for Bess to eat. Even though a carcass was her favourite food, Bess refused to approach her meal. As I was standing between Bess and our front door, I repositioned myself between Bess and our driveway. Bess then ate her meal, but only as long as I remained at that intervening position. Years later we would find a cache of dog bowls under a dense bush[46] across from the driveway on our neighbour's property. As Ian explained the matter from his perspective, "The beastie was standing over the two dogs."

The availability of food and its continuity of supply are important factors in the survival of any animal. For Fatfoot, being able to access an occasional chicken carcass or a can of Pal Meaty Bites was probably a welcome change from her regular diet but it would not have been essential for her survival, as the swamp is a rich source of animal protein with the surrounding human settlements providing a wide range of alternative food choices.[47] However,

---

[46] Laying prostrate, particularly under thick bushes, seems to be a very common behaviour that is used to reduce the possibility of detection. See 'blink comparator' photographs. Rather than the discarded dog bowls simply being stored under the bush, it seems more likely that this was where the food was consumed and the empty bowl left. Similarly, laying prostrate during the middle of a pursuit allows the chaser to overrun the pursued, which also gives the impression that the animal has miraculously vanished!

[47] A few well-known alternative food choices include frogs, bandicoots, sugar gliders, cats, small dogs, other domestic pets, and many types of fruit, including apples, pears, citrus, stone fruit, berries and reintroduced native fruit, such as any of the variants of *Syzygium australe*.

for a well fed and intelligent animal, entertainment might assume a higher priority than food on some occasions.

AN IMPORTANT PART of our research involved speaking to neighbours and wildlife experts about their experiences with, and knowledge of, the local animals and environment. One weekend I walked down to a neighbour's house a few hundred metres to our north. Their house was isolated to the same extent as ours and they had been living there for almost as long, so I thought that they might be having similar experiences. After introducing myself, I indicated where our house was and made various comparisons based on our similar locations and environment. By using a modified form of the 'standard question', I asked the owners if living down the back, like us, resulted in them experiencing anything unusual around their house. They said that, apart from the wallabies eating their grass lawn,[48] they were not experiencing any worrying problems and suspiciously returned the question to me. Without launching into any seemingly hidden issues, I simply said that we had been hearing unusual noises around the house during the night. They agreed but also added that they had been hearing the sound of objects being moved about under their house during the night, particularly their glass recycling crate. After confirming that we had had similar experiences, along with other residents, I left the matter at that and returned home. Although these residents didn't seem to be aware of the broader issues at play in the valley, they did share an under-house experience that was starting to seem more common than a few isolated incidents.

---

[48] With large areas of grass lawn replacing bushland, swamp wallabies are attracted in larger numbers to eat and breed in urban habitats. As with chicken coups, the wallabies attract Dooligahl, increasing the incidence of encounters with these predators.

8: Dooligahl

We also had the good fortune to host a stay at our home with Gary Opit and his daughter Lowanna. Gary is a wildlife talkback radio broadcaster for the Australian Broadcasting Corporation, Radio Station Northern Rivers, 2NR. During their week-long stay, we discussed many things and went for lengthy bush walks. It was a rare and excellent opportunity to run questions past Gary who has extensive bush knowledge. I remember driving with Gary along the dirt fire trails to our north, for as far as we could travel and then proceeding the remaining distance on foot. One thing that was apparent was how far certain noises can and cannot travel. In particular, how low-frequency sounds travel farther than high-frequency noise because they are absorbed and dispersed less by obstacles. Having become complacent with our ambient noise, hearing what remained of these sounds at a distance of about five kilometres was a new and enlightening experience that highlighted how sound is filtered over distance and what sound generators still dominate and are therefore, likely to gain attention. This experience refreshed my memory from when I bushwalked and camped as a teenage Scout, orienteering through remote bushland for up to three days in areas that are now heavily settled and how we experimented with Aboriginal bull roarers and cooee calls, used for long-distance communication.

From a distance of about three kilometres, most common urban sounds have been filtered out of the bush environment because the sound energy has been absorbed and dispersed by objects lying along its path. Depending upon the terrain density and the observer's location, the few remaining sounds at our position were typically derived from railroad transport, highway traffic, building and home maintenance, and unusual acoustic events. When Gary and I were standing on a hilltop about five kilometres from home, virtually no urban sounds remained. The occasional sounds that we could barely detect at that point were the remaining low-frequency sounds from trains, truck exhaust when used for braking, a commercial wood chipper and hammering. Of the few remnant sounds, it was the solitary and staccato hammering from a very distant home builder that was the most fixating because one could easily imagine that the carpenter was a lone survivor repairing a shelter after the apocalypse had removed all other audible traces of civilisation.

Some circumstances seem to gain the attention of these animals, possibly because the sound that is made is understood for what it represents and can also travel a long way. Some very disturbing examples of this, which Elder Merve initially warned us about and was frequently reported to the Octopus, are where babies are heard crying or children are heard playing. Using the example of the low-frequency sound made by a hammer strike heard five kilometres away,[49] this provides an audible coverage of seventy-eight square kilometres. It is probably not surprising, therefore, that a Dooligahl visited the fifteen-thousand-acre or sixty-square-kilometre farm where I was living in 1981 after farmer Ned replaced the water tank using a variety of heavy equipment. It was also not surprising that Fatfoot visited our building site in December 1983, after it was excavated by a fifteen-tonne dozer Drott.

A short distance to our north live neighbours Phil and Helen. When we first moved into the area, they had already been living here for some time. Like us, they built their house, but first, they constructed a large metal shed at the back of their land as temporary accommodation. Alternatively, we chose to rent a house nearby. After

---

[49] Foot thumping, used by many species of macropods, is a low-frequency method of long-distance communication using infrasound.

8: Dooligahl

introducing ourselves, we exchanged narratives about our respective builds, although Phil and Helen held back on a few unusual aspects of their story.

After we became Yowie-aware, we specifically spoke to Phil and Helen as we did with many others, about the other wildlife. To our surprise, they told us about the encounters that they had experienced, almost from the very beginning of their house construction. Phil said that the problem with living onsite, apart from the similar issues related to camping, was that the metal shed was not thermally insulated. The inside of the shed would become very hot during the day and very cold at night. However, Phil also said that the metal shed was not insulated acoustically either.[50] Anybody on the outside could hear everything happening on the inside and vice versa. Consequently, after spending time building during the day, they would hear disturbing sounds on the other side of the metal walls during the night. In particular, they would hear walking, breathing, branches braking, objects being picked up and moved about, tapping and occasionally, growling and grunting. Phil found the nightly task of checking the bush surrounding their temporary accommodation very challenging, but he never saw anything.

During this early research period, Phil and his two sons, Tom and Michael would contribute significantly. Phil would often telephone to keep us informed on the outside activities of Fatfoot and his two sons would help by searching the bushland for any evidence. Like us, Phil tended to contribute mostly on Friday nights when Ian and I would be about, trying to engage with Fatfoot in some manner.

---

[50] Our timber house, with its low-density construction, allows most low-frequency sound to pass through the walls.

The evening of Friday, 6 September 1993, started like many other pre-weekend pursuits, with noises being heard across from the driveway. With the camera charged and, more importantly, loaded with film, I headed off in the direction of movement, trailing behind the disturbance by about twenty metres. As I moved in a northeasterly direction towards Phil and Helens' house, I was very surprised to intercept their youngest son, Michael. With this extra help available, I got Michael to return home and ask his father for assistance. After Michael returned and with the two of us standing in the bush, we heard further movement as Phil arrived. He commented on hearing a large branch breaking deeper into the bush, which we interpreted as being a deliberate distraction from some other activity that we weren't aware of. Also, Phil commented on how strange the large branch break was, because he didn't hear it fall to the ground. This was further confirmation of Fatfoot's intelligence by employing a 'Theory of Mind' technique to intimidate Phil—something that Fatfoot had long abandoned using on Ian and me.

I returned home and telephoned Ian, suggesting that he might be able to move ahead of us by travelling along the public road before entering the bush and then, try to ambush Fatfoot as she was pushed in that direction by the three of us. However, Ian soon realised the futility in the attempt and came down to where we were standing. We soon heard a change in movement towards the swamp, so Phil and his sons returned home and Ian and I retired to the verandah for a refreshing beverage. Around 9 p.m., Ian and I heard some provocative movement from the verandah, which we followed down into the swamp and then south onto Ian's land. We followed a very strong lead that was between twenty and thirty metres ahead of us and pushing through some very difficult terrain. However, it was becoming patently obvious that we were being

8: Dooligahl

Fig. 8.11 Unaided sketch of a three-second encounter: my observation of Fatfoot, as seen from behind, walking diagonally away at a range of ten metres, using backlight from Phil's low-wattage verandah light, 6 December 1994. When I heard movement in this area I would phone Phil and ask him to turn on his verandah light so that I could see Fatfoot. (Drawing: Neil Frost)

led astray because the sound source was very deliberate and there was no hope of chasing the fugitive down in the thick vegetation.

The following morning Phil called out to me from his verandah. He walked over to our house with some very interesting news. He said that he had stayed outside that night and had seen both of us move off the verandah and head down towards the swamp. Phil then said that Ian and I had missed Fatfoot who was elsewhere. I tried to convince Phil that we weren't chasing a phantom, because we were following a very strong lead. Phil insisted that we had missed the action because he could see the outline of a hominoid-like figure standing beside a large tree, on our northern neighbour's land, halfway between our two houses. Phil said that he had been standing on his verandah, with the outside light on. He said that the large figure seemed to be looking at Ian and I as we were heading towards the swamp. The figure then turned back towards him as if it was monitoring his presence as well. Phil said that during the time that it was looking at him, "I saw the yellow-red eyes for thirty seconds." The figure then briefly looked away before looking at him again. He also said that "the eyes were at a height of about six feet".[51] Although we suspected for some time that we were dealing with more than one individual, this was substantial proof that we had, at least, two individuals. Over the following months, we paid closer attention to the location of additional or quieter background sounds and thought that we had a minimum of three individuals, probably keeping

---

[51] Phil's estimate of six feet to the eyes was the same as the measured height of the red eyes seen in the photograph, being 1850 mm or slightly less than 6'1".

8: Dooligahl

together as a family group. The dominant source of any disturbance was always in our foreground, with the others tending to hold back in the far distance.

Another interesting aspect of Phil's observations was the colour change of the eyes. When I spoke further with Phil, he said that the eyes first appeared as yellow discs, which quickly changed colour to red as the head slowly turned towards him. When the Dooligahl looked away, the eyes briefly changed back from red to yellow, before disappearing altogether. This was the same pattern of eye colour change that Sandy and I had also observed from our verandah for many years, without truly knowing what animal was responsible. With these cryptids, the reflected light from the tapetum lucidum as it exits the eye is coloured by the chemical compounds that form the red photo-amplifying receptor protein found in the retina, or rhodopsin. Animals with photo-amplifying receptor proteins with a different chemical composition have a different coloured tapetum lucidum. It seemed, therefore, that this colour shift was simply the result of chromatic aberration, where the light of different frequencies or colours passing through an optical medium is refracted or bent by different amounts. The colour shift seen simply depends upon where the observer's eye intersects the light cone along the optical axis containing the range of outgoing spectral colours. An extreme example of this is the full-colour spectrum, red, orange, yellow, green, blue, indigo, and violet, produced by light passing through a diffraction prism. In this case, however, the distortional variation is less severe, with the outgoing red light only being shifted towards orange and yellow, the next two colours in the spectral order. No other colours can be seen because they lie too far from the optical axis, where vignetting results in a rapid loss of light at the edges, followed by the eventual loss of the image.

Consequently, the ability to see the nighttime 'red eye' phenomenon is very critically determined by the viewing circumstances, something that I learnt from grinding and testing telescope mirrors. The three major elements are the observer, the light source and the subject. Optimally, all three should be in alignment. The light source can be any form of illumination, including faint sources like the moon. The observer can be a camera. The subject's eyes must face directly towards the observer. When the three elements are aligned along a common axis, with the subject's eyes furthest from the other two, red eyes should be seen. As the observer and/or light source moves off-axis, a position will be reached where the eye colour should shift towards orange and then yellow, before finally disappearing due to vignetting. Similarly, when all three elements are centrally aligned and the eyes are looking along the optical axis, red eyes should be seen, however, if the eye's gaze is rotated left or right of the centre, the colour should shift to orange/yellow until it is lost altogether. This is the same principle employed by 'red eye reduction' systems used by some cameras, where a small light, positioned near to the flash, but furthest from the lens, prematurely comes on to distract the gaze of the subject, as the shutter button is belatedly engaged. For a researcher, therefore, the ideal situation when attempting to photograph a cryptid would be to have the flash positioned close to the optical axis of the camera and for the observer to make a noise in order to direct the subject's visual attention towards the lens. Additionally, when the red eyes are visible to the observer, this indicates that the animal can see you and is looking directly at you—as seen in the red eye photograph. When the eyes appear yellow, it is a very good indicator that the animal has an approximate idea of your location, but has not precisely determined your position. When no red eye is visible, the animal is not looking at you or there is no longer an optical alignment along the axis.

8: Dooligahl

In addition to supporting their father during nighttime investigations, Tom and Michael both carried out their own independent searches of the bush on a regular basis, looking for any prospective evidence. A landmark discovery was made by Tom during one of his afternoon searches. There were many things to look at across a very wide area of swamp and sclerophyll forest and many different aspects of investigation to consider. When Tom found hairs wrapped around a fence barb, he immediately knew of its potential value in helping to identify the animal and excitedly believed that they were most probably from the Dooligahl that we had all experienced. When Tom came to give me the good news, I felt similarly excited by his find.

Tom took me down to the swamp and showed me what he had found. A wire fence with an open strand construction passed through the swamp, bisecting the main track.

Fig. 8.12 Orange hairs caught on the barbed wire. (Photo: Neil Frost)

Fig. 8.13 Tom pointing to hairs captured by the barbs on the top strand of the wire fence. (Photo: Neil Frost)

8: Dooligahl

Fig. 8.14 The main swamp track in early 1993, before it became heavily trampled, showing the typical vegetation density and the northern barbed wire fence boundary, where the hairs were later found. (Photo: Neil Frost)

Wrapped around the top strand of barbed wire were several, predominantly orange-coloured hairs. The hairs were about three feet above the spongey swamp base and were located near to the centre of the main track. I photographed Tom pointing to his find and then took some close-up shots and other photographs of the immediate area. As the hairs were very well snared by the barb, I returned to the house to get broad and long-nosed pliers, tweezers and several new, sealable plastic bags. It required the two of us to untangle the captured hairs. Tom held each hair securely with the tweezers while I used two pliers to unwind the barb. Tom placed each hair inside the plastic bag and sealed it.

With the hairs collected and bagged, the next step remained uncertain. The only knowledge that I had of hairs, was what I had picked up listening to sheep farmers and shearers at the pub. Applying this knowledge to the samples told me that these hairs had a very fine crimp that would make the yarn from this wool very resilient and elastic.

Also, the hairs were predominantly orange, which in some animals is a spectral adaptation that allows more UVB light to pass through the fur without being absorbed, which increases the photosynthesis of vitamin D. Alternatively, some animals photosynthesise vitamin D from secreted oils on their fur, which are then ingested when the fur is groomed. These adaptations and other variants are typically found in animals that live at high latitudes, or with animals that normally have a low exposure to sunlight.

Over the next few months, we cautiously asked around for advice. We felt secure enough to ask Michael, our family General Practitioner for advice on whom to get assistance from. Despite what we told Michael, he remained open-minded but was unable to do much more than discuss the process with us because this science was still in development and extremely expensive. Consequently, finding someone to do DNA analysis was very unlikely and far too expensive for our budget even if it was feasible. (Nine years later we obtained contacts within the Commonwealth Scientific and Industrial Research Organisation [CSIRO] that were willing to help.)

In the meantime, several other researchers suggested that we should consult experts involved with microscopic hair analysis. In these studies, an analyst examines some different characteristics of the hair using a microscope to arrive at a conclusion, which in this case, would be the animal from which the hair samples came. After discussions amongst ourselves, we chose an Australian expert who seemed to mainly specialise in the analysis of native animals. I telephoned the analyst and discussed the samples and the circumstances surrounding them. The discussion went well up until the end when I was asked what I thought the animal was. Rather than not disclosing our suspicions by keeping the sample blind, I naïvely mentioned the 'Y-word' to the analyst. The change in the tone of the reply should have been a warning but I proceeded with sending the samples to the analyst with the payment. Sending all of the hairs was my second mistake!

On receiving a postal reply several weeks later, the responsible animal was identified on a mainly blank piece of paper, as a 'cat'. None of the sample hairs was returned with the reply, despite having been stated in our letter. In multiple ways, this was a shocking result! I immediately telephoned the analyst who was adamant that the hairs were from a cat. I was also told that the hairs had been 'binned' and were 'long gone'. Thinking about this a lot, as I have unfortunately done, I didn't put enough thought into this problem because no one could have had any idea of what Yowie hairs are supposed to look like anyway.

In our opinion, accepting that a cat was the source of the hairs didn't make sense. How could a cat get its hair forcefully wrapped around the barb when the fence was open-wired? Instead, it had ample opportunities to simply walk between the wires lower down. Getting impaled on the isolated top barbed wire strand, three feet above, was hard to believe. The only explanation that we could think of was that perhaps the cat, if it was a cat, was dead and its hairs became caught on the barb as it was being dragged over the fence by the hunter. This was the only way that we could rationalise the choice of animal, given the circumstances. Not having the sample hairs returned after being studied, caused our morale to sink and I felt terrible having to apologise to Tom.

About a decade later, things became clearer, unfortunately. Another cryptozoologist was also of the opinion that hair samples were being unreliably identified by using this technique. To test his fear, he went to the trouble of travelling to Taronga Zoo Sydney and acquired some black panther hairs from the helpful zoo keepers. He sent these known hairs as blind samples to an analyst who misidentified them. The inaccuracy of this method of hair identification is now widely known. "It is still acknowledged as a useful technique for confirming that hairs do not match. But DNA testing of evidence has overturned many convictions that relied on hair analysis. Since 2012, the [U.S.] Department of Justice has conducted a study of cases in which hair analysis testimony was given by its agents, and found that a high proportion of testimony could not be supported by the state of science of hair analysis" (*Wikipedia,* https://en.wikipedia.org/wiki/Hair_analysis, accessed 11/2023).

Having very supportive neighbours has been a major element in the success of our local research group. Many people made themselves available without notice to assist

8: Dooligahl

with any unforeseen event, with Phil and his sons being good examples of this cooperative spirit. On several occasions, when the conditions were favourable, I remember being able to telephone Phil and ask him to turn on his low-wattage verandah light after a timed interval, so that it could be used as a subtle backlight when we were attempting to sketch a silhouette of Fatfoot, provide additional light for night vision equipment or create a diversion when we were attempting an ambush.

OBSERVING AND RECORDING behaviours is a relatively easy task. However, as we were dealing with an intelligent and highly capable humanoid, it was becoming increasingly difficult to acquire specific physical information about the Dooligahl. Through frustration, Ian suggested that we should split up and try and herd Fatfoot towards one of us, who would be concealed in the bush, waiting to pounce with a camera. The plan was exciting for some and fortunately, Ian volunteered to be the cameraman.

As we tried this technique several times, we spaced out our attempts across several weeks so that the intent would not become too obvious. As the barking of the dogs was heard moving up the valley, Ian would hide in a carefully chosen spot. Finding the ideal position was usually my job during the week. It needed to be an active path, so many tracks needed to be tested using cotton thread to determine if they were being currently used. Sometimes there were too many active options that would make the nighttime muster too difficult to manipulate, so some of these alternatives needed to be blocked off with 'fallen branches', in an attempt to persuade the taking of another route. When Fatfoot arrived, I would conspicuously move to a position that might cause her to move down a prescribed path. Typically, this would result in some early controlled deviations, towards the desired destination, but eventually, Fatfoot would avoid Ian's hiding place. Sometimes, though desperation or frustration, Ian would launch himself regardless, believing that he had gained enough advantage for a successful intercept.

When we discussed this failure, we both agreed that Fatfoot had better senses than us. Other than having superior hearing and night vision, Ian raised some interesting points based on his extensive knowledge of herpetology. He proposed that, like snakes, perhaps the animal may be able to see very small differences in heat. Whether this was achieved through the use of a specialised pit organ or, eyes with sensitivity shifted towards the infrared spectrum was unknown, but Ian thought that this was a real possibility. Over the following years, this issue would be raised many times by other researchers as well.[52]

A variation of this technique involved both of us staking out positions on either side of an established track further down the valley. Having already tested the track for activity, we positioned ourselves before the sound of the dogs could be heard in the valley. On arrival after the barking, the gentle sound of twigs breaking underfoot could be heard as the immediate swamp wildlife suddenly became very quiet. Despite our optimism, these attempts also failed as Fatfoot's progress would falter and then halt, a short distance from us. As usual, we had been detected and most probably by Fatfoot after using multiple senses.

Also around this time when Ian and I were still undertaking an occasional bush pursuit, I experienced behaviour that we had not fully seen before. After a lengthy Friday

---

[52] The complex micro-structure of guard hairs in certain small mammals, including the mouse-like marsupial *Antechinus agilis,* may act as infrared antennae. Baker, Ian. M. 2021. Infrared antenna-like structures in mammalian fur. *Royal Society Open Science* 8: 210740.

8: Dooligahl

evening chase we were becoming tired from repeatedly initiating a pursuit and then searching for the method of escape or means of concealment that Fatfoot was using. Over the many months that we had been chasing Fatfoot, we had noticed several changes in behaviour. The principal difference was the distance that Fatfoot was travelling away from us after we had engaged her. Sandy and Cheryl had initially made us aware of this after they had heard Fatfoot mirroring our movement, a short distance behind us as we returned home. This behaviour modification made sense because there is no benefit obtained from overexertion. Consequently, it was starting to seem more probable that concealment was the main method used for escaping our detection. One thing was certain when surrounded by bush, Fatfoot would never stand still in the open and allow a torch to be used to observe her!

After this long night, Ian needed to go home and sleep so that he was ready for work the next day. As I was talking to Ian, we could hear that Fatfoot had arrived back at the house and was standing directly across from the driveway. As an experiment, I suggested to Ian that I should stay up and see how she might react to me sitting outside and ignoring her.

In preparation for a long night, I moved the outside table and chair towards the edge of the verandah where it was illuminated by the light of the full moon. From inside the house, a got my watch, a six-pack of beer, and a new packet of cigarettes. On returning outside, I sat on a chair facing Fatfoot's position. Ian went home. As he left I could detect movement in the bush at a distance of about thirty metres.

After half an hour of silence spent sitting on the verandah, I got up and moved the table and chair back into the moonlight. As nothing else was happening, I opened another beer and lit a cigarette. There was no reaction from the other side of the driveway but I knew that Fatfoot was still there. Half an hour on, I continued to follow the moonlight with the furniture. By now I was starting to think that I had grossly underestimated the quantity of provisions that I needed for this job. For the third time, I got up and moved the table and chair, although I was starting to run out of verandah space and knew that I wouldn't be able to keep seated in the moonlight for much longer. I opened another beer, lit another cigarette and sat down at the table. By now we had passed the ninety-minute mark and the time was nearly 2 A.M. Suddenly the bush opposite my chair exploded. Looking across the road in the moonlight I could see that bushes were being violently thrashed about, with some being uprooted. Then, a small stand of tall bloodwood trees began to be whipped about and it seemed that Fatfoot was moving around in a tight circle, trying to keep them all moving as a group. In the dim light, the display was extremely impressive. After about twenty seconds, there was a momentary pause as this rage was transferred to a solitary bloodwood tree that was being rocked back and forth, through an arc of about ninety degrees. The rocking was so violent and powerful that I thought that the tree was either going to snap or become uprooted! Ten seconds later, Fatfoot turned and very noisily walked at a brisk rate, diagonally towards the valley neck, avoiding the existing tracks and breaking a large number of branches, trees and other foliage in the process. Waking the dogs from their sleep, the valley erupted with loud barking focused on the biped as she moved away towards the swampland. Calling out to Fatfoot made no difference to her withdrawal. I turned on the spotlight and shone the 'Bat-Signal'[53] into the night sky with no result either. There was no stopping her loud and angry walk down to the swamp, which continued to be punctuated by the noisy snapping of branches and trees.

On inspecting the site the following morning, the solitary hardwood tree had an estimated height of about fifteen metres and a trunk diameter of two hundred millimetres. In comparison I attempted to shake the tree, but was only able to work up a small wobble of a few degrees. The grass around certain parts of the tree bases were heavily compacted and the tree rootballs had substantial cracking in the partially raised soil surrounding them. As they were bloodwoods, a considerable amount of fibrous bark had been removed from high up where the hands had been holding onto the tree. Clearly, the 'marsupial hominoid' was extremely powerful and clearly, a very cranky beastie. Talking to Ian later that afternoon, we discussed the physical and emotional response that Fatfoot had made. We thought that she displayed some human emotions, particularly frustration and anger. For me, the major concerns were the explosive anger and very formidable strength. These were traits that I subsequently remained very mindful of. After making strength comparisons with chimpanzees, baboons, gorillas and other primates,[54] in his usual fashion Ian simply stated, "If anything happened, it wouldn't hurt for very long."

ON REFLECTION, IT is very interesting to compare the Blue Mountains community attitudes towards these cryptids from the early nineties, with just a few years afterwards.

---

[53] Ian and I used the spotlight to mimic the communicative ability of the 'Bat-Signal' to summon Fatfoot when she was not in the immediate valley. It was successful in summoning Fatfoot about a third of the time that it was used. On its inaugural use, Fatfoot could be heard coming from a long way to the north at high speed, clearly in anticipation of a chase and play.

[54] Ian had a very powerful dislike of Olive Baboons (*Papio anubis*), because of his lengthy association with them in research institutions.

When the Octopus was just starting up, it was quite common to be ridiculed in the street with distant calls of "Yowie" from the sceptical wannabees, just as Troy had experienced a decade earlier at school and also in public. When Yowie-related discussions based upon what we were doing started to percolate through the community, a small number of historical accounts freely rose to the surface without much resistance. The uncovering of these belated reports was the result of candid discussions between locally-involved residents and past witnesses responding positively to the 'standard question', resulting in all participants feeling comfortable with the disclosure. However, a certain degree of hesitancy initially remained. The two newly-uncovered historical accounts each had a caveat. They had to be anonymous.

One afternoon in mid-1993, I received a telephone call from a local solicitor. He called to speak to me about our experiences with the Hairy Man, because he had an experience when young that continued to disturb him, several decades later. He firstly explained that he owned a local legal service and did not want this story to become associated with him and consequently, he wanted to remain anonymous, though he felt that the local community should be aware of the incident and its potential implications.

The witness said that the incident occurred twenty years earlier, when he was sixteen years old. He had been walking through bushland north of the current location of the Warrimoo Bushfire Brigade when he came upon a body of a middle-aged man in the bush. After returning home, he contacted the police who accompanied him to the site. He said that the police were very puzzled by the death, because the man's head had been forcefully removed and was found about fifty feet away in the bush. The police said that the man had died 'in situ', but they could not find any evidence of how the head could have been pulled off the body by using any available mechanical means from the area. When he inquired later with the police, he was told that they had no motive for the murder and concluded that he had been 'killed by persons or animals unknown'. As we were about to have a police stakeout, I told this account to Graham. He said that he would have a look at the records but doubted that he could easily track a paper trail to find the details. When I followed up on this, Graham said that he was unable to find anything. From my perspective, having witnessed the strength of Fatfoot, I have no doubt that these 'marsupial hominoids' have the physical ability to achieve this—removing wallaby heads and dismemberment are certainly typical methods of killing used by these powerful animals.

Also around this time, I received a telephone call from a curious Blue Mountains guesthouse and restaurant owner. He said that he ran the establishment with his wife and they were very interested to know what we knew about the Hairy Man. In particular, they wanted to know how these animals interacted with people. He told me that, having seen their red eyes on many occasions, they were aware of their nightly presence, but said that they had never seen one, despite having had interactions with them for many years. As they ran a restaurant, most nights they would put suitable leftover food out on a platform away from the guesthouse for any of the local animals to eat. Their interest, however, was piqued after they woke one morning to find a wallaby body laying across the entrance stairs at their front door. Since then, there had been a small number of incidents involving offerings of other dead animals at the door. Since they operated a highly successful food and accommodation business, they asked me not to mention the establishment's whereabouts or refer to them by name.

In response, I thought that they may have had more initial experience than us regarding this particular feeding matter. However, using Aboriginal knowledge as a guide, providing food for these animals is not recommended. I recounted our story about feeding the carnivorous birds at our house and how that compounded issues, leading to our current situation. Food is always a very powerful attractant, so I further emphasised the issue by highlighting the problems associated with domestic chicken coups, which function to a large degree as food distribution centres, by providing water, chicken pellets and other food scraps for rats and possums, foxes, feral cats and dogs, eggs for snakes and so forth along the food chain. Also, the less well-maintained the coup, the worse the problem tends to become.

Brushtail possums also compete with others for the opportunity to live in the roof cavities of residents' houses, to eat their ample supplies of fruit and vegetables as well as the eggs and young of nesting birds. After capturing and releasing eight possums in eight days, mostly brushtails, in a small hundred-metre by hundred-metre area of bush on our land, it was obvious that we had a bigger problem than we originally believed.[55] Most of the captured brushtails were large, highly-aggressive males with long and very sharp, slashing claws.[56] We caught these possums in an attempt to stop their damage to our domestic electrical wiring

---

[55] Many years later, after purchasing a thermal camera, the number of possums in the area became frightfully obvious. On any evening it was possible to scan the treetops and observe multiple possums moving about on many properties. I think that most New Zealanders would appreciate the seriousness of this problem.

[56] Inserting a wooden rule across the cage provided a safe opportunity to determine the IPD of captured possums after they had settled down, which was measured on average to be 40 mm. The maximum pupil dilation measured was 13 mm.

8: Dooligahl

and prevent our house from burning down.[57] On a simple basis, this gave a combined brushtail and ringtail possum population density of eight hundred per square kilometre, where the brushtails dominated by a ratio of three to one. We were supporting a potential possum population that greatly exceeded their typical numbers in the surrounding wilderness and was about fourteen times greater than the human population density of the Blue Mountains, which was only fifty-five persons per square kilometre. However, if someone were to walk a short distance away from the valley, as Gary Opit and I did during his visit, there were no possum nests in hollow trees and no dreys visible in the immediate area. It was only after travelling a long way out into the scrub that dreys, the most conspicuous type of possum nest, started to become visible as they clung to the trees. Checking for brushtails in hollow logs was too much work. The obvious conclusion was that our high possum density was the result of migratory pressure pushing inwards, towards the productive and highly desirable valley settlements.

In spite of the advice and our experience during this early period,[58] we still continued to use food in order to achieve an advantage. At first, I used a banana suspended high on a cotton thread along a track intersection near the

---

[57] Despite having a corrugated metal roof, brushtail possums made their way into the small cavity at the top of our roof. After unfastening the roof ridge capping, the dry nesting material was removed, which revealed the chewed and bare electrical wires. How our house didn't catch fire remains a mystery. After repairing the damage, the void was filled with a large quantity of compressed strips of chicken wire. Contrary to popular belief, a family of ringtail possums will occupy a previous brushtail nest in a roof, seemingly unperturbed by the powerful scent.

[58] See Log entry for Wednesday, June 16. / Jerry and Sue O'Connor similarly used peanut butter bread rolls as an enticement, which ultimately concluded with the 'Bucket Experiment'.

swamp to conspicuously gain Fatfoot's attention in order to audio record her efforts to obtain the food. Similar to the restaurant owners, we found part of the banana skin, dried and with a short length of cotton still tied to the stem, several days later on the side of our front steps.[59] We regarded this find, eighty metres from the swamp, as a 'thank you' gesture.

In a similar way, I built a permanent 'feeding platform' in a medium-density section of bush on the edge of the swamp, made from a short timber post with a round wire barbecue rack fixed to the top. The rationale behind its construction was that it would be a raised area for food that Fatfoot could frequently eat from or could be a centre for hunting those animals that came to feed from there instead. It was never successful, probably because its real purpose, especially considering our frequently demonstrated *modus operandi*, was too obvious for a highly intelligent predator.

IN LATE NOVEMBER 1993, I talked to Russell's[60] father, who surprised me with a historical report that had taken place immediately across from and including our property, covering a period of fifty years. He said that Roy had lived across the road from our land since the 1940s and had a long, detailed story to tell. Lamentably, Roy had died a week earlier from lung cancer, but a group of his

---

[59] John thought that the other possibility was that a possum carried the skin back to the house, but considering the inaccessibility of the hung banana high up in the treetops, the distance from the house and the location of the find, he thought that the circumstances seemed too deliberate to be a random event.

[60] Russell was a drinking mate at the club. We regularly spoke about the research and Russell would pass on encounter contacts that he came across.

Fig. 8.15 'Feeding station' located in medium-density bush near the swamp. (Photo: Neil Frost)

mates were having a private wake at the local club later in the week. It was very disapointing that we had waved and said "G'day" to Roy on most afternoons for nearly a decade, as he drove out of his front gate in his light green, early 1980s Land Rover Defender and headed off to the shops, without knowing that Roy was the highly experienced witness that we had longed to speak to!

Speaking to Roy's mates at the club was very interesting and a valuable education that I paid for with multiple schooners of Resch's Draught[61] with successive shouts at the bar. Unfortunately, at this early stage, I did not fully appreciate all of the implications, nor did I have enough knowledge to ask as many specific questions as I should have. According to his mates, Roy would occasionally relate a recent local encounter or would simply explain his theories. They admitted that they would have found his stories hard to believe if Roy had not been a mate.

From what they could collectively remember from around the table, I was able to make a simple list of characteristics and advice. According to Roy, they are mostly seen and heard at night, they are most active in winter, their eyes 'glow' red at night, they are very fast, they are big and powerful, they have the same outline as a person, but larger, they are clever, they have favourite trails that they use but this has changed over the years as houses have blocked their habitual routes, they cross between the major valley catchments which are separated by the main western highway and rail link at an Aboriginal sacred site,[62] "they

---

[61] Resch's Draught is a NSW lager beer that is sold in hotels and is a fruity, golden nectar that is highly recommended by the author. A schooner is a hotel glass size of 425 ml which the author finds to be an adequate size for this beverage. A 'shout' is a person's turn to buy a round of drinks.

[62] Being a Sacred Site, it is better that it is not named or identified.

8: Dooligahl

will leave you alone if you leave them alone", "if you come across one in the bush at night, don't stop, keep walking".

In addition to this information, Roy's mates were also able to provide a crude map drawn on the back of a Resch's-soaked beer coaster. The basic purpose of the map was to highlight how their routes had changed over the years. In essence, it illustrated how Dooligahl had free range throughout the area when Roy first came to the area. As houses appeared and fences were erected, these free paths were progressively closed off, forcing them to create alternative tracks. As the area was originally zoned as rural, clearing further restricted their movement.

When Roy moved into the district, sometime during the 1940s, the area was sparsely settled with a small central village and a dispersed population of about a thousand. In those days he needed to walk to the shops and would pass through our land and valley along the bush tracks. On many occasions, he had encounters along these paths during the late afternoon or early evening. During this time, many tracks could be found along the valley floor and also along the ridges, with direct connections between them. With the growing population, the settlement developed along the ridges which were the most desirable building locations. This expanding housing barrier restricted access along the ridges and around the periphery of these narrow settlements, while still leaving the valley floor relatively open. According to his mates, Roy described this as being "similar to how a glove fits a hand". Instead, these barriers increasingly tended to redirect the majority of movement along the unrestricted valley floor, between the ridges.

As the evening discussion and bar tally lengthened, Roy's mates took turns at dredging up many previously lost stories. One story, in particular, was remembered by them all in detail, mainly because of its obvious effect on Roy.

On that occasion, Roy had just moved to another house, not far from his original residence. It was winter, very cold and misty. Roy had taken his chainsaw down into the valley in the late afternoon to cut some firewood.[63] After a lengthy period of work, he stopped the chainsaw to have a break and when looking into the heavy mist, he made out the very large outline of a 'humanoid' figure standing close by. Staring at the silhouette, it very slowly and noiselessly began to move. Watching the figure, he saw it walk behind a large tree and then back away into the mist until it disappeared. According to his mates, Roy was noticeably affected by this encounter. It was several weeks before Roy had enough courage to recover the chainsaw and then, only in the middle of the day. Shortly after this, he returned to his old house.

TOWARDS THE END of 1993, we were all very tired. We had achieved several things, but we were also frustrated by our inability to achieve certain goals, like taking a detailed photograph of Fatfoot, or simply being able to see her clearly without any obstruction for even a brief moment.[64] For Ian, a very considerable frustration was not being able to get close enough to Fatfoot so that he could either obtain a hair sample or take a clear photograph up close. However, for the rest of us, we thought that Ian just

---

[63] Making loud noises, particularly in, or into a valley floor increases the probability of initiating an encounter, just as when Damien cooeed into a misty valley. Similarly, endlessly riding trailbikes inconsiderately along valley tracks tends to irritate every biped, placental or marsupial.

[64] Over several decades we facilitated opportunities for many interested people to have an ethereal encounter with Fatfoot. We always found it very interesting after a novice would claim to have seen her and then attempt to provide a detailed description based upon the compliant Dooligahl's behaviour!

8: Dooligahl

wanted to see Fatfoot with his own eyes so that he could gain some closure from the ordeal.

Undeterred by our long series of pursuit failures, Ian hatched another bold plan during one of our recovery coffee sessions. As the evening had currently been uneventful, even though Fatfoot was watching us from the bush across the driveway, Ian suggested that he and Toby would push Fatfoot down the valley, while Cheryl and I would take up concealed positions along a cleared easement at the valley neck, hoping to observe whatever emerged from the dense undergrowth. According to Ian, as Toby had "recently grown some testicles", he might add a new element to the pursuit.

Arriving at our picketing location at the neck of the valley after a long and slow journey, Cheryl and I stood on a rise that gave us a good view along the easement clearing in both opposing directions, covering a combined distance of approximately sixty metres. Cheryl and I varied the spacing between us to maximise the picket coverage but we were limited by the many undulations in the terrain. By now it was approaching midnight and we continued to wait in position for about half an hour for the much slower arrival of Ian and Toby. The moon was overhead and full or very close to it. After listening to an approaching sound for several minutes, Ian emerged from the undergrowth without Toby. He said that he had heard movement ahead of them for most of the journey until they began to approach the easement. Up until then, the other dogs in the valley had been barking ahead of them[65] and Toby had been walking quietly at Ian's left side in line with his training.

---

[65] During our many nighttime pursuits, we noticed that the local dogs would bark towards Fatfoot in preference to us. In this way, we could take advantage of the greater sensitivity of the dogs to accurately locate the position of the 'marsupial humanoid'.

Meanwhile, there was no sound from Toby, not even from his barking. Ian said that his dog had disappeared shortly before he had arrived at the easement. At this point, we readjusted the spacing amongst ourselves to maximise the intimidatory presence of the picket. Cheryl stood on the rise, with Ian to the west of her and me to her east. Although it was not planned, we hoped that Toby's presence and approach from the rear might force Fatfoot to adopt an uncharacteristic and risky strategy.

The easement was a cleared space beneath some high-voltage power lines. In the twenty-metre-wide centre of the clearing, the trees had been completely removed to prevent high voltage arcing and reduce the incidence of preventable bushfires. The undergrowth had been mowed to the width of a hitched tractor implement to allow four-wheel drive inspection vehicles to travel along its length. On either side of this, the bush had been crudely slashed for a distance of about thirty metres or more, leaving longer grass and some minor regrowth trees. To the north of the easement was the neck of our valley, which further descended into very dense vegetation that drains towards the impenetrable Grose Valley.

After a further wait, there was a sudden and loud disturbance as Ian ran down the slope and mostly disappeared from my view, beneath the rise. Without warning, Fatfoot had commenced her high-speed run from the southwestern edge of the clearing, as Ian was attempting to intercept her from side-on. With Fatfoot running across in front of Ian at a distance of about twelve to fifteen metres, he had an excellent view of her in the full moonlight. As she was nearing the northern extremity of the slashed easement, Fatfoot had jumped to clear a deep ditch and had then vanished into the dense cover, stopping silently after a few seconds.

Standing at a distance of about fifty metres from the activity and behind the intervening rise, I had missed my

8: Dooligahl

golden opportunity. My experience of the incident had been limited to the very loud timpani-like bass beat of Fatfoot running, with the lesser audio accompaniment of Ian's marimba-like response subtlety superimposed over it. Cheryl had been more fortunate in witnessing the run, being at a closer distance and from atop the rise. Ian, of course, had the best view, witnessed during his intercept in good light, without obstructions and lasting in the vicinity of eight seconds.

As Cheryl and I reunited with Ian a short distance down the easement, Ian immediately got me to stand next to a sapling that had been in the immediate background when Fatfoot had run past it less than a minute before. Comparing my height to a limb growing from the trunk, Ian was accurately able to determine that the height of the Dooligahl was seven feet. This was an excellent result as we had been variously estimating Fatfoot's height from numerous sources over the past year and had refined the range to between six foot ten inches and seven feet. Similarly, in comparison to myself, Ian thought that its body mass was extremely large, about three times bigger than mine. At that time I weighed eighty-four kilograms, which took Ian's body mass estimate for Fatfoot to two hundred and fifty-two kilograms. Ian also said that the animal was dark in appearance. As Ian would later explain, determining colour is extremely difficult in the moonlight because people can not rely upon their cone vision, resulting in most objects appearing as sepia. As an additional comparison, Ian asked me to move into the bush where Fatfoot had entered the clearing and run as fast as I could along the same route. After my laughter had subsided, I trusted the worthiness of his request. From this simulation, Ian confirmed two obvious facts. Firstly, the inadequate sound that my eighty-four-kilogram mass had produced during my run was insignificant compared to the noise made by the biped. Secondly, it moved very much faster. Overall, the speed at which these massive bipeds are capable of achieving over long distances, somehow seems incongruous and therefore, they must be hiding some aspect of their biology which is uniquely different compared to other bipeds.

Five minutes after these experiments were concluded, we decided to continue the pursuit past the valley neck and into the dense bush beyond. Knowing that Fatfoot had stopped soon after entering the scrub on the northern side, we cautiously proceeded into the narrow valley. As Cheryl was new to these pursuits but having heard Fatfoot use our walking to mask her movement, we reminded Cheryl about our timed halts, where we would suddenly stop after a set count and try to catch out Fatfoot's continued steps. Following a track that Fatfoot would have regularly used, it was very difficult not to make noise and the progress was very slow. Despite the moonlight, visibility was very poor under the dense canopy. Counting down to the halt, the three of us were making a considerable racket but, on jointly stopping, nothing was heard in any direction. This procedure was repeated many times and still, nothing was heard afterwards.

Having travelled less than a hundred metres, we decided that any further progress was pointless and that the probability of some misadventure occurring seemed unreasonably high. Leaving the bush we headed towards the road as Toby quietly appeared from somewhere behind and joined us.

Arriving back at the house a short time afterwards, we sat on the verandah and had a cup of coffee as we told Sandy of our successful pursuit. By now it was past two in the morning. Despite his tiredness, for Ian this sighting provided a certain degree of closure, while I would need to wait much longer until I would get a lesser opportunity.

8: Dooligahl

As we concluded our evening on the verandah, we could hear the subtle sound of Fatfoot's approach towards our home, as Ian and Cheryl returned to theirs.

On the Sunday morning the three of us returned to the power transmission easement, because Ian worked on Saturday. The first task was to verify the height of Fatfoot at seven feet using a tape measure. After that, Ian made an attempt to run as fast as he could across the easement clearing and was further amazed by how quick Fatfoot was in covering the same distance. We walked around the area, but there was no other physical evidence found. We checked the area where Fatfoot had jumped across the ditch for a foot imprint but without success, probably because the ground was too hard. However, Ian did find the track that Fatfoot had last used on Friday night, before abandoning it in favour of crossing the easement and noticed that it turned down into the lower creek area under the power transmission lines and then started up the steep incline towards the top of the next ridge. The visibly worn track was positioned close to the southern edge of the cleared easement as it moved up the very steep hill. It didn't seem to be the ideal terrain to support a track that wallabies could use, particularly downhill, and similarly, it wasn't the safest path for people to regularly take in any direction, assuming that they had some worthwhile purpose in travelling along it. Regardless, Ian spotted a tree on the steep hill across from him, that had been snapped off low to the ground and was trailing downhill along the track at the edge of the easement. For all of us, it seemed that the snapped tree was serving the purpose of a handrail. Looking further up the slope, other snapped trees on the track edge could be seen that were also, for all intents and purposes, handrails that could be used to assist a biped with grasping hands, in climbing the steep path. This was not the first time that we had seen this. Six months earlier, Basim[66] had found similar 'tree rails' that he thought were used to assist in climbing the very steep embankments at the back of his property in the valley to our west. Additionally, these power transmission easements and similarly cleared paths like hiking and bicycle tracks, seem to be well utilised by these 'marsupial hominoids' for the obvious transportation benefits that they provide.

ASIDE FROM SOME obstacles in the human environment like fences and cleared areas which hinder the free movement of these hominoids, occasionally the natural environment throws up its own barriers. On Friday, 7 January 1994, fires broke out in isolated bushland not far from the Bells Line of Road in the Blue Mountains. At that time we were holidaying with friends in Brisbane and we monitored the news on the Australian Broadcasting Corporation (ABC) radio with some trepidation. Over the following days the fire seemed to worsen and after telephoning John, he advised us to return home as quickly as possible, because he thought that our house might not survive the approaching fire front during the next day or two. We left Brisbane first thing the next morning and drove steadily south along the Pacific Highway, only stopping for fuel and food. The journey was remarkably fast and hassle free, despite the emergency and having the majority of the eastern seaboard on fire. As it turned out, we seem to have driven through the main police road block without knowing it. The total lack of traffic, in both directions, should have been the clue that alerted us to the fact[67] but we pressed on through

---

[66] Basim is a marine biologist and teacher who I also knew as an ICT client. Basim assisted us with a variety of research tasks during this early period.

[67] We had been travelling on the Pacific Highway, the main road link between Brisbane and Sydney, also known as Highway #1. For a very

8: Dooligahl

some dangerous situations regardless. Confirmation was only made after we were stopped by the police at the other end of the highway road block, outside of Newcastle. We were prevented from proceeding any further. The next morning we made a very long detour out through central NSW, before turning back towards the coast and the Blue Mountains.

On arriving home we had a very good understanding of the extent and severity of the east coast bushfire emergency. From our home, the main fire threat was coming from the northwest, along the tablelands and through the main Grose Valley system. During the day the smoke was very thick to the north and at night the red glow from the fire was very intimidating. Each day the fire front moved closer with the path of the fire being reassessed with every wind change.

From our early work, we had generally determined that Fatfoot usually approached our valley each evening from the north. There were exceptions, particularly during bushfires and extended periods of heavy rain. Later on, we established a better understanding of the route taken, because I had spent several nights sitting on local hilltops, listening for her approach.[68] As Fatfoot would often be present at our house before sundown, I would drive via the lengthy scenic route along the dirt backroads to a northerly location well before sunset and then climb the hill and get into a comfortable, overseeing position. On the best of these occasions, I had arrived at the hilltop more than an hour before sunset and had been seated for only a few

---

long distance, we had travelled the route during a bushfire emergency, without seeing any other cars, including police and emergency service vehicles.

[68] About a decade later I improved this listening technique by buying a parabolic dish microphone. In addition to increasing the observable audio range, it greatly improved the discriminatory audio quality.

minutes when I started to hear distant and uninhibited movement. Coming from the sunlit wilderness below, I could loudly hear the sound of a large tree falling or being pushed over. Using cupped hands to my ears, a similar sound of a falling tree was heard a short time later, which was ample verification for me that this was not a random event. Both trees were sufficiently distant to hide their fall from view. Using my local topographic map, the distance to the sounds was approximately five hundred metres and the distance to our valley neck was slightly more than a kilometre. Over the next quarter of an hour, various sounds that were reminiscent of loud bipedal walking and the occasional tree break were heard, which made progress along a common isoheight that led to the neck of our valley system. In 2002, about nine years after this planned reconnaissance, I was making a last-minute effort to record the GPS coordinates and details of a treebite before dark, when Fatfoot walked parallel to the road that I was working from. Her path on this occasion was much closer to home, but she was still coming somewhere from the north.

Like the wallabies, cockatoos and most other wildlife, Fatfoot's nightly presence was noticeably absent as the fires were approaching. The reliable indicators of her presence had been the foot thumping and the sound of the dogs barking throughout the valley. Everything was unusually quiet.

Over the three weeks that the fire was active, more than fifty thousand hectares of bushland were destroyed in the Blue Mountains. Twelve kilometres to our north, the destruction had initially been confined to the Grose Valley, until wind changes resulted in outbreaks that brought the fires to within a few kilometres of our valley, before rapidly moving on to destroy the settlements to the east of us.

8: Dooligahl

The destruction was further complicated by back burns[69] that moved into some of the smaller valley systems in an attempt to prevent future secondary flareups.

For some time after the bushfires, the valley was a much quieter place. Many species of bird had left the area, simply because they had been driven out of the area by the smoke and there was very little vegetation cover and less food available to their north. The wallabies had left very early on during the disaster, moving rapidly south and creating their own problems as mobs chanced their way across the highway, looking to find an escape route through the maze of housing and transport infrastructure.

With many animals having departed the valleys, it would not be surprising that Fatfoot was most probably following the food supply. If the many examples of predation discovered by local residents over several decades are considered, wallabies and joeys are an important source of protein for these ambush predators. For Ian and I, an early example of this predation was finding the partial remains of a dismembered swamp wallaby further down the valley. Eight years on, Michael found a large wallaby that had "been cleaved in half, or [having] one leg ripped off" and in 2019, Anne found several wallaby heads lying in the bush at various locations at the back of her property. It seems that the common killing method involves breaking the neck and back or the removal of the head. Having witnessed the impressive strength of Fatfoot, it would seem that these are the definitive methods of dispatch used by these 'marsupial humanoids'.

---

[69] Throughout the Blue Mountains there is an established pattern of back burning that has been determined through best practice, using past bushfire experience. This accepted practice is known as 'The Black Line'.

The photograph below was taken by John Appleton who found the adult swamp wallaby remains in bush near an isolated house at the end of a Blue Mountains' street. The otherwise normal suburban house had a high security fence surrounding the property with four motion sensing lights on each roof eave corner.[70] Our neighbours, Murdock and Julie, had an eight-foot high-security fence installed around their house because Fatfoot would typically go underneath their house and tap on their bedroom floor.[71] Similarly, an isolated house near Jerry and Sues' property, where a Quinkan frequented that valley, had a high-security fence with multiple motion-sensing lights located around the perimeter of the house. In addition to 'marsupial hominoids', the other natural predators

Fig. 8.16 Swamp wallaby stomach, intestines and left foot. (Photo: © John Appleton)

8: Dooligahl

of swamp wallabies are dingoes, hybrids, wild dogs and wedge-tailed eagles.[72] Alternatively, another cryptid like a black panther, sometimes sighted within the region, could have been responsible for this particular kill. In this case, a significant difference is that the wallaby was disembowelled and butchered at the site, which would suggest that wild dogs or a panther were most likely responsible.[73] Regardless, the important point to consider is that predators follow the food supply. When the swamp wallabies were not seen in our valley, Fatfoot was also missing.

During early 1994, the typical indicators of Fatfoot's nightly presence were absent. For many weeks we started to believe that Fatfoot had moved on and we thought that our opportunities for further interactive research had ended. However, some of the local residents had been talking to whoever would listen to them and a few historical encounters had emerged as a consequence, which provided

---

[70] Clearly, the owner was experiencing some difficulty.

[71] As the brick sub-floor of their home was high off the ground, Murdock said that Fatfoot would follow their movement around inside the house at night, tapping on the floor and moving stored objects about. They were particularly concerned with the attention that Fatfoot was giving to their son Jeffery. Julie was concerned at nighttime because Murdock would not return home most nights until after midnight. Fatfoot would frequently be waiting in the bush across the road for Murdock to arrive home. Murdock would frequently see the red eyes and occasionally heard movement in the bush.

[72] Dingoes, hybrids and feral dogs are common, but they are not often seen by people. During the forty years that we have lived here, I have only seen one wild dog, in the valley at 2 A.M.—a striped, inbred juvenile dog of doubtful origin. These dogs are variously controlled by 1080 poison baiting that is carried out by National Parks and Wildlife. Shooting and trapping are other minor measures undertaken. Wedge-tailed eagles are not predators in the Blue Mountains.

[73] Most encounter reports involving kangaroo and wallaby kills by 'marsupial hominoids' involve the bodies being carried away, relatively intact.

just enough information for a follow-up interview to be possible. The first of these historical encounters involved a witness who had sort assistance in the past, more than a decade earlier and was still awaiting follow-up information. I can't remember which neighbour had made the referral and what the name of the witness was. However, I drove to the address that I was given and uncomfortably spoke to the lady who answered the door. This was the first Hairy Man interview that I had made where I didn't already know the basic circumstances, the witness's name, what had happened or anything else of relevance. Trying to establish a connection with the lady was extremely difficult and attempting to achieve it without causing alarm was complicated.

The address that I had arrived at was on a side street, off Hawkesbury Road, Winmalee. The lady whom I spoke to was quite welcoming after I had stumbled through my introduction. We went inside and sat in the lounge room where she told me about their encounter history. She said that the protracted encounters started with their young daughter complaining about seeing an intruder's face at her bedroom window on many occasions. With her husband they both carefully explained to their daughter that her window sill was slightly more than eight feet off the ground[74] and that no one would be able to see into her room. Despite this rationalisation, their daughter continued to report seeing a face at her window. After further discussion, it seemed that these incidents had been occurring for several years, but only during the summer months. After being woken by the sound of something

---

[74] The raised subfloor of the house was built on steel and brick stumps which was mostly unenclosed on three sides. The slope of the house site increased towards the rear. The mezzanine area beneath the kitchen was used as a handyman workshop with a minimum six foot head clearance.

8: Dooligahl

moving about under their house, the father got out of bed and turned on the under-house light from a switch located in the kitchen above. The response to this had been immediate. Something under the house had accelerated rapidly, resulting in a collision with one of the vertical steel posts used to support the floor bearers, causing the house to shake and resonate afterwards. The sound of running continued towards the perimeter of the building where a second collision with a mature golden pine tree caused the tree to be snapped off at ground level.

The crucial point about where this interview was heading was the minimum required height of the 'marsupial hominoid'. At more than eight feet from the ground, it was obvious that a Dooligahl like Fatfoot would not be tall enough to adequately see into the girl's bedroom. She would have been at least a foot and a half short. However, this was only the beginning, where conflicting evidence from this encounter interview was going to severely challenge our establishing framework, which had been developing for nearly a year. Consequently, it seemed reasonable to assume that there was another 'marsupial hominoid' type that we were not fully aware of or perhaps, we simply got things wrong. This particular story will be further explored in the Quinkan chapter.

DURING THIS EARLY period of research, some of the puzzle pieces weren't falling into place as easily as they should have if we were on the correct discovery path. It seemed that the scaffold that we were using was either incorrect or it was missing a significant number of important keys.

At the building site in 1984, we had undoubtedly found the best set of Fatfoot tracks that we were ever going to find but, unfortunately, we had no detailed record or recollection of the foot morphology of our resident Dooligahl because, at that naïve time, we didn't appreciate the future need to cast the tracks or take measurements and photographs. Other than having a large and unusual foot size, we remembered next to nothing. We couldn't recall how many toes there were, nor the overall shape of the unusual feet. Back then, there were not many people to speak to about this. One historical report said that controversial evidence was either ignored or, in the case of divergently-shaped castings, erased with a hammer. However, the prevailing dogma amongst the Australian few was that these animals were either hominids or at least primates, which by default, meant that they must have five toes. Similarly reasoned, they had five toes because our Australian 'hominoids' were regarded as close placental relatives of the North American Bigfoot or Sasquatch, where such basal matters amongst collegial researchers are considered to be incontrovertibly true. Consequently, being presented with evidence suggesting that fewer toes might be closer to the norm, required some getting used to, but after all, this is the land of the platypus!

There was also the question of height. As stated in the previous case example, the height of the window sill was slightly more than eight feet, something that I was able to confirm. Although we were aware of the high incidence of this window-perving behaviour, we had not previously or subsequently come across a reported encounter that was this far off the ground. Additionally, this high threshold only suggested a minimum height of the observer, with the actual height to the top of the head probably being substantially more, maybe by as much as eight inches.[75] Subsequently, reports of these frequent encounters occurred at

---

[75] Although Australia recognised the legal use of the Metric System in 1947 and adopted it as the Standard in 1970, most Australians continue to measure height with the Imperial System using feet and inches.

8: Dooligahl

the standard window sill height of four feet, where verandahs or favourable external conditions provided sufficient elevation to allow for a comfortable observing height.[76]

With the probable height of this 'marsupial hominoid' being around eight-and-a-half feet, this placed it in an upper category that we were only just beginning to comprehend. Over the previous year, we had been struggling to obtain an accurate height for Fatfoot, which ranged from more than six-and-a-half feet to slightly more than seven feet. From around this time, however, we had refined our estimate to between six feet ten inches and seven feet. Knowing that Fatfoot was a dominant and mature female, the height disparity between the two individuals seemed to be easily explained by sexual dimorphism, where the sexes of the same species have different characteristics, in addition to those associated with reproduction. Consequently, the sex of the taller individual appeared to be male, where this difference in height is simply a common dimorphic example found in many animals, which is typically explained by the need for males to effectively compete with each other in the battle to reproduce. The simple way for a male to achieve an advantage is to become bigger than the opposition.

Or, so we thought! During the following six years as our knowledge consolidated, we would eventually realise that the explanation for the differences in height was not solely because of sexual dimorphism, but broader speciation instead, as some other traits were also different. Also during this learning period, other encounter reports from

---

[76] The O'Connor's window encounter allowed the height of their Quinkan to be measured at 8′4″. The top of the observed head was the same height as the upper edge of their bedroom window and the underside of their external boxed eaves.

this valley system would be received that confirmed this early account and the unique characteristics of these 'marsupial hominoids'.

After speaking to the Parental School Council Representative at our daughter's preschool, Julie told me that they lived three kilometres further up the road and along the same valley system as the previous encounter witnesses. They had just completed building their sandstone house on an acreage, not far from Yellow Rock. Their new home was situated on flat land, south of the main valley, with few houses surrounding them. She said that the area was normally very quiet, day and night, and that they had not experienced anything untoward for most of their short time there, except for a few nighttime experiences during their first summer living at the house. One of the early morning incidents was very disturbingly and surprisingly loud.

Julie's sandstone house was extremely impressive and very expensively built. Sandstone houses, mud brick and other heavy masonry buildings, like castles, tend to be very cold during most of the year, except during summer when they can become comfortable if left sealed. In addition to the thermal qualities of thick masonry walls, they also tend to be acoustically insulating as well.

Julie recalled a few nights when they could hear unusual noises and a few very loud calls coming from within the valley behind them. Julie said that neither she nor her husband could identify the animal that was responsible for the unusual noises or calls, but they knew that it had to be big and strange. A few nights after this, Julie and her husband were woken during the early morning by very loud calls coming from the direction of the valley, but clearly from on their land. As they listened intently, the loud calls began approaching their house. A short time after this, they began to hear other noises which increased in volume. Julie and her husband both believed that they were hearing low-frequency noise as the animal was approaching the house. Julie told me that, "As it approached the house it started to sound like a tank or heavy truck moving past us." They were too scared to look out of their bedroom window as the animal continued past the corner of the house. She also said, "We could clearly hear the heavy walking. Normally we can't hear anything through these walls." The animal walked up the slight rise at the front of the house towards the road, without stopping.

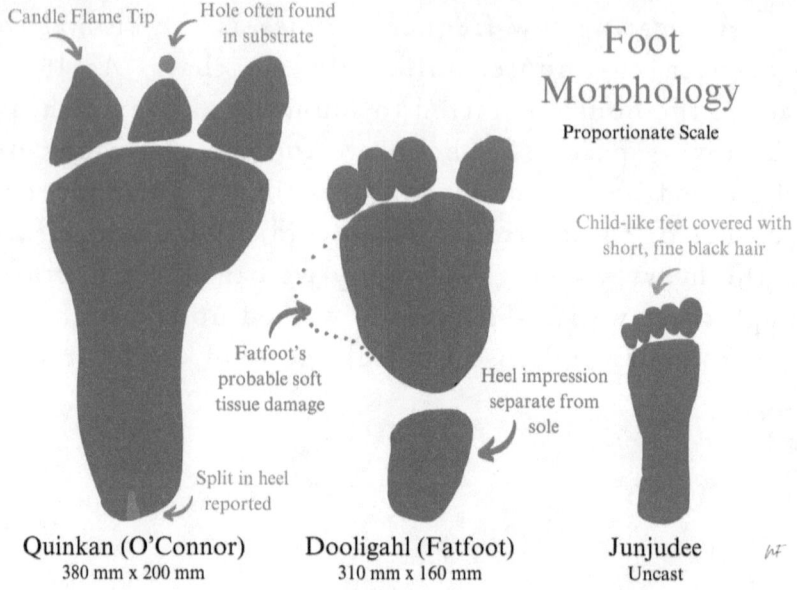

Fig. 9.1 'Marsupial hominoid' foot morphologies

Fig. 9.2 Small macropod tracks (© Damian Herde)

9: Hominoids

THE FIRST EIGHTEEN months or so of our research could be expected to form the foundation of learning for our small community group. Having started from a very narrow and scantily-lined knowledge base, our collective intelligence steadily grew over this brief period. However, it was becoming apparent that some of the imported and entrenched basal knowledge was misinformation when compared with the local evidence and within an Australian context. Consequently, some housekeeping tasks would be required along the journey—specifically a late paradigm shift away from an imported North American, placental Bigfoot model, to a more biogeographically-relevant convergence theory, based upon the evidence derived from Aboriginal knowledge of marsupial megafauna and what we had observed. Also, much of what we had learnt together with the increased volume of Octopus reports improved the resolution of our understanding to the point where conspicuous gaps in our newly developing perception were starting to open up. For example, the height estimates collected from many early encounter reports clumped together but there were some that lay significantly outside the main band, which could not be simply explained by sexual dimorphism alone. This seemed to suggest, therefore, that there were more than one 'marsupial hominoid' species. Also, there was inconsistent and poorly understood evidence in Australia regarding the expected shape of the feet and the number of toes, which was incompatible with the established five-toed primate model unless these different morphologies were somehow the common result of inherent injury or runaway mutation. Such egregious predicaments were impossible to resolve using the adopted North American framework. Once again, the observational evidence was pointing towards more than one species, just like the sequel to the 1979 science fiction horror movie, *Alien*.

It was very difficult to find knowledgeable people who were seriously willing to discuss the work that we were doing. Typically, any mention or hint of the 'Y-word' would immediately result in some sort of visceral response from the audience. On the best of occasions, when a seemingly open minded person was found, their objectiveness only tended to last until the first cataphysical speed bump was reached. Academics were no exception to this situation, in fact, they were probably worse than the average person because they thought that they exclusively knew what they were talking about! You know when you have lost when they give you a 'knowing smile'. Also, there was a high profile crypto-author who regularly wrote about this subject in the newspapers, lumped together with other paranormal topics, who was well known for raising the level of scepticism amongst the general public. Unfortunately, these writings were at odds with what we were uncovering.

After reading an alternative newspaper article on the topic, I wrote a letter to Tony Healy and several others who were mentioned in the story, outlining our situation. Tony replied enthusiastically. Shortly afterwards, Paul

9: Hominoids

Cropper arranged to stay overnight,[1] hoping to have an experience, so before sunset I prepared the unfinished wine cellar as an observation platform and shelter for him and another colleague. The following morning I got out of bed well before sunrise, but unfortunately, Paul had already returned home. When Paul telephoned me, he said that they had heard a 'loud, hollow thump', which seemed to be the same foot-thumping sound that we regularly heard around the house most nights. Unfortunately, it would be another six years before we would cooperate further.

However, we were extremely lucky during our early work to find several other researchers who were also, knowledgeable and supportive. Steve Ruston was one of these people. He was a Queensland builder and an Australian megafaunal researcher. After many lengthy telephone conversations, Steve was very interested in our work and very keen to help us in any way that he could. After discussing areas of weakness in our research, we expressed a desire for an extremely long stereo audio cable that we could use to remotely monitor sound activity in the swamp and elsewhere. Within a week, we received a courier delivery from Steve of a custom and professionally made, one-hundred-metre-long stereo audio cable. With this cable, we were able to passively listen to and record live activity from a remote location. This cable was the second piece of equipment that we could use to passively study these 'marsupial hominoids'. The first was the cotton thread that we used to track movement.

With this in mind, Steve thought that having night vision equipment would be the next logical step towards using passive equipment. To say that we were all excited

---

[1] I should have suggested that Paul arrive early so that he could have spoken to local witnesses and neighbours.

by the possibility of having night vision, was an understatement, especially considering that in Australia, at this time, this equipment was both prohibitory expensive and typically unavailable.

As the Soviet Union had collapsed less than two years earlier, Steve thought that he might be able to obtain some ex-Russian military equipment. He went to the docks at the Port of Brisbane and found a Russian freighter where the crew were selling a range of high-value items. In particular, he found a sailor who was selling night vision binoculars and a light vision video camera adaptor for a total of three thousand Australian dollars. The night vision binoculars were first generation and had been standard issue to Soviet tank crews. Selling off Soviet equipment was a common practice by Russian sailors visiting foreign ports after the collapse of the USSR because the Rouble had nosedived and the best way to make money was to sell items for stronger foreign currency like the Australian dollar and then convert this amount into very large amounts of their domestic currency on their return home. With the night vision equipment in hand, Steve drove to Sydney to try it out for himself.

The Russian binoculars were large, heavy and battery operated. As first-generation night vision (NV) binoculars, they are passive devices that amplify the intensity of the available light. Their image intensifiers increase the light by up to a thousand times and need moonlight or some other external light source to function effectively but, unfortunately, gave off an audible low-level buzz during their operation. The image produced was green in colour and variable in quality and detail, depending upon the nature of the light. If the light contrast was poor or there was excessive light noise, the amplification produced an image that was mostly unusable. Unlike forward looking infrared devices, the depth of vision tends to end at the first obstacle visually encountered. For example, if insubstantial fog is present in the foreground, that will limit or obscure the visibility of objects in the background, whereas an infrared device will be able to see further because of the longer IR light frequencies used which have greater penetrating power.

With Steve unpacking at our house, we made him comfortable for the night ahead at the YOP (Yowie Observation Platform), also known as the children's cubby house.[2] I ran a power supply to the shed and conservatively cleared some of the forward vegetation between the YOP and the swamp. We set up the video camera with its attached NV video adaptor on a tripod, partly closed the window shutters on the two windows to restrict any inward observation and swept the area outside the rear door of leaves and other noise-producing debris. Steve set up a VOR audio recorder in the swamp.

After physically setting things up, we spent the majority of the afternoon getting Steve primed for the upcoming night. It was vitally important to give Steve a 'dog tutorial', so that he could understand and anticipate Fatfoot's movements as announced by various dogs with their characteristic barks. We gave him an understanding of the current track layout and took him for a walkabout. He was introduced to the typical sounds that he could expect to hear and their meanings, where possible. Specifically, foot thumping, stick and branch breaking, rock throwing and other behaviours. We highlighted the importance of tracking, combined with the necessity of having a long attention span, by remembering the position of the last movement heard and updating it, possibly over a very long

---

[2] It was very good to have the cubby house repurposed. It took me eleven days to build it from scratch while our children only played in it for eleven hours!

time. There was also the visual confirmation provided by seeing the red eyes, that indicates that his position has been found and yellow eyes, meaning that the Dooligahl has a very good idea of his position, but is still actively searching. With this and more downloaded to him, I told Steve that we would leave the back door open and showed him the way to my bedside, in case he needed help during the night.

Sandy and I went to bed, having left Steve in the YOP several hours earlier. Sometime after midnight I was woken by Steve standing at the side of my bed. I asked him if he was OK and he simply said, "It's down the back." I replied, "It's always down the back. Have you seen the eyes yet?" Steve replied, "Yes." I got out of bed, dressed and followed Steve down to the YOP.

As a result of Steve having returned to the house and both of us having moved back to the YOP shortly afterwards, there was no need to further maintain any subterfuge. We both sat in the YOP while Steve quietly outlined the sequence of events leading up to this. As was typical of Fatfoot's approach to our house, Steve said that, starting in the early evening he heard the dogs fire up down at the valley neck and he followed their progress, one by one, as the barking moved slowly up the valley. When the barking started to approach our land, the dogs became quiet. He said that there was a very long pause afterwards, for over an hour or more, where nothing was heard. Spread over a considerable period of time, he heard subtle walking with the occasional twig snap, adjacent to the YOP on the far side of the block, along the northern track. After Fatfoot had reached the high point on the track, above the YOP and adjacent to the house, he witnessed very loud and fast running back down the northern track, which terminated in the swamp. A few minutes after this he saw the red eyes. It seemed certain to me that Steve's presence had been spotted, or possibly verified through the open easterly door of the YOP, because everyone underestimates the superior sensory capabilities of these animals by failing to take proactive and extreme precautions.

By looking through the gap in the window shutters, we could see the unresolved red eyes at the edge of the swamp, about forty metres away, which indicated that Fatfoot knew our location exactly. As we continued to observe with our naked eyes, we could perceive slight changes in colour from red to yellow and back, which further indicated that Fatfoot was cautiously looking about.

With the NV binoculars powered up, Steve could easily resolve two green eyes, but little else. He commented on how large the eye dilation appeared. While Steve set up the NV video recorder, I had a turn playing with the binoculars. This was my first time using night vision equipment and it was disappointing. Perhaps this lack of performance was because we were only using ambient light. If we could have turned on a few verandah lights behind us, our vision might have been better. After observing the eyes for a few minutes, it was interesting to see one, or the other eye disappear behind a swamp branch, as the head moved about in the vegetation. An exciting moment occurred when both eyes disappeared, which suggested that Fatfoot had blinked! Overall, the vision was very noisy which made any detail very difficult to see.

With the NV video recorder powered up, Steve had a lot of trouble obtaining focus. When he was successful, the image through the viewfinder was a circular field of view, rather than the normal rectangular shape. The image quality was not as good as that obtained through the binoculars and there was much more noise. Steve left the recorder going until the tape ran out, while we took turns using the binoculars.

By now it was about two in the morning. Fatfoot's eye position hadn't changed very much and so we discussed options for the remainder of the night. I said to Steve that, as long as we could see the eyes, we should continue to observe from the YOP because if we moved towards the swamp, Fatfoot would simply run away and we would waste observing time until she returned. If we maintained our position, we might see her change position. About two hours later, the eyes gradually disappeared, either because Fatfoot had cautiously left the swamp, or the lighting conditions had changed. We didn't see the eyes again, nor did we hear the dogs bark.[3]

As the dawn light started to appear, we left the YOP. Steve collected the audio VOR unit and we took the video recorder and tape back to the house. When we played the video tape, the recording noise was so excessive that we could not see anything of significance apart from a brief period of recording at the beginning of the tape, where we could imagine seeing the eyes. The audio recording had captured the barking of the dogs and little else.

In spite of the recording failures, Steve was very excited. As an experienced researcher, he had only recorded other witness encounters and had never had one himself. Later in the morning before he returned to his hotel room to sleep, Steve said to Sandy and I that, "This has been the best day of my life." In response, I suggested that he should not share that opinion with his wife, Carol.[4]

Before Steve returned to Queensland, he left the NV binoculars with us to use. As we had learnt, it was very difficult to see anything with the NV binoculars, unless there was an unrestricted view and there was sufficient

---

[3] Ian and I had noticed that Fatfoot was well aware of the role that the local dogs played in disclosing her location and regularly changed avenues of approach to confuse us. The eastern side of the swamp had fewer dogs.

[4] When I met Carol, she told me that she hated how much time and money Steve spent on his research.

9: Hominoids

ambient light, or some other unobtrusive light source. With these limitations in mind, I searched for and found a relatively cleared bush corridor that bisected the land leading towards the swamp and its numerous tracks. The narrow, central portion of the corridor provided the best chance of seeing something, because it offered unobstructed, full-height views. The vegetation on the outer edges of the corridor were not very thick and tall, although I had sparingly trimmed down anything that might cause a significant problem,[5] so there was an additional chance of viewing the upper part of Fatfoot as well. Over many evenings, I stood in a partially concealed position, knowing that my presence would eventually become known over successive evenings. I kept the binoculars pointed towards the bottom of the land, in anticipation of some inevitable activity.

During one of these evenings when scanning the area towards the swamp, I became aware of something moving towards the south along a track at a distance of thirty metres. This perceived direction of movement was further reinforced by the sound of a rock, or some other heavy object, crashing ahead of Fatfoot's current position. My presence had obviously been discovered. After continuing to scan to the south, I quickly rescanned back the opposite way and caught a confused image of Fatfoot bent severely over and possibly changing gait as she ran in the opposite direction. Although the quality of the NV image was very poor and furtive, there were sufficient visual clues to suggest that Fatfoot was possibly running as a quadruped, or was about to, in order to travel below the height of the vegetation cover. This made further sense

---

[5] We always exercised caution when trimming or removing vegetation because we knew that Fatfoot would avoid any area which seemed to be recently disturbed by us.

because Fatfoot's seven foot statue was not subsequently seen towering over the adjacent, four foot high vegetation, that dominated the area past this observation point. Ian and I had sometimes speculated that Fatfoot was quadrupedal, because we never caught sight of her upper body, or even her head, when pursuing her through intermediate bush at a relatively close range and when using a torch. This was similar to the circumstances involving the superb fairy-wrens taking flight from bushes after being disturbed by Fatfoot at this same location two years earlier, when I was unable to observe her potentially towering presence as she ran away, despite the encounter happening during the late afternoon. Additionally, there are two photographs, also from this location, that show Fatfoot in a prone position, hiding under the vegetation and from the overhead PIR motion sensing camera.

Another suitable area for observation using the NV binoculars was the bush to the northeast of the house. This location had been highly successful when making naked eye observations, particularly when using Phil's verandah light as a backlight. When these opportunities were combined with the use of NV binoculars, the clarity and resolution of the images were higher on most occasions. Shown (Figs. 9.4, 9.5) are two examples of NV sketches using backlight that show the range of obtainable NV images. The sketch showing 'Fatfoot moving rapidly', was very poorly drawn by me. The general appearance of the biped should be considered as illustrative only, because the subject was in motion during the short time that it was observed. The only significant aspect of the observation was the forward leaning stance of the individual. The other sketch, 'Fatfoot standing', was made by Sandy and shows more accurate detail. Accompanying her sketch were the following notes that she wrote after making the sketch. "Saw—head to lower leg. It was very big, with

9: Hominoids

Fig. 9.3 A vague representation of Fatfoot transitioning from bipedal to quadrupedal locomotion along the edge of the swamp track at a distance of thirty metres, using NV and briefly viewed along a narrow bush corridor.

square shoulders and a gap was visible between the [right} arm and the body. Could not tell the height, but it was some distance from the house—about forty to fifty metres. It then bent forward to the side. Saw it standing there for about two minutes, before it disappeared." Another drawn aspect of the sketch which Sandy did not mention in her notes at that time, was the merging of the head with the shoulders.

Apart from his very strong and cooperative support for our research and that of others, Steve Rushton was the first non-indigenous researcher that we came across, who believed that our 'hominoids' were native to Australia. His research was in total alignment with the empirical evidence that we were accumulating but his work was further advanced in other ways. From his studies, Steve believed that there were two main 'marsupial hominoid' groups: one small and the other big.[6] During this formative time, however, we were mainly learning from the solitary and local experiences gained from Fatfoot and the recent revelations provided by her accompanying family group. For us, Fatfoot was in the 'big' category. Additionally, there were other external influences on our research, which were increasingly coming from broader reports obtained by the Octopus. A few of these early encounter reports seemed to suggest that there was a 'marsupial hominoid' group that was bigger than the archetypical Fatfoot. For a long time we assumed that these larger individuals were part of the same species, with the size difference simply being the result of sexual dimorphism, where the large dominant males (including bachelor males) lived apart from the female family groups. However, apart from their size,

---

[6] Steve commissioned artwork of a bush diorama showing many of the megafauna present during Aboriginal times that also included the 'marsupial hominoids'. After his visit to our house, Steve got the artist to repaint the eyes of the larger 'marsupial hominoid' red.

Fig. 9.4 My crude sketch of Fatfoot moving rapidly through the bush to the NE of house. Sketched after using NV and Phil's verandah light as back illumination.

the significantly different foot morphologies[7] of the two seemed to suggest that they were, at least, separate species. Similarly, during this explorative period, there were few identifiable reports of little 'marsupial hominoids'[8] for several of these early investigative years, until reported by the Pendlebury family. One questionable exception was a report from a near neighbour who suggested that the little fellas were gnome-like. Most importantly, unlike other Australian researchers, Steve believed that our 'hominoids' were endemic to Australia, marsupial and a remnant population of extant megafauna. His opinions also contrasted sharply with the prevailing, imported paradigm from North America and were therefore, very appealing to us.

As a consequence of Steve's many contacts with witnesses, I received a long-distance telephone call from a man living in Far North Queensland, after our phone number was passed on to him, sometime after Steve's last visit. He didn't want to be identified (I'll call him Anton) or have his precise location known, because he was concerned that someone might attempt to shoot the animals for personal gain. Similarly, the deliberate withholding of Fatfoot's various locations over the decades, we thought, was more than a reasonable request that we made unconditionally to all outside visitors to our valley, for identical ethical reasons.[9]

---

[7] Dooligahl, like Fatfoot, have four toes, whereas the larger Quinkan have three toes. Junjudee have five toes. There are other differences.

[8] For Steve, these little hominoids were widely known throughout Queensland and northern New South Wales (NSW) as 'Brown Jacks'. On the South Coast of NSW, the Aboriginal word 'Junjudee' is used to identify these small hominoids.

[9] Dr. Helmut Loofs-Wissowa from the Australian National University strongly encouraged this ethical behaviour as well. He wrote submissions to the United Nations seeking to have these unrecognised hominoids covered by 'The Declaration of Human Rights'. However, there is always someone, usually with an axe to grind, who thinks that ethical behaviour in all of its forms is unnecessary and non-compulsory.

9: Hominoids

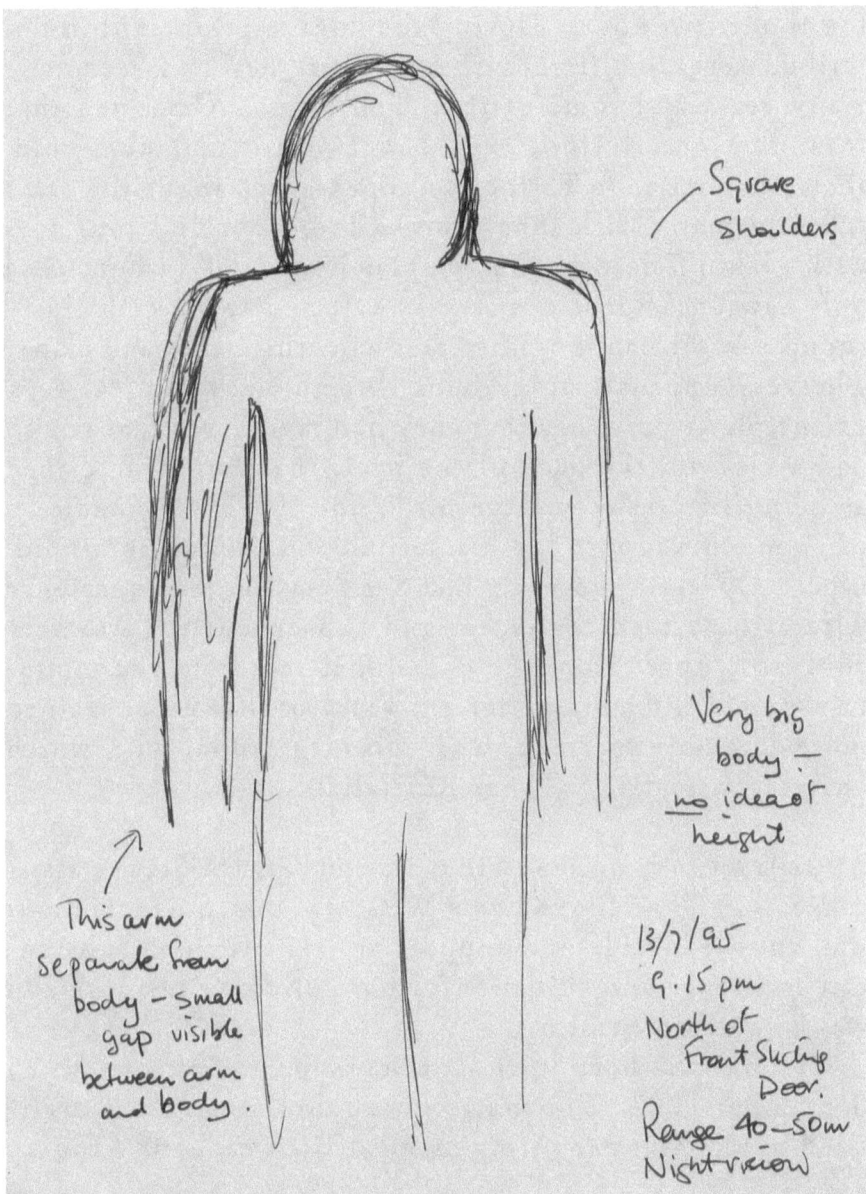

Fig. 9.5 Sandy's sketch of Fatfoot standing to the NE of house using NV and Phil's verandah light as back illumination.

I was extremely pleased that Anton was paying for the Subscriber Truck Dialling (STD) call because we spoke for about two hours. During the conversation, Anton described several different species that they had seen over many years and some of their behaviours. It seemed that from one description, they had a 'marsupial hominoid' that was similar to Fatfoot in appearance, about the same size and mass, with long arms, a long foot and four toes with claws. The other 'marsupial hominoid', whom they only saw occasionally, was very tall and thin, with fluffy round ears. It had long thin feet with three toes and claws. The very unusual thing about this animal's foot, as seen on the plaster castings that they had made, was that it also had a claw on the back of the heel, that Anton described as being like a "very large bird's foot."[10] I was unable to offer much support for his second 'marsupial hominoid' type, only having recently had one very limited experience with a three-toed encounter and consequently, I was very confused! Apart from the occasional report in the mountains, it would be another six years until we had gained more detailed experience with this megafaunal type, which characteristically, was very Australian!

ANOTHER EARLY CONTACT that was made with Steve's assistance, was Dr. Helmut Loofs-Wissowa. Like Steve, Helmut was knowledgeable and supportive. He provided substantial assistance and guidance for our collective research and was a valued mentor.

Helmut was born in Halle near Leipzig, Germany on 25 September 1927. Whilst still at school, Helmut was drafted into the German Army as an officer cadet in January

---

[10] An excellent representation of this three-toed, clawed foot is shown in the painting of 'Turramulli the Giant Quinkan' by Percy Trezise, as used on the cover of Healy and Cropper, *The Yowie: In Search of Australia's Bigfoot*.

1945 and sent on foot to the Eastern Front to fight the advancing Russians. As Helmut would frequently say when recounting his army experience, "The front came much quicker to us than we got to it, which saved on marching." After reaching the relative safety of the American Zone, the MPs sent him back, which required Helmut to make the journey a second time. In 1948, Helmut enlisted in the French Foreign Legion for seven years. When asked why he joined the French Army, Helmut would jokingly reply using the typical Legionary cliché, "Of course, I did it to forget." When serving in Algeria, his drawing skills and intelligence were noticed and he was posted to French Indochina as a Captain, serving in the Press Corp of the French Army. Whilst on patrol in a Vietnamese rice paddy, Helmut's group was ambushed. He lay in the fields for three days until he was found—he was the only survivor. As a result of long periods of contemplation when on sentry duty during his earlier service and through contact with the culture of many local inhabitants, Helmut developed an interest in the origins of Mankind, so he left the Legion in 1954 and began an academic career instead.

At the age of twenty-nine, Helmut began his studies in archaeology. After gaining a scholarship to the Sorbonne, he obtained his doctorate in archaeology in 1960 by studying the mountain tribes in southeastern Indochina. Helmut found employment as a lecturer in the Department of Asian Studies at the Australian National University in 1961. During this time, Helmut undertook extensive anthropological and archaeological work throughout Southeast Asia. It was while carrying out these overseas studies that he heard about local accounts of the Hairy Man, or Wild Man across the Asian region.[11]

---

[11] Adapted from his obituary, written by his son: Jean-Jacques Loofs. 2018. Pursuing reason and dignity at all cost. *The Canberra Times* (29 September).

Like the introductory scene from a Cold War spy movie, we first rendezvoused with Helmut and his wife Ziggy in a dimly lit and rundown motel room, located on the freezingly cold and heavily fogged, eastern side of Northbourne Avenue at the northern entrance to the nation's capital sometime in the mid-winter of 1993. We had chosen to meet them at our motel room because we had two young children that needed to settle in and sleep.

During our joint introductions, Helmut[12] outlined his background, including his war and broader military experience, interspersed with his humorous antidotes. Helmut's strength and courage were very obvious, as were his strong anti-war, pacifist beliefs. He also spoke of his Anthropological and Archaeological experiences in Southeast Asia but particularly their Archaeological excavations at the prehistoric burial site called Khok Charoen in Northern Thailand, between 1966 and 1970. From his broader Asian studies, Helmut collected stories of the Wildmen that he said were broadly reported across the region. Although Helmut was aware of our Australian Yowie, he had no personal experience with them and was extremely excited to hear of our community research. On hearing about this, Helmut stated that the existence of Wildmen was not a rare, local occurrence, scattered amongst a few regions but was actually a 'worldwide phenomenon'. Consequently, Helmut gave us his contact details and we paid both of them a visit shortly afterwards when visiting family in Canberra. He also regularly sent information and relevant research from the Australian National University and we kept Helmut up to date with our work. He often posted academic articles to us and kept in touch by email.

---

[12] Loofs-Wissowa, Helmut Hermann Ernst (1927-2018). Obituary at: https://www.canberratimes.com.au/story/6002457/helmut-loofs-wissowa-obituary-pursuing-reason-and-dignity-at-all-cost/

Additionally, Helmet possessed an alternative European perspective on this matter, which had been heavily influenced by his academic friend, Bernard Heuvelmans, a Belgian-French scientist who wrote *On the Track of Unknown Animals* (1955). This book is regarded as the foundation of cryptozoology.[13]

Unfortunately, Helmut did not have a prior and detailed knowledge of our Australian 'hominoids', because of his European background and with his early experience and focus being centred upon Southeast Asia, where he researched the Nguoi Rung Wildmen in Vietnam and Laos. Therefore, it was not surprising that Helmut's initial view of our Australian Wildmen was similarly shaped by his knowledge of Asian anthropological models, as well as the Bigfoot phenomenon, where such hominoids are believed to be either hominid or pongid.[14]

As our work continued to progress, it became increasingly apparent that our Australian 'hominoids' were very different. Even though Fatfoot was neither hominid nor pongid, we continued to seek Helmut's help, keeping him up to date with our controversial findings, while he continued to send relevant anthropological articles to us and occasionally made personal visits to our home with his wife Ziggy. Helmut continued to be a friend, supporter and mentor.

Helmet maintained very strong ties with his German homeland and the French culture that he loved.[15] As a

---

[13] Bernard Heuvelmans and Ivan Sanderson are considered to be the founders of Cryptozoology.

[14] 'Pongid' was a historical term from the 1940s that Helmut frequently used to describe the great apes: chimpanzees, gorillas and orangutans. It is an obsolete primate taxon. The great apes are currently classified as Hominidae because of their close relationship to humans.

[15] In 1983 he was awarded the Legion of Honour, which is the highest French civil and military order of merit.

result of his well-known work on Wildmen, Helmut would frequently give interviews to television and film producers, many of them European but mostly German and French. In response to one of these requests, Helmut organised a German documentary team to interview him at our home, which we were very glad to assist with. We have fond visual memories of Helmut sitting on the grass and leaning against a scribbly gum tree near to the swamp, holding his sketch book while he was being interviewed. We had no idea of what was being said, but later Helmut told us that it "went well". The other thing that ended up being extremely well done was the meat-heavy barbeque that we organised for our German visitors. While we were being interviewed, the German producer decided that she would cook the kangaroo fillets, lamb backstraps, crocodile tail[16] and beef steaks for us, to a uniform charcoal consistency. Helmut also arranged for an overnight visit from two German 'Men's Health' magazine reporters, who were lucky enough to hear walking and see the red eye shine.[17] However, not all of Helmut's international contacts were German film producers or journalists. Helmut also had a lot of cryptozoological contacts and so he put me in touch with a Russian scientist that he thought would be interested in our crypto-work and might be supportive. Unfortunately,

---

[16] For non-Australian readers, kangaroo is an extremely lean meat which dries out and toughens past the traditional 'rare to medium rare' stage of red meat cooking. Similarly, crocodile only requires brief cooking, otherwise the moisture runs out of the meat, rendering it hard, dry and flavourless.

[17] Shortly after this, at this same location, a new researcher was lucky enough to have his first encounter. Fred was very determined to have an experience and was prepared to put in long hours of solitary observation. He maximised his opportunity by spending several very long nights at this location. He was rewarded by hearing running down the remote bush road towards him, before sharply diverting into the scrub next to his hidden location, probably because he had been detected by one or more of Fatfoot's senses, at around 4 A.M.

after sending an email summary of our Australian research to him, it was strongly rejected because it was inconsistent with the established dogma.

DURING THE MID-TO-LATE-1990s, the number of encounter reports received by the Octopus steadily increased. Nearly all were local in origin, mostly historical reports initially, together with some currently active incidents, collected by community members as they fished for responses from family, neighbours, work colleagues and casual acquaintances. Sometimes the 'standard question' was used and at other times chance opportunities arose. Rather than unifying 'marsupial hominoid' characteristics, the collective responses received highlighted considerable variance in population traits found across the mountain region. For some, these inconsistencies were viewed as proof of observational errors or some fundamental problem with the research. Some local areas seemed to provide ample examples where the resident animal was very big and tall, aggressive and violent. Other investigative groups gave conflicting reports of small, monkey-like 'anthropoids', whilst a number of areas seemed to provide examples that were more like Fatfoot. Consequently, these collective reports made the study more complex and confusing for a long time, which only served to further downplay the veracity and thoroughness of our research during the ongoing battle with sceptics.

Though surprisingly uncommon,[18] there were sufficient footprints found throughout the broader area that

---

[18] The number of foot impressions found was far less than might be expected. This shortfall sharply commenced after we became Yowie-aware when Fatfoot noticeably started avoiding or minimising any evidence of her presence. The best example of this was when Fatfoot began walking around sand laid along the main track, which we had placed there with the specific intention of obtaining her footprints, despite obtaining fox, wallaby and other wildlife footprints.

had been photographed or cast from time to time, to allow for meaningful comparisons of foot morphology and draw some basic conclusions. The problem with this physical evidence was that it covered many morphologies that exceeded and simultaneously contradicted everyone's expectations. From our narrower perspective, however, we had become extremely confident of the shape of Fatfoot's foot and how many toes she had. In having only four toes, including a large offset hallux, it was clear that Fatfoot was not human, nor was she a variant of any of the other great ape possibilities. It was also very clear that the foot is prehensile because the hallux is offset like the thumb on a human hand, which means that it is capable of grasping objects, which is particularly useful when climbing a tree. This would further suggest that the evolutionary origin of Fatfoot is arboreal.[19]

Also discovered in the mountains and other parts of Australia were three-toed footprints, like those found in the sandpit at Winmalee in the 1980s, with the case involving the 'marsupial hominoid' that looked through the girl's window. The plaster cast was reported to be long and thin with three large toes. Similarly, the footprints that were cast by Anton from Far North Queensland in the early 1990s were three-toed, with the foot being long and thin, but were further complicated by also showing claws on each toe and one on the heel, like a large bird. Other witnesses, like Ray and Anna from southeastern Queensland confirmed finding footprints showing each of the three large toes having 'flame-like' or 'candle-like' impressions at their extremities that could have been made by claws. They reported the presence of two heel variations, one where a rear claw impression was found and another, where an open 'V' in the heel was impressed into the soil.

---

[19] Aboriginal Elder Merve told me, "They live in the trees."

9: Hominoids

Additionally, they found left and right tracks that showed digitigrade and plantigrade stances, indicating that running is carried out using the toes—when resting or standing, the heel is used for support.

The best preserved three-toed footprint that I saw and photographed was located in soft, fine moss at Jerry and Sue O'Connors' house in 2000. The imprint was very well defined because the delicately fine green moss had been sharply compressed along its edges. (See Chapter 10.)

Not surprisingly, reporting on the physical evidence of the three and four-toed footprints is extremely problematic, or should have been, for Australian Bigfoot Theory dogmatists. Whereas Bigfoot can be typed from many biological traits, such as having five toes with flat and pink nails, having three- or four-toed feet with thick black claws are not defining characteristics of primates.

Aside from the alternative foot morphology, having five-toed footprints had been discussed and photographed for a long time by other researchers across Australia. For us, such footprints had not been verifiably seen for the majority of our early research history, except for some uncertain clues coming from the Pendlebury encounters and during two other distinct periods. The first period occurred from February 2002 to around 2004 and the second, generally between 2015 and 2017 when we had Junjudee active throughout the local area and a small but poorly defined child-like foot imprint was seen along the creek's edge in the swamp. In appearance, the feet of Junjudee have five toes and are the size of a small child. I would estimate the foot to be about 175 mm in length. The feet are covered all over with very short, fine, black hair, with underlying black skin.[20]

---

[20] Details were obtained from a close, late-morning encounter with a Junjudee on 5 May 2017. A foot was clearly observed as the small biped pivoted on its left foot as it turned away.

Simply based on the foot morphology, it would seem that there are at least two separate 'marsupial hominoid' genera, but probably three. Considering the large number of minor and major description variations, each genus must have a number of species. For example, our Blue Mountains Junjudee have a wide, yellow annulus surrounding each eye, which as far as I am aware, has not been described elsewhere in Australia. Taking this further, based on some obscure observational evidence, with Quinkan having a rear bird-like claw, for example, it is likely that there are other Australian conundrums awaiting us, like those provided by monotremes and marsupials. The only problem is that we need a couple of bodies in order to be definitive—I think that we are making excellent progress regardless.

As THE REPORTS accumulated across the Blue Mountains, certain areas became associated with an individual who had a particular, recognisable persona. One of these areas was centred around the southern side of the mid-mountains region, along Magdala Creek and Sassafras Gully, Springwood, and the surrounding valleys, from 1993 to late 2010.

An early daytime encounter report from the Magdala Creek valley was of bushwalkers who became aware of something that was following them along a bush track from behind. After several minutes the animal rapidly overtook them through the thick bush alongside the track. They knew that the animal was very large and powerful, even though they could not see it, because it stopped along the way to shake the trees. The bushwalkers rapidly left the area as the animal continued to escort them.

This encounter report was similar to more than several dozen others that were received over the next decade or so from the Magdala and Sassafras areas. Usually, on knowing the location of the incident, it was typically possible to predict the report contents, as they were being received. All witnesses reported being followed at an angle from behind by something that was big and powerful. Some mentioned that the animal followed them from the bush alongside, at an angle of about forty-five degrees, with some mentioning that the bush or trees had been shaken. Most witnesses believed that the animal was deliberately trying to drive them out of the area.

Another early encounter report involved a builder who lived in Martin Place at Linden, about four kilometres west of Magdala and Sassafras. I interviewed the man in 1993 at his home where the incident had recently occurred, about a week before. This was one of my first 'outsider' encounter interviews. He was a large muscular man who enjoyed a long bush walk into the valley most Saturday afternoons, sometimes travelling as far as Magdala. He said that he had walked to the point where he usually turned around when he suddenly felt a terrible need to stop and immediately turn back. He said that as he was walking home, he got the impression that someone was following him, so he began listening more intently to the sound behind him. As he became convinced that this was the situation, he stopped and faced the direction of his pursuer, saying, "I know that you are there and I am not scared!" He told me that he could not see anything and whatever was following him seemed big and that it was not using the same track as himself. Instead, it was moving parallel to him, about forty-five degrees to the rear. As he continued to walk home, the animal behind him was slowly narrowing the gap and he was becoming genuinely afraid. In response, he picked up the pace and whatever was behind him matched his speed. He said that the beast was obviously big and heavy because of the sound that it made. He then started to jog and he could now hear the sound of something crashing through the bush behind him. Jogging turned into running with the builder running as fast as he could, while the pursuer continued to cut its own path through the adjacent vegetation and having no difficulty in doing it. Glancing behind, he still could not see what was chasing him but he could see the bush moving violently about. On reaching his rear property fence, the builder said that he "vaulted over the fence, turned around, but nothing was there!" He then told me, "I am never going bushwalking again!" Encounters like this often falsely guide witnesses into believing that these animals are supernatural.

Also in 1993, in a similar manner to the above, two close colleagues, Pam and Brian, who also lived along Martin Place at Linden, did frequent nighttime bush walks to their south and east. As Pam and I had previously worked at the same school, it was she who had told me about the encounter with the builder from her street. On their occasion, Pam and Brian were returning from their long midnight walk and were nearing home when something approached the two through the bush at high speed from behind. Fortunately, they were not far from home but this did not stop them from running! Over the following years, Brian would sometimes hear heavy walking and running through their interconnecting, double property at night, which directly linked with the street and valley behind them.

This next encounter report is unique and extremely significant because it was a large group encounter that occurred overnight in two stages, over a period of about eight hours and involving the participation of more than thirty-five witnesses, ranging in age from fourteen years to adult parents. The report was made shortly after I had returned to local teaching in 2002—a student telephoned me on the Saturday morning of the encounter, requesting my presence. After getting the address, I asked if I needed any plaster for footprints and was told that it would be a good idea.

9: Hominoids

The location of the encounter was a street located within a residential area on the southern side of the Springwood town centre. As I drove down the main thoroughfare adjacent to the side streets, I overshot the address but as I did, I noticed a large number of people milling around on the road. After reversing and turning into the street, I grabbed the crowd's immediate attention, like a solitary human gaining the awareness of ravenous zombies during an apocalypse. I recognised a student amongst the mass of people and he approached my car, along with the crowd. He gave me directions on where to park. I asked the student why there were so many people in the street. He replied, "They all saw it!"

The residential land associated with the street was landscaped and well-maintained. The houses seemed to have been part of a recent property development because they were all very similar in style and age. Like most residential housing in the Blue Mountains, each house had been built on the high ground and backed onto sloping bushland at the rear of the properties. Each property had minimal fencing on most lateral boundaries, having low to medium-high hedging instead and generally having no fencing or hedges at the rear. The back of each block collectively overlooked a cliff with an average height of about ten metres and a slope of about fifty degrees or 120%. The steepest sections of the cliff at the back of some houses had rough steps cut into the soil at various locations along its length. Looking from the cliff top at the back of one of the houses, the remains of a campsite could be partially made out below, in the far distance.

After speaking to a number of children and parents on the street, the student ushered me away to his house where I parked the car. Leaving the entourage of neighbours behind, I was guided through the house to the back garden. Another boy of about the same age then appeared in his neighbouring back garden with his parents and took up a position along the border hedge. From the garden, I was given an overview of the situation by the student, supported by his parents' interjections and supplementary comments from the neighbours.

The background to the story provided significant insight into the circumstances leading up to the encounter. The street had the typical adolescent demographic structure that is typical of new housing estates with a common starting date. By now, the teenage cohort living in the street had obtained their peak destructive capacity during their mid-teens or at about fifteen years of age. By some chance, the majority of children in this age group were male and if this community was a small country hamlet, I would say that their fathers had all been ingesting too much zinc from their ageing, corrugated water tanks. Since the boys, all nine of them had similar conception dates, the local community maintained a common yearly observance ritual, that often took the form of a street party that all community members, regardless of their fecundity status, could unashamedly rejoice in. I was beginning to have some sympathy for the animal. It seemed that in addition to the annual street festival, the boys had decided to extend their celebrations into the local amphitheatre, which was the shared bush valley below.

The boys had been making additional preparations for their valley birthday festival well in advance. Some had cleared a formally pristine area within the environmentally protected Crown Land Reserve, whilst others had started accumulating firewood for their hazard-reduction bonfire and food-related camping activities. The steps and track into the valley were upgraded so that various camping and entertainment items could be more easily transported into the valley in anticipation of the increased participant traffic.

On a Friday afternoon, the party was ready to be unleashed. The boys started the fires, while the local adult

9: Hominoids

residents maintained their courteous indifference by not calling the fire brigade. A short time after it became dark the boys noticed movement in the bush below their campsite. As the evening progressed, the movement in the bush became increasingly violent and some of the boys insisted that they could see red eyes in the bush. Understandably, many of the boys became afraid and returned to their homes, leaving the more hardened boys at the campsite.

As the remaining boys contemplated holding their ground, the level of acoustic and physical violence continued to grow, reaching a peak when rocks and plant material began to rain in on the campsite. In response to these retaliatory threat displays, the boys retreated to the section of clifftop belonging to their respective houses, whilst other teenagers and parents joined them. Those parents who were numb to the preceding raucous behaviour were woken or were alerted by the sudden and unscheduled arrival of children into their homes. By now most neighbours had assembled along the southern cliff boundary and they all witnessed the activity in the valley, with occasional displays of a red glow being seen in the scrub below.

The bush disturbance subsided and eventually ceased. Despite this, many teenagers and parents remained at the scene until after midnight, hoping to experience more activity and optimistically wishing to see what the animal might be like. By early morning, all of the residents had returned home to sleep.

Before dawn, a few residents were woken by the sound of various noises coming from their back gardens. Uncharacteristically rising with the Sun, some teenagers got out of bed and reassembled at the cliff with a few parents. After much discussion from a growing number of assembling witnesses, the student from school suggested that I might be someone who was able to explain the matter to them.

As I stood in the rear garden, I could see obvious signs of recent damage. I spoke at length to the student's parents, who were very proud of their former garden. To the rear left of their property, several grafted eucalyptus trees had once been planted—now visible some distance down the embankment. These trees were not mature, having an estimated height of around six to eight feet, but they were of sufficient age and size to prevent me from thinking about pulling them cleanly out of the ground and hurling them some distance away. On the opposing side of the garden was a mature border hedge, probably Japanese box, that was about two feet high, with smaller plants arranged neatly along the flower bed. A section of the hedge had been trodden flat with a large depression found in the composted bed that was most probably caused by a large foot. The 'imprint' was very poorly defined without any recognisable structure and was not worth casting or even committing to the expense of a new roll of film.

After much discussion, the other neighbours who had been patiently standing by called on me to step over onto their property so that they could show me their garden damage. They mostly pointed out specific examples of the destruction that only they would have been aware of, although some larger broken branches and uprooted plants attested to the possibility of an angry 'hominoid' attempting to make its retaliatory presence felt within its territory. Similarly, I was beckoned onto the next property with an awaiting queue of inquisitive witnesses looking on. With the initial inspection completed, I was taken down the steps and track to the newly developed camping area. There seemed to be a lot of damage inflicted upon the bushland, although it was difficult to differentiate between the indulgent destruction brought about by the teenagers and the reactionary responses of an apoplectic pseudo-anthropoid. However, there were plenty of torn plant parts and uprooted vegetation that might

9: Hominoids

be more indicative of a powerful animal's rage. Like most well-compacted bushland environments, no footprints were visible.

Having sated the southern crowd's curiosity, we moved back onto the street. With some people still patiently waiting for personalised answers, the remnant group barricaded around my car door where I gave a simulated press conference. Like the other witnesses, these residents asked a variety of questions, mostly centred upon what the animal might be. I told them what I knew and didn't know. I said that the beast was most likely around eight feet tall, extremely powerful, very fast, had red reflective eyes for excellent night vision, had three toes, dislikes loud noises and intrusion into its territory, was most probably male, was a marsupial, an indigenous Australian omnivore and though like the North American Bigfoot in appearance, was not the same animal. I also told the concerned group that this valley had a history of violent activity by this marsupial omnivore and that bushwalking should not be encouraged, particularly for unsupervised children. I undoubtedly mentioned other things in response to their specific questions, with the above statements being the main issues that I usually covered. I particularly remember that there was a great deal of interest in the reflective eyes, as most witnesses had seen them at some stage through the night and had noticed their large size and intense red 'glow'. One thing that was certain amongst the group when I left the street—none of them were sceptics.

After many investigations, the reasons which instigated this encounter were fairly obvious. The activities of the boys and the loud noises that they made, were probably well known to the animal. With the large bowl-shaped local terrain and many connected valleys, the sound would have penetrated a long way into the remote southern wilderness, spreading across many tens of square kilometres. There was also the bonfire and the aromas from the burnt meat which should have further broadcast the boys' presence, however, I am certain that their ruckus would have been sufficient by itself. When arriving at some convenient vantage point, the Quinkan would have found the source of the disturbance and also would have seen its territory being violated. Over the years, a few Aboriginal Elders have said that the Hairy Man is 'the protector of the environment'. It is a naturalist belief common in a variety of forms in many traditional societies, as with the North American First Nation tribes that respect the culture of the Sasquatch. They similarly regard their Wildmen as 'the guardians of the forest'. When Sasquatch see land that has been cleared, it is said that they become sad. On the contrary, our Australian Wildmen become very angry. I think that the message being conveyed during this incident was, if you invade and destroy my territory, I will invade and destroy yours.

What struck me about this mass encounter most of all, was that afterwards, no one was afraid to speak their mind about anything that they had experienced. There was no reservation, hesitancy, doubt or apparent fear of ridicule from anyone. I assume that this was because the encounter was a shared experience and as with a herd, there is safety in numbers. This tended to contrast with my frequent encounter debriefing with individuals, where more often than not, people weren't initially prepared to speak as openly, or venture as far, in case I might think that they were barking mad. This tended to be the case if the witness had already attempted to speak to someone else. Imagine a situation where you confided in a friend, only to have them tell you that you had 'misinterpreted' the situation.[21]

---

[21] Nearly every time, regardless of the circumstances, a witness will be told that they had an encounter with a kangaroo. Most Australians seem to believe that they possess innate bush skills and knowledge.

9: Hominoids

That was why interviewing witnesses was a very important task to perform because it is cathartic and many traumatised witnesses need exorcising. Having had an encounter is a big thing and it involves a lot of personal adjustments! This was what Sandy and I were actively seeking after we became Yowie-aware. Having someone to talk to about the experience should be considered essential therapy.

It is also interesting to reflect on how we viewed the over-represented and violent encounters from this area, before the end of the century. During that early period of research, we didn't know a great deal about the very tall, three-toed Quinkan. Our only contact had been the encounter interview where a Quinkan had been spying on a girl through her bedroom window, but we knew very little and weren't even sure if that Wildman was a different genus. The only disclosed information regarding its potential violence was reported by the witness and their neighbours when large rocks were thrown up against the cliff face and loud vocalisations were heard coming from out of the valley. Consequently, we applied what we knew, thinking that the 'marsupial hominoids' from this region were actually four-toed Dooligahl, whose overly aggressive traits could be principally attributed to their male bachelor status and their attempts at gaining dominance or a mate.

However, by 2000, the footprint evidence found by the O'Connors determined that their resident 'marsupial hominoid' was a three-toed Quinkan. Over their lengthy period of encounters with the Quinkan, Jerry would be roared at on several occasions and had rocks and other objects thrown at him.

Of all the reports received by the Octopus, the most poignant account received involved a girl on her first day at high school in 2015, when she came to the staffroom at the end of the day to speak to me about her father's recent encounter. She was advised to speak to me after telling another teacher her story. The girl said that her father was a professional photographer. He had been commissioned to photograph the recently refurbished Hydro Majestic Hotel, a grand tourist destination that runs for a kilometre along the escarpment at Medlow Bath, at an altitude of 1050 metres. The resort sits on a mountain spur that overlooks the sweeping Megalong Valley, about 500 metres below. The Blue Gum Forest in the Grose Valley lies on the other side of the spur, six hundred metres below and ten kilometres to the northeast of the Hydro Majestic Hotel.[22]

The girl's father had driven to the Megalong Valley on the lower western side of the hotel using the local roads to find a suitable shooting position. She said that her father had taken their four-wheel-drive car so that he could go anywhere in the bush alongside the road. By special arrangement with the management, the hotel had organised to have all of the building, garden and guest room lights left on for her father. After driving around the area searching for a suitable location, he eventually settled for a spot on a dirt road that gave him a good view of the hotel above. He took some time to set up his tripod and camera. As he was doing this in the quiet of the night, he started to hear the very distant sound of something approaching from the bush alongside the road. The approaching sound progressively became louder and after a little while, he thought that he could hear high-speed, heavy running heading straight for him. Continuing to listen, there was no let up in the approach of whatever it was, as it came closer to his car. Fearing the worst, she said that her father

---

[22] See also from Chapter 1: Who made the strange cry. 1976. *The Macleay Argus* (28 September). The 'marsupial hominoid' featured in this newspaper article was most probably attempting to move between the Megalong and Grose Valleys, across the spur.

threw all of his expensive camera equipment into the back of the four-wheel-drive, "something that Dad would never do normally" and waited until the very last second before speeding off down the road. She said that her Dad didn't see anything!

After driving around for a little while, her father found a new location over a kilometre away, at the other end of the hotel property and pulled the car over. Again he set up the tripod and camera. As before, he started to hear high-speed running approaching his car from a long distance away. He put his camera gear away and just listened. According to the girl, Dad said that whatever the animal was, it was "very fast and very heavy". Her father waited until it was on top of him before driving off again.

The girl said that since that recent event, her father had not been the same. Not wanting to traumatise the girl, I didn't explain the circumstances to her, despite her asking. Instead, I wrote out my name, address and phone number which I gave to the girl, telling her to get her father to phone me. I told her that I would talk to her again later in the week to see how her father was. When I followed up on the concern with the Year Seven Advisor, I was told that the girl had changed schools.

DURING OUR EARLY period of research, we had been told that there were also small 'hominoids'. We had no substantial knowledge of these small bipeds other than what we suspected when working with the Pendlebury family. We were told that they were about three feet tall and had five toes. However, we unwittingly came across some inharmonious oral testimony from many hundreds of early interviews that we had conducted throughout the mountains, that suggested that we were probably missing, or misinterpreting certain evidence. Most Aboriginal tribes had a name for these small Hairy Men. 'Junjudee' was a commonly accepted name from the South Coast of New South Wales. European Australians and many Aborigines living from northern New South Wales to Far North Queensland (FNQ), commonly refer to these small bipeds as 'Brown Jacks'.

Apart from a few early reports, the first time that we detected other 'hominoids' that deviated from our physical and behavioural expectations of medium and large bipeds, was in 1995. At that time, our gregarious daughter Avril was attending a local preschool and had made friends with Zoe. After talking to Zoe's mother, Lynn told me how Avril had become a very close friend of her daughter and asked if the two could play together after school. When we arrived at Lynn's property, Avril and Zoe ran off to play and left Lynn and me to talk. Lynn confided with me about Zoe's problems with sleeping and how she claimed to have something that regularly knocked on her bedroom window and looked inside. From personal experience and having received many encounter interviews that specifically involved children being contacted by 'hominoids' through their bedroom windows, I became alarmed. Many children also reported seeing a face or red reflective eyes through the window.

Lynn and Gordons' property was a very large acreage, capable of pasturing many horses and providing plenty of riding area. Horse riding was a major interest of Lynn. The land also possessed several caves which contained Aboriginal instructional drawings inscribed in charcoal and were recognised by local Aboriginal tribes as a traditional camping site, providing running water, tool-sharpening grooves and good hunting, with reminders to always burn off the area when leaving. At that time, there were few other properties in the area and the landscape was still largely native bushland.

9: Hominoids

When talking to Lynn about Zoe's sleeping issues, it was blatantly obvious that it was going to require a significant disclosure of what we knew about these animals and it was going to be a very difficult task to introduce sensitively. After a very long and cautious introduction that was going nowhere fast, I asked Lynn to trust me, because I was going to predict what we would find outside Zoe's bedroom window. I told her that we should also trust Zoe because she wouldn't be making this up. I asked Lynn where Zoe's bedroom was and she told me that it was on the far side of the house. Before we went there, I told Lynn that we should find evidence of a flattened area outside the window, with a visible track leading to it. After walking around the house, we found that the far side of the building was very close to the land boundary, like a normal residential property, so we went under the wire fence onto the neighbour's land to make the walking task and approach easier. On arriving at Zoe's bedroom window, we could see Avril and Zoe playing inside and Lynn could see a well-worn path leading up to her daughter's bedroom, with a large flattened area below the base of the window. Lynn was impressed and wanted to know how I knew this. I said that I had personal experience of this phenomenon when living on the farm out west and had spoken to many witnesses in the Blue Mountains over the past two or three years who had experienced the same. What surprised her more, was me telling her that Avril and Drew had both had similar experiences at our house and that Drew had been sufficiently traumatised by what he had seen that we had to remove his previously favourite soft toy from ABC Playschool, called 'Big Ted', mainly because the stuffed bear had red eyes.

Many things would happen at Lynn and Gordons' property over the following decade, that will be discussed in a later chapter. One of the identifying behaviours of these 'hominoids' was the damage that they did to their trees, ripping long sections of bark off the trunk. However, many years later, the best evidence was Lynn seeing a four-foot-tall, upright monkey analogue run across their road as she was walking her dogs at dawn. Shortly after this, when Lynn and Zoe were riding their horses along the same road at sunrise, they both saw a 'mischief' of short monkey-like bipeds run in front of them and down into the gully. It seemed like Junjudee were mostly responsible for a number of these common behaviours, but it would take a few more years of listening to witness reports and our encounter comparisons, to sort out the apparent confusion.

For us, our neighbourhood encounter experiences continued to centre solely on Fatfoot and her joeys until February 2002, when we noticed that her regular behaviours suddenly ceased to be observed. Consequently, we suspected that Fatfoot had either died or moved on. We suspected that she had died because she would have been at least thirty years of age by now, with the survival conditions in this harsh environment being extremely brutal.[23] If she had moved on, we were bound to recognise her distinctive behaviours from reports from other areas in the mountains at a later date. By March 2002 and for some time afterwards, it was obvious that the valley had gone quiet. Whatever the cause, there was now a 'predator vacuum' in our local valley system.

---

[23] On 9 September 2001, I was bitten on the left hallux by a yellow-striped swamp leech which became a cellulitis infection that rapidly spread up my leg to the thigh. Suppressing the cellulitis required two high-dose intravenous antibiotics for a week, followed by oral antibiotics for two weeks. Consequently, I have become very impressed with the robustness of Fatfoot's immune system.

9: Hominoids

As with Aristotle's 'horror vacui', the predator vacuum seemed to be filled over the following months. As usual, the dog's behaviour was the first indicator of the changing patterns of activity, both day and night, followed by the emergence of new sounds and different behaviours. Finally, on the evening of Saturday, 25 September 2004, the new apex predator in our valley was revealed when Sandy and I investigated a new behaviour and were confronted by three very hostile little 'marsupial hominoids'.

BY 2004, SOME semblance of order was developing in our understanding of the 'hominoid' hierarchy. At the simplest level, we now confirmed that there were three different 'anthropomorphic' types, which in order of increasing height were Junjudee, Dooligahl and Quinkan. Our knowledge of Dooligahl was reasonably complete—similarly, the Quinkan. Apart from encounter reports, Junjudee were relatively new to us and we had a lot to learn. For all of the 'marsupial hominoids', the mystery of the toes had disappeared. Using the same height order, their number of toes was five, four and three. The number of fingers, however, remained uncertain.

Despite the progress in accumulated data and improved order, much remains to be learned. We are also certain that there are many more surprises awaiting, most of which lie ahead of this long-term study.

10: Quinkan

DURING THE FIRST week of September 2001, the Octopus received a typical number of encounter reports from the local community. Two reports centred on a valley to the west of our house, so I went for an evening walk during the weekend. I also had a very interesting telephone conversation with Sue, the mother of a former student from several years earlier. Sue recounted a recent incident where she and her husband Jonno, had observed and listened to a 'terror'[1] of Quinkan at the back of their property late one evening. There were more than two very tall 'marsupial hominoids', most probably Quinkan, 'talking' in the bush below their house, early in the morning. She described their 'utterances' as monosyllabic, consonant-vowel combinations that were not human. Together with grunts and other deep sounds, she thought that the animals had big chests because of the apparent high volume of air being exhaled.[2] She

---

[1] We coined 'terror' as the collective noun for a group of Quinkan because Aboriginal accounts frequently described encounters as 'terrorising'. It seemed that Aboriginals were more afraid of Quinkan than Dooligahl because of their greater size and aggressive attitude, though both were regarded as dangerous.

[2] Dr. John, a trauma neurosurgeon, after listening to a Dooligahl breathing, thought that its lung capacity was large—around 5 litres or twice the volume of an adult human.

also saw two pairs of greenish-blue eyes[3] reflecting out of the bush that she estimated to be more than eight feet from the ground. When their son Ben heard that I was thinking of investigating a nearby valley, he asked if he could come along.

Ben had been studying to become an actor at the National Institute of Dramatic Art (NIDA) for the past few years and had taken a minor role in an exportable and well-known Australian soap opera television series. He had become interested in cryptozoology after his mother's previous encounter a few years earlier and because of what he had heard from other students in the local area when at high school. During Sue's first encounter, she had been woken by sounds at the back of their house after 3 A.M. She then heard a very loud 'bellow' with a 'sustained pitch' for three seconds, followed about a second later by the loud sound of a rock striking another. The morning after his mother's first encounter, Ben accompanied her down through the swamp towards the waterfall where the sounds had been heard. Sue saw a large rock sitting on a cliff edge that she thought looked like it had very recently struck the rock beneath it. After Ben had climbed down the cliff, he lifted the large rock and could see bright yellow sandstone detritus under it, which suggested that the two rocks had recently collided.

The following weekend, I drove around to speak to Sue in the afternoon, seeking her permission to be telephone interviewed by Paul Cropper and Tony Healy.[4] This was my usual practice as it provided another opportunity for the encounter to be studied and recorded by other trusted researchers, whom I knew would be ethical and not take

---

[3] Eyeshine is mostly described as red or yellow, with this colour combination having never been reported previously.

[4] Authors of *The Yowie: In search of Australia's Bigfoot*, 2006.

alternative ownership of the research. Likewise, I arranged for Mike Williams to conduct a video interview for similar reasons, as well as providing an audio-visual alternative.[5] While I was there, I drove Ben to the valley and we went for a walk through the thick bush. Ben was very interested in learning about all aspects of these Australian Wildmen and it was good to provide an opportunity to mentor another potential researcher, as I had frequently done with varying degrees of success, since 1993. Ben was also very intelligent, a quick learner and ethically sound, which made the task much easier.

As we slowly inspected the valley we found several well-worn and regularly-used game trails. Unlike typical wallaby and other animal tracks, they reminded me of the paths at home. Some possessed specialised features whilst others seemed to be disused, with alternative routes evolving around them. The upper edges of some paths had broken branch tips clearly visible, hanging down.[6] The starting height of these twig snaps was above my waist and variously continued upwards to approximately six feet. Having studied the development and progress of these twig snaps at home for a long time, we concluded that Fatfoot was slowly opening up access with each passing, possibly to minimise the sound of movement along the track or to make travel less restrictive. Whatever the reason, Ian and I referred to this twig-snapping behaviour as 'track maintenance'.

---

[5] Mike Williams' YouTube video, 'Yowies talking', https://www.youtube.com/watch?v=eFSHLaON6Yw

[6] As a Scout, I was taught to snap the branch tips along a track on the right-hand side, at about waist height and every so often along the path when bushwalking. The purpose of this was to allow our path to be retraced if lost. If the snap was on the left-hand side, then we were heading the correct way back.

On the far south side of the valley, we could see a flattened area of grass that could either have been wallaby or 'marsupial hominoid' beds.[7] As it was too late and too difficult to investigate the area, we don't bother on this occasion.

High up in the trees, we found several damaged branches and trunks made by the much-maligned yellow-tailed black cockatoo, as well as several treebites lower down made by non-avians with 60 mm / 55 mm upper and lower canine separations. We also found steep embankments with paths down them, that had trees snapped a metre or so above the ground with the trunk laying downhill, acting like handrails for the near vertical track. These deliberately constructed 'tree rails' had been originally found by Basim,[8] a marine biologist who lived nearby and had been providing us with activity reports from his immediate area. After finding the treebites, these steep paths and the twig snaps, I was convinced that the two recent reports were accurate and the area was active. Since it was becoming dark, we decided to stay in the valley for a little longer.

After it became dark we started to hear possums and sugar gliders moving about in the tree canopy.[9] The sugar gliders, in particular, seemed to be a very active family group of about six identifiable individuals that were gliding from tree to tree in search of a variety of food. Shortly

---

[7] Sandy had originally found these apparent 'beds' many years earlier. In this instance, the flattened area was on the southern side of the valley, beneath a relatively clear area. It looked like a place where wallabies might sun themselves in the early morning, but we were also aware of similar behaviour by our 'marsupial hominoids'.

[8] See Log entry for Monday, June 21, 1993, and footnote.

[9] A major predator of possums and gliders is the Powerful Owl, *(Ninox strenua)*, the largest Australian owl with a 1.4 metre wingspan. No 'woo hoo' call was heard on this occasion.

10: Quinkan

afterwards, we heard the unmistakably slow and subtle movement of a biped entering from the opposing side of the valley. The noise and activity in the valley started to subside and for the next ten to fifteen minutes, Ben and I weren't exactly sure where the biped was. Suddenly there was a very loud scream coming from the last position where we had heard the walking. More screams were heard, interspersed with several calls which eventually ceased after ten seconds or so. It sounded like one of the sugar gliders had been taken. After that disturbing hunting demonstration, Ben remained focused. Having had his own experience, he wanted me to drive him home so that he could tell Sue and Jonno.[10] When I arrived home and removed my shoes, I found that I had the remains of a yellow-striped leech that had become blended into my sock, between the big and second toes of my left foot. In nature, any bright colour, particularly alternate yellow stripes, tends to be a warning to any potential predator. In this case, it certainly was a portend of future ill health.

When recess came to school the next morning, my foot was extremely hot. When I removed my shoe, the skin from the ankle down was starting to turn black.[11] I could not replace my boot. There was a medical centre across the road, so I hobbled into the building and showed my foot to the triage nurse who immediately called a doctor. As the doctor started to write up a letter, he told me to get to the nearest hospital as soon as possible and present the letter to their triage, telling me that there was no time to waste. I was immediately admitted and the Professor in charge

---

[10] Ben attended a number of other field excursions until his parents sold their house and the family moved out of the area.

[11] The medical professor at the hospital told me that wearing thick hiking socks and steel-capped boots only served to incubate the bacterial infection caused by the leech bite.

of infectious diseases examined my foot and immediately summoned all available interns to attend. He said that the current border of the infection needed to be outlined and regularly updated with an ink pen, dated and timed. He prescribed heavy alternate doses of intravenous flucloxacillin and intravenous penicillin, saying to the interns, "If we don't get on top of this infection quickly, the patient will die".[12] Three nights later, during the early morning hours, I watched the first plane fly into the World Trade Center whilst having an intravenous injection.

THE NAME 'QUINKAN' is primarily used in northern Queensland to refer to the largest 'marsupial hominoid' and other Aboriginal tribes would have had their own names. 'Quinkan' is a collective name given to a group of humanoids, large and small, by the speakers of Kuku Thaypan, Koko Warra, Koko Yalanji, Koko Minni, Olkola, Guugu Yimidhirr and other languages. The people of this linguistic group have held continuous ownership of the land for the past thirty-five thousand years and also produced their world-famous Quinkan Rock Art over the past twenty-seven thousand years. Not surprisingly, there are many interpretations of Quinkan existence. Some believe that Quinkan are the spirits of the dead. Others believe that they are entities associated with the Dreaming. Other accounts simply claim that the Quinkan are the largest of the original inhabitants of Australia who were in constant competition with the recent Aboriginal arrivals. As a consequence of their superior hunting abilities, Quinkan are said to have eaten men and women, with a preference for

---

[12] My mother's aunt died from a skin infection caused by a saddle sore from riding her horse, during the early 1900s. My infection response contrasts poorly with the obviously superior immune system of these 'marsupial hominoids'.

young children. In response, Aborigines said that Quinkan were very difficult to hunt because of their great speed, power and agility, with many Aborigines being terrified of them, although Elder Merve told me that they managed to recover one taken child. They followed the noise of the boy until dawn and showed the 'marsupial humanoid' their raised spears as a guarantee of the tribe's determination to carry through with their threat of war. This suggests that the 'marsupial hominoid' that kidnapped the child from Merve's tribe may have been a Quinkan rather than a Dooligahl. Some stories claimed that Quinkan had their feet facing backwards, which is an illusion, or excuse for not being easily tracked. Some of the definitive attributes of Quinkan can be clearly seen in Aboriginal Rock Art, particularly their night vision ability, enormous height, disproportionate limb size, large feet, three toes, claws and other characteristics.

As a consequence of this research, Quinkan are regarded as an extant, omnivorous, Australian megafaunal predator. However, Quinkan are primarily carnivorous. Having been observed or noted from encounter reports, their main natural prey are kangaroos and wallabies, birds, gliders and insects. Undoubtedly there are many more, unobserved natural prey. Recently introduced prey include cattle, sheep, dogs, cats and other domestic animals and pets. During the sixty-five thousand years of Aboriginal history, humans would have also been included on the prey list.[13] In addition, Quinkan have also been known to eat native

---

[13] "[The Devil-Devil] lives in the tops the steepest and rockiest mountains, which are totally inaccessible to all humans and comes down at night to seize and run away with men, women, or children, whom he eats up, children being his favourite food." Mrs. Charles Meredith. 1973. *Notes and Sketches of New South Wales: During a Residence in that Colony from 1839 to 1844.* Sydney: Ure Smith.

Fig. 10.1 Resting plantigrade footprint (380 mm x 200 mm) from the O'Connor Quinkan made in fine moss. (Photo: Neil Frost)

10: Quinkan

Fig. 10.2 A complete, resting plantigrade Dooligahl (Fatfoot) right footprint cast (310 mm x 160 mm), showing a large hallux, group of three toes, probable outside foot deformity or soft tissue injury, small narrow heel with arch. Cast 22/2/1995. (Photo: Neil Frost)

fruit, blackberries, raspberries, bananas, peanut butter and bread rolls, various food scraps and other unobserved food items.

A definitive and highly recognisable trait of Quinkan is their feet. In having three toes, this trait is so unique amongst the other 'marsupial hominoids' that it can be used as the primary key in their identification. However, as Quinkan and Dooligahl are most likely genetically related to tree-kangaroos and similar tree-dwelling ancestors, Quinkan footprints, in particular, may be misidentified because they can appear similar to kangaroo and wallaby tracks made in coarse substrates where the two feet land together when hopping.

Like Dooligahl,[14] Quinkan also have long, thin feet. Measurements taken from photographs of the O'Connor Quinkan footprint give a length of 380 mm and a width of 200 mm. The feet appear to have an arch, which is a possible adaptation for walking. Unlike the feet of Dooligahl, Quinkan seem to have gained an additional prehensile hallux, which suggests an even greater ability to grip and climb trees. Of the three 'marsupial hominoid' types, the Dooligahl and Quinkan seem to be very closely related, whereas Junjudee are clearly a separate genus.

ALTHOUGH IT IS very difficult to be certain, I believe that the first time I came into contact with a Quinkan, was on the second and third nights of our seven-day camping trip in the Blue Gum Forest, within the Grose Valley, in 1966. Even though a unique Quinkan identifier, like a three-toed footprint, a primary key, was not found in the alluvial deposits along the Grose River Valley the following

---

[14] Fatfoot, as the name suggests, is different to most Dooligahl because of an apparent outside foot injury.

morning, I am confident that this encounter involved a Quinkan because several other secondary keys could be identified and used to uniquely match with the animal. Of all reported Australian 'hominoid' encounters, incidents involving Quinkan are the most violently sustained and varied behaviours of any biped. On this occasion, the animal characteristically behaved very aggressively during the second night, bellowing[15] intimidatory calls for many hours from the cliff top, six hundred metres above. By the third night, a large biped had descended to the valley floor and continued to demonstrate many types of aggression. It proceeded to move around the group campsite, outside of the protective light coming from the central campfire, whilst loudly thrashing the vegetation. The biped then left the main area and moved to our isolated and unprotected tent, where it effectively intimidated our group of twelve- to sixteen-year-old boys with 'Theory of Mind' games, for the majority of the night. We clearly heard heavy footsteps, grunts and growls. I remember hearing very large branches breaking, that could only be broken by something with incredible power and supported by a very angry disposition. In particular, it singled out the young twelve-year-old boy for special attention, growling in his ear that was only separated by three or four millimetres of the khaki canvass tent wall. Even without the primary key, this was clearly a Quinkan encounter.

As BRIEFLY INTRODUCED in the Dooligahl chapter, I first interviewed someone who had an encounter with a Quinkan in early 1994, but we would not become fully convinced of this separate 'marsupial hominoid' type for six more years. The reported events occurred during the

---

[15] See newspaper article, 'Who made the strange cry,' in Chapter 1.

1980s. The encounter witness lived in a house on a side street off Hawkesbury Road, Springwood, New South Wales, an area where many similar reports would be received by the Octopus over the following decades. The woman that I spoke to lived with her husband and daughter in a fibre cement-clad house, built on steel posts because of the sloping block of land. The rear underneath of the house had plenty of headroom and was used as a semi-open storage and work area, which had a wired light that was switched from the kitchen above. The western side of the land had a medium-sized golden cypress pine that had its lower branches removed so that it was easier to freely walk along the side of the house. The entrance to the house was via a small vestibule which faced the driveway on the eastern side. The backyard was small with a high and thick hedge along its northern boundary, that was primarily used to stop anyone from walking off the edge of the rock face, that was immediately behind it. The cliff was very steep at this location with a typical height of about ten metres. The valley behind was drained by a creek that flowed eastward towards Yellow Rock and eventually to the Nepean River, about seven kilometres away and three hundred metres below.

The woman said that their young daughter had sometimes experienced difficulty sleeping in her bedroom over several years, saying that she would be awakened by a tapping noise at the window, followed by a face that would appear at the window's edge. These incidents mostly occurred around late spring or early summer.[16] To comfort their daughter, they would tell her that nothing could reach her bedroom window because its height at that position was well over eight feet from the ground!

---

[16] Southern hemisphere spring and summer—September to February.

At around the same time, usually in November, the woman said that they would sometimes hear the very noisy approach of an animal moving up the valley, loudly calling out as it arrived. These sporadic vocalisations would usually continue over the following months. In addition to this, the unknown animal would throw large rocks from the valley floor that would hit the cliff face and then fall back down. On nights when it was still early, they would meet with other local residents along their property boundaries, listening to the sounds and discussing what the animal could be. Whatever it was, it was capable of throwing rocks![17]

One evening while the family was asleep, she and her husband were woken by the sound of something moving about under the house. After getting out of bed, the husband went to the kitchen and turned on the under house light. The immediate response was the very loud sound of running, followed by the noise of something colliding heavily with a supporting steel post. The house noticeably shook and continued to vibrate in response to the heavy impact. Immediately following this was the sound of something exiting from under the house on the western side and colliding with their cypress pine tree. At this point, the lady took me to view her daughter's old bedroom to confirm the window height from the ground. We then went outside so that I could see the same window

---

[17] Most witnesses do not seem to make the conscious connection between throwing or handling objects and the broader implications of this dextrous behaviour—for example, tool usage. Also, having free hands further implies the possibility of being either obligate bipeds (like humans), or facultative quadrupeds (like kangaroos and wallabies). Kangaroos and wallabies are normally bipeds that hop but can walk on four (quadrupeds) or five legs (pentapeds—by using their tail) when moving at low speed.

from this lower perspective and measure the sill height and also see the sawn-off remains of the tree stump. From this position, there was no chance that an individual could see into the bedroom without the aid of a ladder or a genetically modified pair of legs! The lady seemed to be slightly nervous and so I reassured her by thanking her for her willingness to explain their encounter in such detail.

On inspecting the under-house area the following morning, no damage could be found. However, outside the family found that the medium-sized pine tree had been snapped off at ground level and that there was a trail of broken branches leading towards the rear of the property. Furthermore, the Photinia hedge at the back of the property had a large V-shaped hole through it, where something had clearly continued on down the very steep, kamikaze descent. The reason for planting the hedge many years ago, was to prevent their daughter and any other children from walking off the cliff.[18] The edge was substantially high, which from memory was about ten metres but I have a feeling that it might have been much more. The cliff progressively increased in height as the associated valley dropped away to the east. Found in front of the gap in the hedge was a large footprint in the sand where the animal had launched itself. Now standing close to the hedge at the back of the property, the lady drew my attention to an old wooden frame that was now embedded into the lawn. She said that it was the remains of their daughter's sandpit where a well-defined footprint was found. The solitary footprint was long and thin with three large

---

[18] A major reason for fencing the railway through the mountains was to minimise deaths caused by falling into railway cuttings. A senior student from our school died after she fell into a cutting while taking a shortcut at night. This fencing also restricts the free north-south movement of 'marsupial hominoids'.

toe imprints. As the footprint was so unusual and well-defined, her husband bought some plaster of Paris and cast it. They showed the casting to a number of people but none knew what the animal could be. After having no success, a friend suggested that they should contact a local crypto-researcher who regularly wrote about these matters in the newspapers. They contacted him and he took the cast. Not having received a reply after some time, they telephoned the researcher and were told that it was fake. After asking for the cast to be returned, they were told that it had been smashed with a hammer because it didn't show five toes.

The lady then drew a sketch of the foot on some paper so that I might have a better understanding of what she had previously described. Unfortunately, the only knowledge that I had regarding the foot morphology of these animals was a vague recollection of a line of tracks that Fatfoot had left across our building site in 1984, that we had naïvely dismissed a decade earlier as being a chance combination of synchronised dog and horse tracks. In any case, what I vaguely remembered were tracks from a four-toed Dooligahl and bore no relationship to the three-toed imprint that she had drawn. It was radically different from the five-toed 'hominin' model that everyone was still expecting us to uphold. Consequently, my ignorance and confusion were of no use to her, but the interview was very formative for us.

The lady also told me that this animal had been regularly visiting their valley for many years—their daughter had repeatedly heard something calling out loudly from the valley floor during the latter part of each year over a long period of time. Their neighbours confirmed this long-term, cyclical pattern of behaviour.

Behaviourally, the animal threw rocks and stones, bellowed loudly, tapped and looked through a window and showed repeated interest in their child. There was also a possibility that this 'marsupial hominoid' conducted some type of migration because of the apparent seasonal cycle centred around the month of November. Any migratory path for an animal of this size, however, would probably require thick, contiguous rainforests along the Great Dividing Range, something that the former Gondwanan Rainforests could no longer provide. This suggested that these animals might have become geographically isolated. Alternatively, this apparent seasonal cycle might be better explained by short-distance altitudinal or vertical migration. Australian East Coast altitudes range from zero to 2,228 metres, but in the Blue Mountains, the range is zero to 1,189 metres. Allowing a temperature decrease of 0.6°C per 100 metres of altitude gained would produce a temperature difference of 13.4°C for the East Coast and 7.1°C for the Blue Mountains. Arguably, having a vertical thermal gradient of 7.1°C might be desirable for such a large, well-insulated animal in order to maintain its body temperature and presumably, efficiently-regulated metabolism.[19] Restricting movements to nighttime would also assist in temperature control.

Having evidence of tridactylism, let alone a well-defined long and thin foot imprint, fulfils the unique identification requirements of a Quinkan primary key. Since Quinkan are the tallest of Australia's 'marsupial hominoids', being able to confirm the biped's height at more than eight foot excludes all other possibilities and provides diagnostic evidence of the animal's unique identity, in the same way that having three toes does. The other secondary elements, whilst not as conclusive, further support this verdict. These secondary keys include a propensity

---

[19] In terms of food efficiency, it is very interesting to note that marsupials have a much lower metabolism compared to placentals.

towards violence, where the 'marsupial hominoid' bellowed loudly out of the valley and in particular, threw rocks and stones at the upper cliff face. Having a fascination for a child and looking through a window, however, are typical behaviours of all three 'marsupial hominoid' types and are therefore, unsuitable as keys for identification purposes.

IN LATE NOVEMBER 1999, Steve Crofts telephoned me shortly before 9.00 P.M. He said that his son Brad had a Hairy Man run across in front of his ute on the 'back road' as he was driving home. During the very brief conversation, Steve said that the encounter had only occurred five minutes before. I told Steve to ask his son to wait at the top of our road if he was prepared to accompany me to the location.

Brad's sighting was less than a kilometre from where his parents had their encounter at Cataract Falls in May 1975. The falls are located within a narrow gorge with steep, vegetated sides. On weekends, Steve and his fiancée Doris, would camp underneath a rock overhang below the waterfall, shown in the photograph as a dark area covered in vegetation at the base of the falls and behind the talus. They would light a fire at the entrance to the rock shelter to keep warm and cook their meals. There was a long rope that was connected to an overhead tree, which Steve would swing on, out over the plunge pool at the base of the waterfall. They called this area the 'Tarzan Pool'.

One evening around midnight, they both heard a very loud disturbance coming from the bush, very high above the crest of the waterfall, about ten metres to the right of where the photograph was taken and directly opposite their camping site. The disturbance moved vertically down at a rapid speed, 'like a fireman on a pole', until it reached the same height as the waterfall base. In the pale light of the fire, about ten metres in front of them, they could see the outline of a humanoid figure standing still on the other side of the plunge pool. It was about six feet tall and Steve simply described the figure as, looking like "Cousin Itt". It stood there, motionless for a time, as if it was deciding what to do. The animal had "brown to orange . . . long straight hair" that concealed the biped's facial features, its hands and feet. The 'marsupial hominoid' then turned right and ran off at very high speed, following the track alongside Cataract Creek until it could no longer be heard a few seconds later. By then Doris had relocated to the back of the rock shelter, while Steve began to rake the fire in towards the front of the cave and started to build up the flame with their stock of firewood.

Fig. 10.3 Cataract Falls, Lawson, NSW. (Photo: Neil Frost)

10: Quinkan

Fig. 10.4 Sketch of hominoid seen at Cataract Falls, Lawson, NSW, May 1975, by Steve Crofts.

Like his father, Brad had extensive bush and hunting experience. Having found a fox cub after shooting the mother, Brad decided against all advice to keep it as his 'dog'. Despite the very wise counsel, Brad succeeded in raising the cub and trained it to respond in a manner that closely simulated abnormal dog behaviour. Wherever Brad went, the fox would always accompany him. Whenever Brad was absent, the fox would cry and howl. During walks, Brad would take it on a lead like other dogs. Brad named his pet fox, Dog.

After picking up Brad, we crossed the highway and railway and drove to the site a few kilometres away. During the short journey, Brad said that his Mum and Dad had told him of their experience in 1975. Clearly reflecting upon what he had just seen, he told me that he believes in the existence of these 'humanoids', "one hundred and ten percent!" Driving to the 'back road' required a detour off the normally sealed village roads by taking a short cut away from local residences and into areas of thick bush. As an unsealed track, the 'back road' was very unusual because the two way traffic deviated around a large eucalypt tree growing in the middle of the thoroughfare and considering that this road provided unmonitored passage for local travellers wishing to avoid scrutiny, it was also very unsafe.

Standing on the edge of the dirt road where the animal had crossed about fifteen minutes earlier, we both listened. Immediately we could hear the faint and distant sound of a slow, alternating gait on the far side of the valley. This was difficult terrain. It involved a thirty-metre walk into a shallow valley that, though of medium density, contained many thorny hakea plants. At the base of the valley was a hanging swamp, similar to the one in our own valley. The vegetation here was extremely dense and the ground was boggy. On the far side of the swamp was a steep sandstone cliff about ten metres high with sparse vegetation, that was nonetheless, a difficult climb. Being very familiar with Fatfoot's behaviour, I felt very confident in obtaining a reaction from this biped as well. I told Brad that I was going to whistle across the valley and anticipated that it would stop. Instantly the sound of walking stopped. After a pause of about ten seconds, the source of the sound slowly continued up towards the vertical incline.

Early the next morning Brad and I returned to the site. We accurately determined our stopping point by finding our tyre tracks on the side of the road, where we had pulled over the night before. Brad soon confirmed from memory where the 'marsupial hominoid' had crossed the road. From my past experience,[20] we both scrutinised the soft shoulder mound that had been formed by the grader during the previous road maintenance cycle and we soon found what appeared to be a narrow, but incomplete imprint, covering the hind to mid sections of the foot, that had impacted into the soft ditch mound. The imprint was unimpressive. We then moved to the northwestern side of the road to repeat the process, but Brad pointed out that in the ute headlights the 'marsupial hominoid' had emerged from the treeline with its arms slightly raised and had leapt a considerable distance onto the road, avoiding the grader ditch entirely. Still, Brad climbed up onto the northwestern side of the road and found an established, well-worn track that had a large and freshly broken branch laying across it. We then turned our attention back to the shallow valley and swamp on the southern side of the road. We both climbed down the road embankment and headed

---

[20] In an early attempt to more easily find footprints, we would often walk along the side of remote roads looking at the grader spoil mounds. We also tried a faster version of this method using a passenger as a spotter whilst driving.

off in the same direction that the animal had taken twelve hours earlier. The progress quickly slowed as we entered the swamp. Not wishing to be bitten by leaches and other things, we stopped and viewed the far side of the valley and cliff face from this position. We noticed that the cliff had a prominent fault or fissure that was slightly off-vertical. Directly above this fault was a dominant tree. We turned around and headed back to the road.

Looking back across the valley and swamp, we could make out the very prominent tree on the cliff top and confirm that its location was where we last heard the walking. It was certainly a useful local landmark. Obviously, the animal was going somewhere and the tree marked the point of origin for someone who was about to have an encounter.

Identifying the 'marsupial hominoid' type from this encounter was very difficult because there was not much useful information. The partial imprint from the roadside was missing the important forefoot section used to determine the number of toes, but its overall suggestive size was sufficiently large enough to exclude Junjudee, whatever their feet looked like. Without a primary key, the remaining contenders were either Dooligahl or Quinkan. As for secondary keys, there were effectively none, because Brad was unable to quantify the height of the Wildman. He could only say that it was "big"; whether it was above or below eight feet was unknown under the circumstances.

As THE WEEKS went by, the Octopus received one and then another report from households within two hundred metres of Brad's road-crossing encounter. This was an area of low density housing with large properties and plenty of cleared land. Surrounding the properties to the north was some remnant bushland. To the south, the natural vegetation remained thick and in the future, it would become a site of many treebites. Clearly, this was a route that this 'marsupial hominoid' travelled along on a regular basis, most probably following the clean line of Cataract Creek and deviating from its regular path as its inquisitiveness demanded. It could also get a quick tree larvae snack along the way. It would also seem that the animal was habitually using this route with its nightly curiosity playing the odds against inevitable detection.

The first report came from a senior student who slept with his bedroom window open when the weather was hot. Like so many others, he said that something would approach his window and he would hear heavy breathing. When he got up to investigate, something would run away at high speed. Additionally, when studying near his open window, he would occasionally hear rapid running in the open and along the fence line. After raising the matter with his neighbouring friend who lived sixty five metres away, that friend also admitted to hearing heavy movement in the bush and running through their property, late at night.

As a result of these contemporary reports, other historical accounts began to make more sense and the pieces started to come together. Another encounter report, a few years before these, came from a woman living in a house on the northern side of the Great Western Highway at Lawson. Her story was one of the more bizarre interviews that we had come across. The woman had been upstairs in her house when she heard a noise coming from the lower section of the building. After turning on the lights for the stairs, she started moving down to the lower area when she noticed a long hairy arm protruding through the doggy door. A hand was desperately grabbing at the dog bowl, but had to abandon the attempt. She then heard very heavy running moving at speed towards the northern bush. Apart from the potential link to a track on the other side of the highway, at the time it was amusing to Ian and I because we had also been dealing with the loss of our dogs' food bowls.

Also from around the same time, a Lawson address was given to me by the Octopus so that I could interview the residents. The people lived on the other side of the highway and were experiencing some strange behaviours alongside their house. From the little that I was told, the residents would very frequently hear high-speed, loud running past their windows at night. Despite several attempts to interview the residents, nobody was ever home.

IN EARLY 2000, I received an unexpected telephone call from Jerry O'Connor. Jerry and his partner Sue were local residents having repeated encounters with a large 'marsupial hominoid' in the bush surrounding their house. After a quick exchange of information, I drove to their home and was pleasantly surprised to find that they lived on a property immediately in front of the prominent eucalypt that Brad and I had identified as a local landmark, several months before. After driving Sue and Jerry the long way to the site of Brad's encounter and having recounted the story of the encounter, Jerry commented, "That tree is at the back of our property."

Jerry, Sue and their many cats took me for a tour of their property. The area around the house was cleared and provided little cover to hide behind. There was a large bush towards the back of their land and some small sheds on the rear southern side of their backyard. The sides of their land mostly had inadequate wire fences but they were in better condition than the front and rear boundaries, which had no fences at all. Moving west behind their property boundary was some light to medium-density bush with a very well-established track that travelled two hundred and eighty metres to the edge of a ten-metre drop into the valley. As we approached the cliff we passed the tall eucalypt that Brad and I had identified as a landmark a few months earlier. The track seemingly terminated at the cliff edge, however, on closer inspection, there was a large fracture in the sandstone face that any competent rock climber could have used as a chimney climb by pressing outwards with opposing hands and feet. At the top of the fracture was a mature tree that had grown as a wedge into the crack and could be used as a secure handhold. Looking at the side of the tree facing away from the cliff, the bark showed signs of polished wear on its lower trunk, as if it had been regularly used as a hand-hold.

Like most settlements in the 11,400 square kilometre Blue Mountains National Park, the O'Connor's house backed onto native bush. To their south is a vast wilderness that includes the Blue Labyrinth, an appropriately named area of "approximately 450 square kilometres of deeply dissected sandstone plateau that rises from the Nepean River in the East and stretches West to Kings Tableland; South to the Burragorang Valley rim and North to the main Blue Mountains East-West ridge." "The Labyrinth is a maze of forested ridges and gorges, tangled together forming a puzzle of landscapes, where the unwary are likely to get lost very quickly."[21]

This area includes the southern half of the Blue Mountains National Park but not the Kanangra-Boyd and the Nattai National Parks which together contribute an additional area of 1,175 square kilometres. Of these parks, the most inaccessible is the Nattai National Park which only has four-wheel-drive access, no facilities and requires visitors to register in and out using the logbook. To the O'Connor's north is the northern half of the Blue

---

[21] Bruce Cameron. 2014. *A History of the Blue Labyrinth, Blue Mountains National Park*. Second edition. Sun Valley, NSW: Bruce Cameron.

Mountains National Park including the highly inaccessible Grose River and Valley, plus the equally isolated Wollemi and Yengo National Parks, which contribute an additional area of 6,560 square kilometres.

Thanks to Brad's timely encounter and our broad community-based contacts, we knew that the O'Connors' 'marsupial hominoid' approached their home from the north. From where it had crossed the back road, there was an obvious corridor along a creek which ran towards the highway and railway, that would provide adequate bush cover when on the move. There were also intermediate encounter reports from residents along this corridor that supported this movement. We also knew from past encounter histories that major highways and enclosed railway lines were no barriers to these animals. Typically used to overcome these obstacles are railway overpasses and underpasses, car and pedestrian level crossings, roads with bush corridors on both sides and many other examples. A definitive report dealt with previously, was the encounter experienced by Craig Holmes on 7 April 2005 when travelling along the Great Western Highway, where a 'marsupial hominoid' ran in front of his vehicle as it crossed the road. As is typical of many of these particular situations, there were a number of collateral encounter reports made around this crossing epicentre.

With highways and railways restricting but not preventing the migration of these Wildmen between regions, it initially seemed possible that Fatfoot or one of her family might include the O'Connors' premises within their foraging range. This later seemed highly probable, especially after Susan, the canteen lady from school, saw what appeared to be Fatfoot returning from the southern side of the highway after crossing at the railway overpass bridge at Park Road, Woodford, in May 2000. However, as Jerry and Sue O'Connor would discover over the following months and years, their 'marsupial humanoid' was not Fatfoot but something different. The great thing was, they had physical evidence to help define the new genus.

IN MANY WAYS the experiences of Jerry and Sue were very similar to our own, except that it took a lot less time for them to figure out what was happening around their home. After the O'Connors moved into their new accommodation in September 1997, they started to hear some very disturbing sounds in the bush around them. Jerry asked the neighbours if they had experienced anything unusual, but they said that they hadn't. Notably, their northern neighbour's property was fenced while their land wasn't, not that that would necessarily prevent these animals from stepping over the flimsy wire barrier—it wouldn't. Of course, the O'Connors' house was the first stop for the 'marsupial hominoid' after chimneying up the fault and as the house had not been occupied for quite a while, having someone new moving into the residence and turning the lights on, would have provided more than enough incentive for this curious biped to investigate.

Like us, Jerry and Sue heard loud crashing noises in the bush during the night. Similarly, Jerry described the loud movement as being like that of a bull elephant, whereas I likened the walking to "an elephant on two legs wearing size 20 boots". Clearly, both deliberate auditory movements were intended as threat display: simple, but exaggerated attempts at intimidation where the animals were communicating their displeasure. From the experiences of many hundreds of witnesses interviewed, the common perception was a sense of power, strength and speed, where the overwhelming intention was intimidation.

We also shared other experiences. The unexpected sound of doors being tested is always terrifying, particularly in the middle of the night when no familiar arrival was anticipated. Having rocks thrown at the house or having the walls slapped by an obviously powerful hand. Being stalked.

After several years of experiencing these strange nighttime events, from their rear verandah Sue saw movement behind the large bush in their backyard. Jerry went outside to investigate and was met with a very loud, lion-like roar. Hearing these roars is a very primal experience. Ian Price often described the personal response that these roars and growls elicited as being, "deep and ballsy . . . earthy . . . instinctive . . . predator/prey . . . gut feeling . . . not rational".

After Jerry had contacted us, we spent sometime together comparing our situations. As we previously did with other new researchers, Sandy, Ian[22] and I shared our seven years of experience and communal knowledge with Jerry and Sue. When opportunities arose, we also introduced Jerry to other local participants and witnesses. As the O'Connor's biped seemed different to our own, we were very envious of their opportunity to deal with a naïve 'marsupial hominoid'. However, from an ethical standpoint, we weren't going to meddle in their study without permission, as a few had overwhelmingly done to us. But, even Fatfoot was once naïve and the problem with interacting with an intelligent 'marsupial hominoid' is that they also learn from you. From our experience, we generally found that you could only try a new technique or use a piece of equipment once before they would adapt. They are like the Borg, the cybernetic humanoids from *Star Trek!*[23] Mostly we recommended that the use of light

---

[22] Undoubtedly because of his background, Ian would always evaluate new participants or researchers that he met. Most people passed the interview though a few didn't. Ian always let people know if he didn't like them and why.

[23] More accurately, the Borg would take up to three hits from plasma, phaser or disruptor fire before adapting their shields to the new operating frequency of the beams.

from a torch or camera flash should not be attempted because it would force the biped to become wary and keep its distance, however, carrying a torch was advisable because it would be an effective deterrent, or weapon, should the homanoid decide to become more aggressive. Consequently, as with us, we recommended making any interactions as passive as possible, although at that time the technology that would allow this to happen was still emerging, or was extremely expensive. Commercially available game cameras were still in development as the new digital technology was blossoming, which required us to make our own. Additionally, game cameras may have been effective in taking photographs of wild deer, but intelligent, adaptive predators are a totally different matter. Similarly, digital audio recording was in its infancy, so we had to use inferior and very noisy analogue tape recorders instead.

For many years we had been using cotton thread to subtly track the movement of Fatfoot and her family through our valley and beyond. This was very effective because it was simple and cheap, but most of all, it was passive and guaranteed to produce results if something was moving through the area. Although the technology was primitive, it could be tweaked to enable a number of different measurements. By only anchoring one end of the thread, the cotton would be trailed along in the direction of movement.[24] Having multiple threads, at different heights, but at the same location could act as a simple filter to help differentiate between different animals moving along a track. For example, multiple threads at four and six feet would indicate the probable movement of a swamp wallaby if the only broken thread was the lower one. Additionally,

---

[24] Jerry started by anchoring the cotton thread at both ends. This made the thread more like a trip wire and easy to detect when pressed against, but most unfortunately, it caused a bird to have the thread wrapped around its leg which eventually killed it. Understandably, Jerry never used cotton thread again.

it was always advantageous to use cotton threads of different colours because the lines were not always broken on the first night. Sometimes a thread might persist for weeks before it was interfered with and having a record of the colour gave some idea of the frequency of travel and the track's history. The cotton thread was also used for other purposes. Sometimes we would use it to suspend tempting food items from trees, only because we wanted to record and observe Fatfoot from a known location. After hanging a banana from a tree on the edge of the swamp at a height that nothing else could obtain, we found the skin with the attached cotton at our front doorstep. We also placed food on a wire 'feeding table' in the hope that Fatfoot would approach it. The logic behind this feeding attempt was, either Fatfoot would eat the food or Fatfoot would eat whatever was attracted to the food.

As nighttime film photography was always problematic and unreliable, the next best alternative was audio recording. After Robert had lent us his VOR (Voice Operated Recording) tape recorder, we had some early success in recording Fatfoot's movement, until she was able to recognise it and knew that it was not food or of any particular interest. However, Robert's VOR tape recorder had a number of disadvantages. The primary problem was that it was quite noisy. For me, I could just hear the noise of the motor and the tape being dragged through the mechanism if the recorder was within a few feet. For these animals, it was apparent that they could hear this noise from much further away and even when hidden underneath other obstacles. Clearly, these bipeds have superior hearing compared to us. Additionally, the VOR recorder would only activate when it heard a sound above the detection threshold but the system also had some latency and so it took some time for the motor to reach the required recording speed. This meant that the beginning of a recording would either be lost, or distorted because it was not buffered. This problem resulted in the inability to record any quick action movements, like foot thumping, which were never recorded successfully. There were also issues regarding battery life and recording time. Battery life remains a perennial problem. Being an electro-mechanical device, this analogue apparatus was more prone to battery depletion, a condition made even worse in the low-temperature environment of a swamp. The available recording time was determined by the thickness and length of the tape. Typically the maximum recording time was 120 minutes, but only if you were able to flip over the cassette, so in practice it was limited to one hour, which was usually adequate.

Like many of us, Jerry and Sue wanted to learn more about these 'marsupial hominoids'. Jerry initially placed cotton threads around their land and was able to trace the movement from the track at the back of their property, alongside their house and to their front bedroom. This path was consistent with their many nighttime experiences, where they would hear the electrical switchboard cover on the outside of their bedroom being opened and the sound of a hand slapping the wall. More importantly, it was through their bedroom window that Sue saw the animal's head and shoulders in August 2000, at about 2 A.M. She said that the Wildman had a human-sized head that was sunk into the extremely wide shoulders. It had a narrow nose, a very wide mouth and a "rounded clump of tan-coloured hair on top of its head". Measuring the height of the head from outside their bedroom window, Jerry said that it nearly reached the underside of the boxed eaves of their house and consequently, would have been at least eight foot four inches tall.

Jerry also obtained a number of black plastic plant pots which he suspended on cotton threads from trees in the bush. For food, he filled the 'buckets' with a variety of fruit and also day-old bread rolls that he obtained from the local bakery. He experimented with different types of filling for the bread rolls, mainly peanut butter because it was recommended in posts from a highly regarded North American Bigfoot website. The peanut butter rolls seemed to be popular. I think that Jerry also tried Vegemite, which the biped rejected for some irrational reason. Jerry said that their Wildman also rejected bread rolls that were too old. As a feeding variation, Jerry started to alter the height of the buckets, gradually raising them until they were no longer able to be touched. Jerry said that the Hairy Man had a very long reach, although I don't know what his ultimate height measurement was.

Continuing along with his theme, Jerry decided to move the bucket experiment further towards the solitary access point on the rock face, nearly three hundred metres from their house. On this occasion Jerry wanted to expand his experimental capability with audio recordings of the animal handling the bucket and food. As I had some experience with concealed audio recording, Jerry asked if he could borrow my equipment. I brought Jerry the equipment that he had asked for and familiarised him with the operation of the VOR tape recorder. I also brought him a high-quality, one-hundred-and-twenty-minute cassette tape, four high-quality alkaline batteries, plastic cling wrap and some fishing line. Jerry suspended the black plastic plant pot or 'bucket' from a tree branch that overhung the cliff void, using fishing line rather than the low tensile cotton thread. The fishing line was made less obvious by repeatedly wrapping it around the tree and securing it in a position that hopefully, wouldn't be seen. Apart from the height of the bucket, it was positioned out from the cliff edge making it impossible, for us at least, to reach the pot without a considerable risk of death from falling over the cliff. The recorder was prepared and waterproofed in the usual manner by using plastic cling wrap and Jerry chose a location under a hollow log that was close to the access point and the middle of the well-established track. I left Jerry and Sue to finalise the experiment. They returned to the recording site during the early evening to fill the bucket with fruit and bread rolls. They then started the recording and hid the recorder under a halved log.

As was my habit during the period that I was keeping the Activity Log in 1993, I kept paper and a pencil next to my bed and scribbled down the time of any identifiable auditory event. Jerry followed the same procedure at his house. Even though we were several kilometres from the recorder, it was a simple task to identify and note the time when coal and passenger trains passed through. In reality, everyone residing within the Blue Mountains City Council region could have participated in these temporal notations if they so wished, since the rail squeal would have shocked most residents awake! Sometimes planes and helicopters, or a recalcitrant truck driver excessively using his exhaust brakes could also be heard. When playing back the recording, most of these auditory clues could be identified on the tape, placing every recorded activity on a convenient timeline. The most commonly recorded sound was the distinctive rail squeal from coal trains as their fixed axle wheels attempted to turn a bend without the aid of a differential and to a lesser extent, the quieter electrical noise from the timetabled passenger trains after they began service around 3 A.M., providing another accurate time check on the recording.

More interestingly, the recording revealed the sound of bipedal walking and the sounds of leaf litter being disturbed. Following this are various sounds of the camouflage being removed, the hollow log being shifted and the recorder being discovered. There is the sound of the plastic being torn and its removal attempted. The most significant part of this recording occurs at 2:33 when the unreachable 'bucket' is hit by some baseball-like bat, that launches the pot and its contents on a horizontal journey of more than twenty metres into the valley below. This audio recording is available on Mike Williams' *Strange Nation* YouTube channel, 'Jerry and Neil—Bucket experiment', at https://www.youtube.com/watch?v=8McBSdlIILY.

The implications of this 'marsupial hominoid' hitting the bucket are highly significant, although a better solution to the problem might have been to capture the fishing line and pull in the pot and its contents with the same hitting implement. Like the other extant genera of Australian megafauna, this large biped also uses tools. Tool usage is a defining characteristic of humanity, alongside bipedalism, language and a growing list of other traits, that seem to have become necessary in order to maintain the distinction and separation between man and animals. I suspect that historically this was necessary in order to clearly differentiate humans from the rest of the animal kingdom for religious purposes. However, because of the work of Jane Goodall, we awakened to the knowledge that chimpanzees use tools as well. Taking this further, many other animals are now recognised as using tools and an excellent example is the New Zealand kea, which not only uses tools but engages in sophisticated problem-solving. This is similar to the actions of the O'Connor's 'marsupial humanoid' which has clearly recognised the problem and produced a mechanical solution. This is high-order reasoning! Regarding language, humans use it to communicate and organise complex group tasks. From what has been previously mentioned, there have been occasions where language usage by the three Wildmen appears to have been demonstrated.

10: Quinkan

Jerry and Sue also found treebites in the bloodwood eucalypts along their road, just minutes after we mentioned that they should look for them during their daily walks. They were very similar in appearance to the early morning bites that Sandy had originally discovered across from our driveway in March 1996. After one afternoon, Jerry reported back to us to say that they had found more than thirty treebites, with consistently spaced tears, at a height of about six feet from the ground. After a wider search of the area over several weeks, Sue and Jerry found a large field of damaged trees near the swamp below their house, containing several hundred near identical examples. On inspection, they conformed to the standard pattern and ticked all of the required keys. Of interest in this case, was the upper canine separation that was consistently measured at 82 mm and 55 mm for the lower canine separation. This compares to the upper canine separation of Fatfoot which was consistently measured around our house at 60 mm and 55 mm for the lower canine separation. Clearly, the O'Connors' 'marsupial hominoid' is larger and likely physically and behaviourally different from our Dooligahl.

Perhaps the most staggering physical evidence produced by Jerry and Sue were the cast and photographs of a three-toed footprint. Like the painting by Percy Trezise of 'Turramulli, the Giant Quinkan', used on the front cover of Tony Healy and Paul Cropper's book, the footprint that was made in the thick, green moss alongside their northern bedroom wall, had three clearly defined toes.

The two photographs of a three-toed footprint in moss that I took outside of the O'Connors' northern bedroom wall as shown in Figs. 10.5 and 10.6. Sue also took some photographs and cast the footprint. There is a line around the imprint where Sue placed a cardboard box with a brick on top, to protect it before she obtained some plaster to cast it. The foot was measured at 380 mm in length and 200 mm in width across the toes. The narrow section of the mid and hindfoot are 90 mm wide. Although Sue took great care in casting the imprint, it was embedded in highly porous moss, where the definition and accuracy of the cast were not going to be as good as the two photographs. In maximising my very limited understanding of the foot morphology of these bipeds, I have taken a fifty percent guess that the imprint is of a left foot and the hallux, if it has one, is in its usual position. It could be argued that the footprint shows a second hallux, or more acceptably, an overdeveloped 'little toe'. It was unfortunate that the imprint did not show any indication of claws, as shown in the Percy Trezise painting and very occasionally referred to by a diminishing number of Aboriginal Elders.[25] Perhaps the claws are retractable, as were the large, retractable thumb claws of the Australian megafauna carnivore, the extinct marsupial lion, *Thylacoleo carnifex*? This marsupial version of a placental lion is another example of biological convergence, where similar environmental niches result in the evolution of similarly adapted animals. It can be argued that these 'marsupial hominoids' are Australia's evolutionary response to the segregant mammal, *Homo*. Stranger still, Anton from Far North Queensland said that there was also a rear claw, which is not shown in the painting of the giant Turramulli, or elsewhere.

It needs to be kept constantly in mind that Australian marsupials are not complete analogues of animals from the rest of the world. The over-quoted platypus, for example, is a semiaquatic mammal that has a duck bill and webbed

---

[25] The giant Pleistocene kangaroo *Procoptodon goliah* was the largest and heaviest kangaroo known. It had an erect posture, forward facing eyes, with a single large clawed toe on each foot. Each hand had two long, clawed fingers.

Fig. 10.5 Three-toed Quinkan footprint (360 mm x 200 mm) in moss. (Photo: Neil Frost)

feet, a beaver-like tail, a furred otter-like body and walks like a reptile. The females lay eggs, but feed their puggles with milk. The males have a venomous spur on their hind legs. Early biologists thought that platypus specimens were a hoax. Sounds familiar!

THE CHARCOAL DRAWING (Fig. 10.7) was photographed by Dom Boidin at Dingoes Lair, Wollemi National Park, in 2011. To obtain this photograph, Dom needed to make a fourteen-hour return hiking journey into this extremely remote World Heritage site. Like its other prized namesake, the Wollemi National Park holds a number of other important secrets. The Dingoes Lair Cave site is a collection of 203 rock art images with up to twelve layers, which are dated to 4000 YBP. As an ancient reference, this drawing certainly pushes back the historical record for this particular humanoid type. As with the pine tree, its location is kept secret.

Assuming that the image shown on the left is human, to be used for comparative purposes, then the charcoal drawing shows another figure that is clearly not. Since the two figures are of different sizes and types, it seems reasonable to assume that the artist was attempting to illustrate a realistic, to-scale depiction, where the figure on the right is 'humanoid' and 55% larger. If inputting a height range for the human figure of between five and six foot, this will give a comparative height range of seven feet and nine inches to nine feet and four inches[26] for the 'humanoid'. The mean height from this range is eight foot, six and a half inches. The actual O'Connor height measurement for their Quinkan was eight foot four inches, which is similar to the estimated height of the Quinkan from the Hawkesbury Road valley.

[26] Corrected to the nearest inch.

10: Quinkan

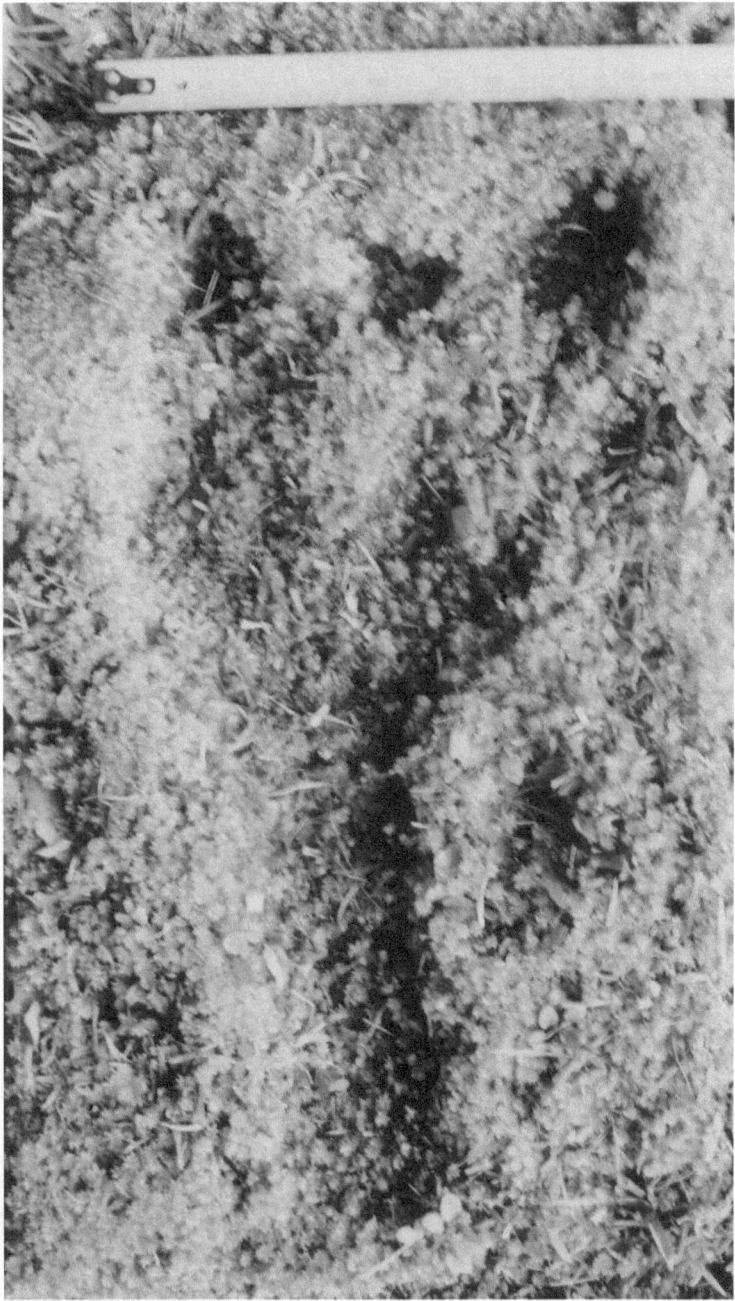

Fig. 10.6 The line around the footprint is from a weighted cardboard box that Sue placed over the footprint to protect it. (Photo: Neil Frost)

Looking at the larger figure, the eyes are very clearly represented and in comparison to the imagined size of the human counterpart, they seem to be very large, which is a realistic comparison. Also, illustrating the eyes as empty vessels provides the artistic license that they are bright or emitting light, which seems to be an effective artistic

Fig. 10.7 4000-year-old charcoal drawing of a Quinkan, from Dingoes Lair, Wollemi National Park, NSW. (Photo: with kind permission of Dom Boiden © 2011)

technique to represent the large red reflective eyes that are commonly seen at night and, from an Aboriginal perspective, from around a campfire.

The top of the head has an obvious point, especially when compared to the rounded human cranium drawn alongside. A potentially similar feature was noticed by Sue O'Connor when she observed that their biped had a "rounded clump of tan-coloured hair on top of its head". This anatomical feature would typically indicate the presence of a sagittal crest, a raised ridge of bone that runs along the joint of the two parietal bones of the skull. Such a bony ridge is used to anchor exceptionally long and strong jaw muscles that supplement the shorter muscles normally present in less powerful animals. Animals with sagittal ridges have an extremely powerful bite force, capable of shearing and crushing through bone and other tough substances, including trees. The drawing also suggests that the head is closely attached to the shoulders, though the same could be said of the human figure.

The feet shown in this drawing are perhaps the most noticeable feature? They are narrow and very long, like those in the three-toed photograph, and appear to be ridiculously exaggerated, compared to the feet drawn on the human. It seems likely, therefore, that the artist was attempting to draw attention to another unusual aspect of this humanoid, other than its feet, because the overall length of the legs are long as some witnesses have commented on.[27] Unfortunately, the artist didn't indicate the number of toes.

---

[27] Sandy's niece Zoe, together with her brother Gabriel and cousins Sharlene and Natalie, had a very close daytime encounter with an eight-foot-tall Quinkan when living at Nerrigundah, NSW, in 1982. She said that it had "disproportionately long legs that took slow but, very long steps, covering the distance very quickly". Gabriel also said that the animal was "very large and powerfully built, about eight feet tall", that seemed to "glide along".

Some disappointing aspects of the drawing are the depiction of the hands, which clearly show five digits and the length of the arms, which should be longer. As with the painting of Turramulli and spoken evidence from Elders, particularly from North Queensland, there should only be three fingers. However, obtaining evidence of the number of digits on any of the Australian 'hominoids' is a problem that twenty-first-century researchers can also identify with. I would like to think that this was an oversight by the local artist four thousand years ago, who may not have been aware of this particular detail at this time and simply copied from the human analogue by default.

AFTER MOVING TO Nerang, Queensland, in 2000, Ian and Cheryl bought property in the area. However, some time afterwards, they both missed having an acreage and purchased land with a partly-finished house in the village of Milbrodale, NSW.

Typical of Ian, he got to know most of the local inhabitants and soon discovered that the village in the Yengo National Park contained an Aboriginal sacred site, known as Baiame Cave, which contained a main painting that is up to two thousand years old and was also the traditional site of a sacred Bora.

Carbon dating estimates for this coastal settlement area are recent and vary within a relatively narrow range. Surprisingly, it seems that this forested eastern coastal strip was occupied much later than the drier western grasslands of the ranges, where occupancy dates of nine to fourteen thousand years have been obtained.[28] If I was to speculate on some of the reasons behind the delayed settlement

---

[28] Moore, D.R., 1967. Archaeological field study of the Hunter Valley, NSW by the Australian Museum: A preliminary report, *Australian Institute of Aboriginal Studies Newsletter* 2(6): 34-41.

of the eastern coastal strip, it might have something to do with the prolonged survival of the eastern Gondwanan Rainforests in the face of increasing aridification in the west and elsewhere. Consequently, the ancient Gondwanan Rainforest habitat in the east, apart from its ancestral significance, would be a more suitable predatory environment for Quinkan, compared to the dry open forest structures typically found on the western plains.

According to the local Wonnarua people, Baiame is an ancestral Great Spirit being, responsible for the creation of the land, animals and people. Baiame had large eyes because "he was all seeing and all knowing". As the sky spirit, Baiame had long arms because he was believed to be "a protector of the area and district". This belief is very similar to other tribes who say instead that, their 'marsupial hominoids' are "protectors of the environment".

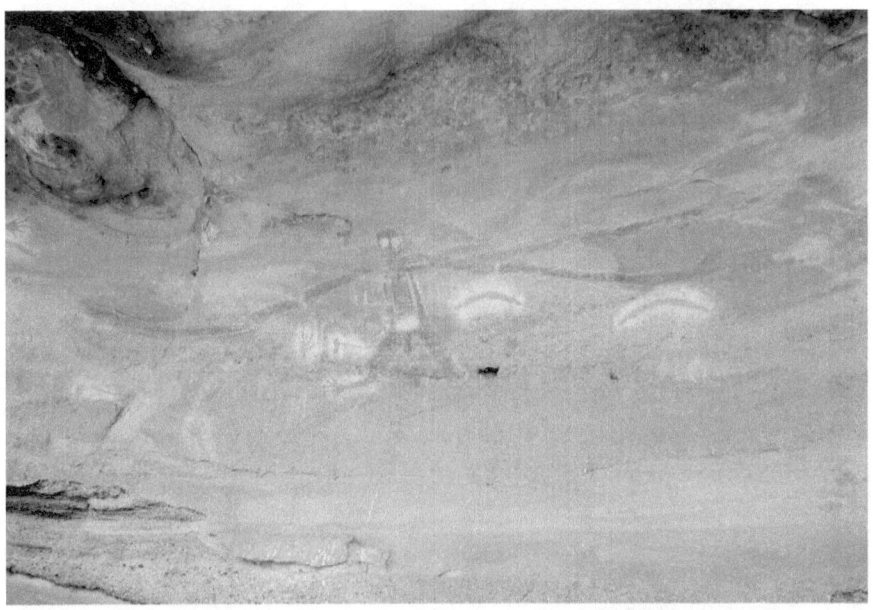

Fig. 10.8 'Sky Father' painting from Baiame Cave, Milbrodale. (Photo: Ian Price)

Although the painting is called the 'Sky Father', Ian immediately recognised the significance of how the eyes had been painted. He said that the eyes were depicted in a style that was typical of large 'glowing' or 'reflective' eyes, as seen in many other Aboriginal paintings and also like the red eyes commonly seen by so many people around a campfire or when in the bush. In other words, the painting was a representation of something that was familiar to the artist. Consequently, Ian thought that the meaning of this painting had more in common with descriptions of the Hairy Man than Baime, a creator sky spirit, and that the story may have been corrupted.

I fully agree with Ian's interpretation of the eyes. There are also other aspects of the painting and the information recorded by an early Australian anthropologist, that are worthy of investigation.

To begin with, Robert Hamilton Mathews (1841-1918) started his career as a licensed surveyor in 1870 and made considerable earnings from it. By 1893 he took up the new study of anthropology, mainly because of his early childhood experiences playing with Aboriginal children and later, from working with Aboriginal people as a surveyor and as a magistrate in local courts. When working as a surveyor in the Milbrodale area in 1893, "my attention was drawn to the existence of some caves."[29]

"Standing in front of the cave with the face towards it, the most prominent object is the grotesque figure of a man about eight feet high, with the arms and legs extended and out of all proportion to the rest of the body."[30] Three years

---

[29] Mathews, R. H. 1893. Rock paintings by the Aborigines in caves on Bulgar Creek, near Singleton. *Journal and Proceedings of the Royal Society of New South Wales* 27: 353-358. P. 353.

[30] Ibid., pp. 354-355.

later, at the Queensland Branch of the Royal Geographical Society of Australasia, Matthews modified his description of the figure: "The most prominent object in this cave is the grotesque nude representation of a man, a little over nine feet tall if the legs were close together, instead of being extended. . . . This is much the largest and most remarkable human figure I have yet met with amongst the cave paintings of New South Wales."[31] Being a surveyor, I would imagine that all his measurements would be highly accurate. When looking for a link, it would seem that the height of the painting is a close match with many actual Quinkan encounter reports, together with the O'Connor experience and a soon-to-be-mentioned Aboriginal Quinkan artwork from Mareeba in Far North Queensland.

The depiction of the arms are also interesting, mainly because of their length. Moving down each arm, there is a white painted line that Matthews simply describes as, "a band around them, about two feet from the shoulder". Unlike Matthews, if I was to look for a meaning for these white lines when there doesn't seem to be an obvious one, I would suggest that they mark the point of articulation for the elbow, whilst also highlighting the disproportional length of the upper and lower arms, as Matthews originally pointed out in his earlier 1893 description. This interpretation is similar to the rationale used to explain Aboriginal X-ray art for animals, humans and spirit beings, where the artist represents the internal workings and structure of the animal, within the context of its external body shape. Unfortunately, the same disproportional length of Quinkan legs cannot be examined on the painting, ironically

---

[31] Mathews, R. H. 1896. The rock pictures of the Australian Aborigines. *Proceedings and Transactions of the Queensland Branch of the Royal Geographical Society of Australasia* 11: 86-105. P. 90.

because of alleged sandblasting involving the lower section of the painting by a state conservation organisation.[32] However, from what might be seen or imagined from the lower section of the painting, there does seem to be some similarity with the charcoal drawing from Dingoes Lair, Wollemi National Park, NSW.

Continuing with the anatomical comparisons, the left hand shown in the painting seems to correctly display three digits, although this part of the painting does not seem to be perfectly preserved. The right hand has no visible detail, although I would imagine that microscopy could provide a simple solution to the original placement of the ochre and a resolution to the hidden features of the image. In his mostly descriptive account of the cave, no details are given regarding the hands and feet.

Usually, the feet and the number of toes are regarded by many researchers as being the most difficult to determine. In reality, clearly defined footprints showing the number of toes, although rare, have been sufficiently found. However, determining the number of digits on the hand would normally require a daytime observation to ascertain. For us, the only source of information so far, has been from Aboriginal paintings and oral accounts.

With this painting showing three hand digits, these could be regarded as a suitable equivalent to the three toes more commonly used as a primary key for Quinkan identification. Also, with 'marsupial hominoid' identification generally, any height greater than eight feet automatically becomes a unique identifier of Quinkan. For Matthews,

---

[32] Brook Boney. 2017. Ancient Aboriginal sites left vulnerable, showing why cultural preservation is necessary in Australia. *ABC*. https://www.abc.net.au/news/2017-12-21/brooke-boney-road-back-home-cultural-preservation-in-australia/9269956

the revised height measurement exceeds this lower marker being "a little over nine feet tall".

Another anatomical feature that is not seen during encounters is the penis. From the perspective of the Octopus, no reports mentioned seeing one and similarly, with mammary glands.[33] Conforming to Victorian virtues, Matthews did not mention or provide a measurement of the erect penis shown in the painting, however, the male spirit being seems to be very well endowed. Having watched male swamp wallabies mating around the cleared grassland near our swamp, the erectile size of their penis is truly phenomenal. Their penises appear snake-like as they seem to take on an active and independent life of their own! I would estimate their erectile penis length at around eighteen inches. According to an encounter report by Mrs. Jean Maloney from the book *The Yowie: In Search of Australia's Bigfoot,* Jean noticed that the penis seen during her encounter looked similar to an uncircumcised human penis that was "quite large: maybe nine inches long" and had a human-like scrotum.[34]

Even though Matthews attempted to be mainly descriptive, he also unintentionally introduced some cultural and spiritual elements. "I have confined myself as much as possible to descriptions only of these drawings and have not attempted to connect them with the myths and superstitions of the Australian Aborigines; neither have I speculated on their supposed totemic or symbolical meanings. I have left these researches for the better qualified to follow them out than I am, or have more time at their

---

[33] As marsupial hominoids, females would not have breasts. However, breasts are reported as being seen three times and a penis once in Healy and Cropper 2006, p. 137.

[34] Healy and Cropper, *op. cit.*, p. 137

disposal." Matthews said that: "Such a figure represented Baiamai, or the Great Spirit."[35] He also quoted, "It is generally supposed by old colonists who have been a good deal among the Aborigines in the early days of the Colony that a figure of a man represents either a good or bad spirit and generally were those who presided over the ceremony of a Bora."[36] Continuing on from this he further mentions, "The figure of Baiamai, or Devil-Devil, or whatever the image represents . . ."[37] As quoted earlier, a nineteenth-century account of Devil-devils taking children in New South Wales was written by Mrs. Charles Meredith during the 1840s: ". . . but they have an evil spirit, which causes them great terror, whom they call 'Yahoo' or 'Devil-devil': he lives in the tops of the steepest and rockiest mountains, which are totally inaccessible to all human beings, and comes down at night to seize and run away with men, women, or children, whom he eats up, children being his favourite food. . . ."[38]

Interestingly, Matthews also wrote, "I was informed by Mr. W. G. McAlpin, who is now eighty-four years of age and has resided in the neighbourhood for the last fifty years, that the figures in this cave were there when he first came to the district; and even at that time the drawings were beyond the knowledge of the local blacks."[39]

IN 2004, OUR family took a flight to Cairns in Far North Queensland for a holiday. In addition to seeing many of

---

[35] Mathews 1893, *op. cit.*, p. 355.

[36] Ibid., p. 355.

[37] Ibid., p. 356.

[38] Mrs. Charles Meredith. 1973. *Notes and Sketches of New South Wales*. Sydney: Ure Smith. P. 95.

[39] Mathews 1893, *op. cit.*, p. 356.

the local attractions, like the Daintree Rainforest and the Great Barrier Reef, we also drove around the area in order to interview some Quinkan witnesses who had been recommended by the Octopus and also to visit some Aboriginal sites.

At Mareeba, we stopped the car so that we could explore the town and I soon tripped over an indigenous store that had a magnificent, full-sized Quinkan painting on display at the front of the store, facing inwards away from unscrutinised inspection. The painting was eight feet tall and would have been a very tight fit inside a house with standard ceiling heights. For amusement, I asked about the price and was told that it was $AUD 20,000. It was extremely beautiful and despite asking politely and immediately following up with another request a few seconds later, I wasn't allowed to take a photograph of it. The sales assistant told me that the artist had strictly forbidden any photography of his painting. After engaging with the sales assistant for about five minutes regarding various aspects of the painting, she told me that she was an Elder, so I enquired about certain aspects of the picture's accuracy. Apart from simply relying on the label, I asked if the picture was of a Quinkan. After her confirmation, I asked her why did the picture show four toes then? I told her that our Dooligahl in New South Wales have four toes and Quinkan have three! After this, the Elder seemed to significantly lighten up and we had a good conversation about these 'humanoids' and our respective encounters.

From my experience in interviewing tribe members throughout Papua New Guinea and the Pacific Islands across many years during the 1970s, sometimes obtaining accurate information is very difficult because it may be sacred, restricted knowledge, or for initiates only and certainly not for general consumption. Sometimes subjects would deliberately provide false, inaccurate or misleading information as a way of circumventing these restrictions or tribal laws, without causing a transgression, embarrassment or personal affront to anyone.

For a further twenty minutes or so, we had a very pleasant discussion in the store about Quinkan. As I did in Melanesia, I spoke the most as I attempted to establish trust by presenting my credentials. I told her about our Dooligahl and then about our experiences with various mountain Quinkan. Around this time, Sandy and the children were making their third circumnavigation of the Mareeba town centre and I was given the wind-up signal.

With Sandy about to make her final approach, the Elder told me that she was responsible for making the Quinkan masks for the traditional dancers. She asked me to guess what she used to represent the eyes. She told me that she used bicycle rear tail-light reflectors, "the really big red ones!" We talked some more about the red eyes and the tendency for them to turn yellow as the alignment changes. She also made me aware of the Laura Quinkan Dance Festival,[40] held at the sacred Bora meeting ground at Laura in Far North Queensland and suggested that I should attend.

ANOTHER OUTSTANDING, MASS encounter involving a recognisable Quinkan-like individual occurred during a two-day school camp near Springbrook, Queensland, in 1977. As reported in the *Gold Coast Bulletin* was the following newspaper story that was further investigated by Tony Healy and Paul Cropper in their book.

"About 20 of us saw it. It was about three metres tall, covered in hair, had a flat face and walked to one side in a crab-like style. It smashed small saplings and trees like

---

[40] See https://www.lauraquinkanfestival.com.au/

matchsticks as it careered through the bush. We spotted it several times and once watched it through binoculars. . . . We first saw the Yowie at 12:30 P.M. on October 22 and last saw it just before we returned to Southport on the afternoon of October 23."

The above newspaper report was provided by Bill O'Chee. Later on, he would become Senator Bill O'Chee, a member of the Senate or Upper House in the Australian Federal Parliament. In addition to being further interviewed by Tony and Paul, he gave an interview to the producers of the *Animal X* Yowie documentary who were making a film of our research on Fatfoot. The documentary, in two parts, can be viewed from the following *StorytellerMedia* YouTube channel. Bill O'Chee's interview is shown in Part 2.

Part 1: https://www.youtube.com/watch?v=EjBw7RDHfVY

Part 2: https://www.youtube.com/watch?v=d81Jf75KPxs

The first part of the documentary deals with two historical cases. One report from the Blue Mountains tells of an encounter by two bushwalkers who found a freshly dismembered kangaroo (most probably a swamp or rock wallaby) on a walking track in April 1979. The kangaroo seemed to have been torn apart. "All the flesh had been ripped away from the back legs." Further along the 'walking' track they found a huge footprint and had a feeling of being watched, a commonly reported encounter emotion that Tony Healy often referred to during his frequent visits to our home as the 'nameless dread'[41] and also known elsewhere by other names, such as 'mountain panic'. From

---

[41] Healy and Cropper, *op. cit.*, p. 172-173

personal experience, the 'nameless dread' is a real phenomenon that is induced under certain circumstances where a person's basal instincts pick up on and recognise the targeted hunting behaviour of a large and powerful predator. This is a typical intimidatory behaviour of our Australian Wildmen, involving 'Theory of Mind' games. After experiencing the dread, the two bushwalkers saw a hairy biped hiding in the bush. It "was a huge beast, at least three metres tall, looking like a man, not an ape and covered in short grey hair". Unfortunately, the number of toes on the footprint were not recorded or were not detectable, however, the beast was at least three metres tall and consequently, a Quinkan.

The second report outlines an encounter involving two Queensland pig shooters in 1979. After stopping for lunch, the two shooters saw a three-metre-tall biped on the ridge above. They shot at the biped and followed its thumping noise along the creek bed. They later found huge three-toed footprints. Clearly, because of the biped's height and having three toes, this beast was a Quinkan.

With Quinkan, height estimates for these megafaunal beasts are often estimated at about three metres and in a few cases, even taller. Such reports will frequently mention an overall hair colour that is predominantly white, grey, or having streaks of lighter colouring, that is suggestive of the animal's seniority and a pituitary gland that won't give up! Such estimates are usually made shortly after an encounter and some might suggest that these measurements are the dubious outcome from a *heightened* hormonal response. However, having interviewed many witnesses over the decades, with some being repeat interviews by myself or other followup researchers like Paul Cropper, there is a tendency for these height estimates to diminish rapidly after the event. I think that most witnesses tend to experience very tough times, where the only worst admission to seeing a Yowie, is to see one that is ten feet tall.[42] Something tends to give, so height drops to a more socially acceptable level, while the remaining and undeniable truth remains intact. In a worst-case scenario, a witness may totally renounce the encounter, especially when not provided with early support and counselling.[43] Although Bill O'Chee originally estimated the height of the 'marsupial hominoid' at three metres or almost ten feet, it was 'more accurately determined' by a few others later in that day at about eight feet or 2.4 metres and was similarly revised by the Senator during his filmed interview. Considering Senator O'Chee's high public profile, it obviously took a lot of courage to speak the truth and to know in advance the inevitable stigma that he would face.

ACCORDING TO MANY Aboriginal oral histories, Quinkan were the most feared of the 'anthropic' megafauna and for good reasons. The mother of Kate, a young Aboriginal woman from Kempsey that I taught, told me an interesting anecdote involving a long history of indigenous cedar

---

[42] During a preliminary interview with a Texan Bigfoot Blogger, he abruptly terminated the conversation and 'hung up' after I mentioned that some of the Quinkan megafauna are reported to be ten feet tall! This proves that the spectrum of narrow-mindedness is much broader than previously believed.

[43] At the time of writing this chapter, it was winter, so we bought a few tonnes of Ironbark firewood for quality heating. The truck driver who was delivering the load was a white-haired sexagenarian with experience and attitude that reminded me of Ian. After asking the 'standard question', he mentioned several Hairy Man-related incidents but most significantly, he told me that he used to be a member of the Rural Fire Service. One afternoon when working in the Megalong Valley, his Fire Captain went into a side valley and was gone for far too long. When he returned to the fire truck, he looked highly disturbed and agitated. Despite repeated attempts to find out what had happened in the valley, the Captain refused to disclose anything to his firefighters. The truck driver said that whatever happened, it had to be very serious!

timber cutters from the mid-northern coast of NSW disappearing without a trace over a long period. She said that the indigenous community had used the services of Aboriginal trackers to find the missing men but nothing had been found on each occasion. She said that, by not finding the men, the incidents were very sensitive and shameful to many in the Aboriginal community who kept the details quiet. It was commonly believed that a Quinkan in the rainforest was responsible for the disappearance of the experienced timber cutters.

Considering the measured physical characteristics obtained by the O'Connors and from many Aboriginal and other witness testimonies, it is not difficult to understand why they were feared. At a height of eight foot and four inches, they are very tall and intimidating, however, there have been some Aboriginal and non-Indigenous reports of very much taller individuals.

There were also more than a hundred treebites in the area surrounding the O'Connor residence that gave upper and lower canine separations of 82 mm and 55 mm, which are significantly greater than those produced by our local Dooligahl, at 60 mm and 55 mm. The savage damage that was consistently inflicted on many of the hardwood trees clearly shows how powerful their jaws are and the vision that Sue had of a "rounded clump of tan coloured hair on top of its head" confirms other observations and Aboriginal Art of the presence of a sagittal ridge that is used to power this formidable ability. The feet were three-toed and very long at 380 mm and 200 mm wide at the forefoot but also very slender along the mid to hindfoot at 90 mm.

There were also aggressive behaviours that Jerry experienced on many occasions where, for example, the biped showed hostility by roaring at his presence. The food bucket was forcefully belted into the valley, plants were

10: Quinkan

uprooted, branches were broken and other actions. This raised aggression was similarly experienced by other Quinkan witnesses throughout the Blue Mountains.

With hindsight, there have been a few recognisable Quinkan-like individuals within the Blue Mountains who made an impression on residents across the decades. Clearly, the O'Connors' experiences were outstanding because they were able to identify and measure a number of specific Quinkan characteristics by studying them. There was also the case from Singles Ridge Road, Springwood, where the large and heavy Quinkan was taller than the eight-foot-plus-high raised window sill and left a large castable footprint that had three toes. Just south of this, in the Sassafras Gully and Magdala Creek area, was an individual that was repeatedly reported to the Octopus, where the dominant

Fig. 10.9 Jerry O'Connor measuring the lower canine separation on a Quinkan treebite. (Photo: Sue O'Connor)

characteristics were size, power and particularly, aggression. There were many other of these local incidents reported to the Octopus; they were mostly very similar where the person believed that they were being hunted from the side by something that was big, powerful and had attitude.

Although it cannot be verified at this time, it seems that Quinkan, possibly due to their size or ecological preference, are restricted in their distribution to the remnant wooded areas of the Great Dividing Range and elsewhere in the Gondwanan Rainforests of Eastern Australia. Unlike other quantifiable and verifiable information obtained throughout our combined research, this notion remains speculative. It is based on the apparent absence of confirmed Quinkan reports outside of the forested areas of Australia and early suggestions that the forested woodlands were not permanently or as heavily populated by Aborigines in previous times, possibly due to fears of predation. This topic would benefit from Indigenous research and is certain to generate discussion.

Since the two largest Wildmen share some similar traits and behaviours, they show signs of some genetic relationship, however, the principal differences between the two are related to their height, weight, foot morphology and their aggressive temperaments. This leaves the third Australian 'marsupial anthropoid', a smaller monkey analogue, with a different set of characteristics.

11: Junjudee

JUNJUDEE ARE THE smallest of the three 'marsupial hominoid' omnivores. Typically, they are about three feet in height or slightly more, with one encounter reporting a height approaching five feet. Junjudee can be considered analogous to monkeys. Junjudee are black-skinned mammals, covered in short, black hair. Their bodies are gracile with a straight back and no tail. The feet and hands each have five digits with black claws. The head is large for the body size and ovate in shape. The head, face and buttocks are covered in what appears to be a fine layer of hair, similar to a horse's coat. The face has a thirteen-millimetre wide brownish-yellow fur annulus surrounding each eye that is highly conspicuous.

Each Aboriginal tribe would have their name for Junjudee, with the appellation in this case having a South Coast, NSW, origin that originally became known to the Octopus from information provided by a tribal Elder from that region. A more generic term used from northern NSW to Northern Queensland is 'Brown Jack'—sometimes 'Little Fellas' is applied as a colloquial alternative.

Almost from the beginning, we received a few reports from Aboriginal Elders, local residents, friends and others that suggested, in addition to large Wildmen, there could be small monkey-like bipeds as well. A very early incident that occurred before our encounter with Fatfoot, involved a neighbour who saw a short 'gnome-like' animal pressing its face and cupping its hands against the glass of her downstairs sliding door when she turned on the lights. When we were later told of this sighting, after becoming Yowie-aware, we misinterpreted it. We initially believed that the small biped was a juvenile Dooligahl, probably one of Fatfoot's offspring[1] and not a different 'marsupial hominoid' type. However, it could have been something else because we knew little; our initial interpretation remains ambivalent.

Occasionally, the Octopus would report a sighting from out of our immediate area, where a small 'marsupial hominoid' had been seen climbing down a tree. The first one of these reports involved German tourists who were walking the very challenging, forty-four kilometre 'Six Foot Track',[2] from Katoomba to the Jenolan Caves. They saw a 'small monkey-like animal' climb down out of a tree and run away. When interviewed, the witness stated, "I didn't know that Australia had monkeys!" Similarly, Miklos, the late Director of Photography of the ill-fated Hairy Man documentary that we were working on in 2010, was inspired to work on the movie after having seen a small monkey-like animal climb down out of a tree in rural Victoria whilst camping with his girlfriend. At that time, these reports seemed to be limited in number and most were restricted to the Western Slopes and Plains of the Blue Mountains, west of Katoomba and generally

---

[1] At this early period, we did not know that Fatfoot was a marsupial.

[2] Alum Creek Camping Ground is the half-way, overnight stay for the two-day, hard-grade hiking journey. The following is a Google review (2019) for this camping ground: "Nice campground, but we have strange feeling about the place. More than one person told me that is something strange during the night on the campground and they will not comeback to sleep there. Anyone out there have strange feeling or encounter in Alum Creek?"

occurring around the farming areas of Lithgow and Mudgee. Some of these reports were very different from what we were familiar with, involving incidents where a 'mischief' of individuals was conspicuously engaged in the disturbance, looking through windows and climbing as a group onto the roof of the dwelling. The most common physical indicator of their presence was said to be their 'small, child-like, five-toed footprints'. When we saw alleged photographs of Junjudee tracks, they looked very much like small, child-like, five-toed footprints and that is what we tended to think that they were! The obvious point of concern was, why would a typical, contemporary parent allow their child to run around without shoes in the bush, considering the high risk of snake and leach bites, together with the chance of contracting various highly contagious bacterial swamp infections? Also, we were very familiar with four-toed prints, while others were convinced that Australian Wildmen were all five-toed primates.

There were many other reported behaviours that were regularly mentioned in encounter reports by others from different regions, that we never experienced with Fatfoot or her family. Similarly, Jerry and Sue O'Connor never mentioned these particular sets of behaviours either, involving their Quinkan. For example, over a decade of frequently reported 'wood knocking' behaviour had us mystified, though witnesses from other parts of Eastern Australia frequently spoke of it. When witnesses reported hearing 'wood knocking', we thought that they were referring to what we had generally experienced—the violent breaking of large-diameter branches and the occasional striking of neighbouring trees with the timber piece. However, we were mistaken—those witnesses were referring to a completely different behaviour. There were other behaviours that didn't initially fit with our localised experience and our nascent and narrow paradigm.

ANOTHER CONTRASTING INFLUENCE on our collective Hairy Man research over a long period was the encounters experienced by the Pendlebury family. In the beginning, their experiences didn't seem to be much different from other reports. From early 1994, their daughter Zoe was attending the same preschool as our daughter Avril and the two were very good friends. During conversations with Lynn, it was apparent that something was conspicuously visiting Zoe's bedroom window most nights and keeping her awake. This behaviour was very similar to Drew's experiences at our house with Fatfoot—it was becoming a common theme throughout the mountains and elsewhere as reported to the Octopus. The immediate suspect was a Dooligahl. Many reports were received by the Octopus on a regular basis, with a few clearly involving three-toed Quinkan as well—for example, the encounters from Hawkesbury Road. Consequently, this behaviour seemed to be common to, at least, Dooligahl and Quinkan.

There is a close association between children and the three 'marsupial hominoids', but Junjudee seem to have the creepiest relationship. Why this is the situation is unknown but it may simply be associated with their closely related size and form. They have the disturbing habit of observing children at the closest distance that they can approach, regardless of the time of day and risk of being seen. Occasionally, Junjudee will tap on a window or make similar noises to gain attention.

Roaring was also experienced by the family. According to Lynn, they never had any advanced warning of this. Sometimes they would simply walk outside after dark and something in the distance would loudly roar from the scrub at the back of their property. Sometimes they could provoke a roaring response by shining a light around. With the O'Connors, Jerry was roared at several times from the back of their property. His very tall stature was probably easily recognised by their Quinkan because Sue never had that experience. For us, Fatfoot or one of her joeys only roared once. That occurred when I found Fatfoot hiding in the swamp using a bright torch. On that occasion, Fatfoot stood up from behind the flimsy bush that was separating us and roared directly into my face. Fatfoot never roared at Ian, Sandy or Cheryl.

However, there were some conspicuously different behaviours. Something that seemed to be unique to the Pendleburys' situation was the damage that was being inflicted nightly on their trees, particularly the large stringybark eucalypts. Each morning after waking, the first task for Lynn and Gordon was to inspect the area at the front of the house for damage. Typically there would be several newly-ripped trees, where the outer fibrous bark had been pulled upward from the base and torn away from the phloem or inner bark layer to a height of about six to eight feet. Lynn described the damage as "looking like they had been shot with a cannon". Other smooth-barked trees like *Angophora*, were also damaged, having deep X- and V-patterned marks cut into their outer layer, which were typically about five or six feet from the ground. This damage was undoubtedly caused by powerful hands with robust and sharp claws. We positively knew that the intruders had grasping hands because Lynn had been deliberately monitoring the security of the chicken shed, with the prevention of a devastating animal loss in mind. The chicken shed had been purposefully and expensively built to resist most feral, if not all, animal attacks, with heavy-gauge side wire embedded beneath the soil surface, heavy-duty hinge and lock fittings and twin outer and inner doors. The sliding bolts on both doors were sufficiently intricate to thwart any known predator—Lynn had placed a twenty-kilogram bag of chicken feed against the inner door to make matters even more difficult. On inspection one morning, Lynn found that the outer door had been released but somehow it was still closed. The heavy bag of food, however, had been carried away and placed against the sidewall. The chickens remained unharmed.

Dealing with this type of harassment can be very difficult to cope with and resolve successfully. For Gordon and Lynn, the best way to achieve a rapid and successful outcome was to hire professional security specialists. We also lent them the Russian night vision binoculars and some VOR audio tape recorders for their use, plus a long-play surveillance VHS video recorder that Ian had been given by his mate Jacko,[3] a witness formerly in the SAS. Later, I was invited to their house to view the installation of the surveillance equipment and was very impressed with the operation, despite its inherent limitations. In those days, effective surveillance technology was still in development and very expensive. The unfortunate thing about the apparatus being installed was that it was an active system, a methodology that we had consciously avoided using across all areas of detection and observation because the subject is aware of the mechanism's operation and consequently, the data obtained are inherently unreliable and affected in some way.

At the centre of the system was a five-hundred-watt, wide-spectrum infrared illuminator which was not dissimilar to an oversized filament light bulb with a rear

---

[3] Jacko was a former soldier in the Australian Special Air Service Regiment (SAS). I met Jacko briefly with Ian. Jacko was a large man who didn't say very much but he wished us the best of luck trying to obtain video evidence of Fatfoot. According to Jacko, he came across a Quinkan when on military manoeuvres in Queensland. He said that the biped was around ten feet tall and could have easily ripped him apart. Ian also said that Jacko made the mistake of telling the other soldiers in his squad. Apart from his overseas military service, Jacko was clearly suffering from PTSD.

reflector. The radiated infrared light was directed towards the front of the house where the majority of tree damage had been located, with the bulb mounted on top of the house roof on a short mast. At a distance, the IR light was highly conspicuous and looked like a medium-sized red ball in the dark. I can only imagine that for a night-vision-capable animal, with greater sensitivity and eyesight that is probably shifted into the infrared as well, the ball must have appeared exceedingly bright, together with the strangely illuminated front yard.

With the surveillance equipment operational, the attacks on the trees at the front of the house ceased. Shortly afterwards Gordon phoned me asking for my assistance regarding two reflective eyes seen at a distance on a recording. To make the task easier and more precise, Gordon obtained a pair of two-way radios to guide me towards the location of the eyes as seen on the video monitor. We were both surprised at how far away the eyes appeared to be.

After exiting their property through the front fence, I was guided by radio across the road and into a stormwater drainage ditch on the far side. The depth of the earthen drain steadily increased in fall from about three to four feet along Gordon's property frontage, with this small gradient accounting for the gradual upward movement of the eyes as they travelled in the opposite direction along the trench. At certain points along the ditch, the eyes also appeared to make an additional effort to briefly pop above the crest of the drainage mound, as if taking a sneak peek at the illuminated surroundings from within the safety of the trench. For these reasons, we both thought that the animal was not overly tall and was wholly contained within the drain.

With the IR illuminator at work on the large front section of their land, the damage to the trees had ceased and no other reflective eyes were seen on subsequent surveillance tapes. The damage to the trees now commenced at the back and sides of their house. They both took turns scanning their property with the night vision binoculars but saw nothing. Unlike our land, their property was clear of undergrowth and ideal for this type of open surveillance. However, early one morning at around 2 A.M., after their dogs had started barking and become highly agitated, Lynn went outside with the NV binoculars and walked towards the stables where she heard "an almighty roar, like a tiger's or lion's". She ran back inside.

They then had the IR illuminator and camera turned towards the back of the house, whereupon the behavioural pattern reversed. When I next visited them, I was horrified to see that the trees immediately in front of their main house entrance, which had been previously spared, were now similarly damaged, as if the Wildman was making a bold statement of defiance.

Many years later, Lynn and Gordon were still having ongoing encounters with these animals and we continued to follow their interactions. Their experiences with the local 'marsupial hominoids' were very different from ours, the O'Connors and most other witnesses. At dawn, while walking her dogs along the road not far from their house, Lynn saw a small monkey-like animal run across in front of her at a distance of about twenty metres. She said that it looked like an "overgrown monkey" that was screaming as it ran away on two legs. It was black and about four to five feet tall. As it ran, its back was perfectly straight and it may have had a tail, although she was not very certain because the dawn light was so poor. The dogs chased it a long way down into the valley while it continued to make "a very high-pitched squeal". Sometime after this, while riding their horses, Lynn and Zoe had an encounter a bit further down the road, where a small biped ran across in front of them and down into the valley. Having known the family for about a decade, we knew that their encounter reports were highly reliable and built on lengthy experience.

FOR SOME REASON, we tended to have significant encounters immediately before going on holiday. Much of this could be related to Sandy's inability to sleep before departure because, once again, she was sitting on our verandah having a cigarette when she called me outside during the evening of Saturday, 25 September 2004, to say that she was hearing tapping in the bush about fifty metres to our north. After standing on the verandah for a short while, we could both hear faint tapping. This was highly unusual because we had never heard this type of tapping noise before and activity around our valley had been minimal for the past few years.[4] Perhaps thirty seconds later, we heard the tapping again but coming from a new location, about thirty degrees further to the west. After another pause, we heard the tapping coming from the original location again. It is difficult to remember the exact sequence of tapping heard, but we soon identified three clear sources that were separated from each other by thirty degrees and about fifty metres range.

As we continued to listen, the three sources were converging on our northern boundary, just below the septic tank, with the apparent purpose of the tapping being to communicate position and progress to each other. It was now very obvious that this was the 'wood knocking' that everyone else had been talking about! After about five minutes, the wood knocking ceased in the bush. I collected

---

[4] In February 2002 we first noticed that Fatfoot's nightly behaviours had abruptly stopped. By March 2002, there was still no activity and we suspected that Fatfoot had either died or moved on. This inactivity lasted for about two years, remaining uncertain for many more years after this.

my new digital camera and mobile phone. With Sandy, we started walking down the block on an intercept course and as we walked, I phoned Mike Williams, who was living close enough to us to be able to arrive in time and possibly share the encounter. I asked Mike to leave his car on the street and walk down without a torch. As Sandy and I continued to move down towards the boundary, we had to correct our direction of travel as the source of the sound was now moving away from us, to the west. We arrived at a path that Fatfoot had regularly used as an entry and exit point onto the 'northern track' from our land, below the septic tank. I told Sandy that I was going to enter here, although Sandy strongly suggested that I shouldn't.

I turned on the digital camera. It was a Fujifilm S7000 with video and an inbuilt flash that we had recently bought for our holiday to India. I then entered the bush along the track, while Sandy remained on the relatively cleared side of the bush interface. The immediate response was the very loud sound of the bush being thrashed at multiple locations, some distance ahead of me. It sounded like sticks were being used to beat the undergrowth. I continued to walk along the track, passing a large stringybark on my left. The loud thrashing continued ahead of me but was now clearly resolved into three separate locations, one directly in front, with another on each flank at about forty-five degrees. I walked further in and the thrashing became extremely violent. I remember on several occasions, Sandy yelling out for me to leave the bush. I briefly thought about what Ian would do and just as quickly, decided that charging headlong into the fray was not going to be rewarding or beneficial.[5] By now the two outlying sources of sound had moved further around onto both

---

[5] In similar threatening circumstances like this, Ian would reassuringly say, "It is going to hurt but only for a little while."

opposing flanks, so I fired the camera at them, knowing that Fatfoot would have been blinded by the flash and would have been repelled by the bright light. Nothing changed, although fortunately the new camera was much quicker at recharging and so I fired off several more photographs. I must admit that I was becoming slightly concerned because this encounter was different in many ways, so I thought that I would try loudly shouting at my invisible opponents as I fired the camera. I imagined that this was making a slight difference in their attitude, so I tried lowering my voice to a more testosterone-rich bass tone, with a ton of extra vocal weight. My demonic voice seemed to make an impression on them, or so I thought, as I noticed that Mike Williams was arriving at our house. The three assailants backed away and made a quick escape towards the swamp, as the local dogs began to fire up. We went to see Mike who only noticed that the local dogs had become very vocal. He then gave me some worthwhile advice. If I had selected video instead of photo, I would have recorded the audio component of the encounter. The photos taken only showed close-up shots of the bush and the flash was inferior to what I was familiar with because there was no depth-of-focus exposure.

The next morning we got out of bed and went to the encounter site to see if anything of significance could be found. Just like the Pendlebury's, we found that the large stringybark tree at the entrance to the track had been ripped up in the same manner. This damage had been made sometime later in the night after the aggressors had obviously returned to the site and we had gone to bed. Like the trees that had been damaged at the Pendlebury's front entrance door, this tree ripping seemed to be making some kind of defiant statement.

Not only were we dealing with a new 'marsupial hominoid' in our valley but they were also the same Wild-

Fig. 11.1 Ripped stringybark tree at the track entrance, damaged during the early morning hours after the encounter. (Photo: Neil Frost)

11: Junjudee

men that the Pendleburys had been experiencing further down the mountain since 1994. These monkey-like analogues are known as Junjudee and they have a different set of physical and behavioural attributes. However, there was another important issue to consider. What happened to Fatfoot? If she had died or moved on, then had the Junjudee occupied our valley to fill the vacant niche? Or had Fatfoot been pushed out of the valley by the aggressive numbers of a 'mischief' of Junjudee?

SEVERAL BEHAVIOURAL ASPECTS of Junjudee[6] were becoming clear. Junjudee move and function cooperatively as a group or 'mischief'. Junjudee are practitioners of 'wood knocking' and use this tapping to communicate their identity and location to others in the 'mischief' when visibility is poor and/or when participants are separated by distance. This tapping is undoubtedly used for other communicative purposes. Junjudee rip bark and claw trees to conspicuously display their presence in an area or to proclaim dominance or defiance.[7] Junjudee are very aggressive and seem to be very territorial, capable of challenging some Dooligahl, like Fatfoot, for area dominance.[8] They make various high-pitched vocalisations, such as squealing and from an audio recording made of a 'conversation' between two individuals, they have a very wide, but mostly high-frequency range of communication. Junjudee have

---

[6] Junjudee is an Aboriginal word. Like most Aboriginal words, the singular and plural forms of the word are the same.

[7] Dooligahl and Quinkan break many large trees in their territory as a conspicuous display of their strength to all, marsupial or placental, that may be passing through.

[8] Based solely upon the large size of Fatfoot's footprints found in 1983, one of a few early traits that we could reliably remember, Fatfoot was probably a mature female, at least in her early teens. This is because large animals, such as elephants, tend to become sexually

broad and highly adaptive behaviours, being both arboreal and terrestrial dwellers, as well as being cathemeral or being active both day and night without preference, as the need requires.

Several physical aspects of Junjudee were also well-established at this point. Junjudee are black-skinned, monkey-like analogues. Junjudee are small, mostly three- to four-foot-tall bipeds, with a small number approaching five feet. Junjudee have clawed hands with (probably five) grasping fingers and opposable thumbs. Junjudee have five clawed toes, with small gracile feet. Junjudee stand erect with a straight back but no tail. Junjudee have reflective eyes that are capable of enhanced night vision that function differently from the other two large 'marsupial hominoids'.

WITH JUNJUDEE HAVING a significant presence in the valley after 2004, we were experiencing a new regime of behaviours that were initially generating some confusion. The most significant changes involved the behaviour of the local dogs. Since becoming Yowie-aware, we had learnt the very familiar patterns of standard dog behaviour in the neighbourhood, which allowed us to determine the location and progress of Fatfoot and her joeys along the valley, as well as allowing some reasonable activity predictions to be made, on most occasions. We also knew that Fatfoot was aware of the dog barking being used by us for tracking purposes because she would sometimes travel on

(cont.) mature later in life, compared to most other, smaller animals. By 1993 we knew that she had several joeys. Based upon this assumed minimum starting age, Fatfoot was at least in her early thirties in 2002. By 2022, she should be in her early fifties. Since Fatfoot still foot thumps and responds to the 'light signal' on occasions, she is currently alive. If Fatfoot is similar to the megafaunal African elephant, she might live to the age of seventy.

11: Junjudee

the near side of the valley where there were fewer dogs but also fewer opportunities for foraging and human engagement. This savvy behaviour usually resulted in us being either caught off guard or being misled in some way. Most of Fatfoot's movement was at a slow and steady pace with occasional pauses, possibly to observe human activity,[9] or to forage for local food. This barking of the dogs tended to be more active during the evening and pre-dawn, with the tone and intensity of the disturbance also giving clues to how provocative her encounters were with the dogs.

With the arrival of Junjudee, there was a very conspicuous increase in dog activity during the day, with no obvious temporal preference and occasionally, with several noticeable and rapid passes being made through the valley, in both directions. When the Junjudee were moving through the valley, the pace of the barking relay was much faster and the tone was more aggressive, compared to Fatfoot. This was probably because there were more of them and they didn't seem to care. During the evening, the dog activity seemed less intense and of a shorter duration, although indiscriminate sounds could be heard outside of the swamp, usually dispersing at speed across the sclerophyll landscape. In intuiting this different pattern of behaviour, I strongly suspect that Junjudee, like Dooligahl, prefer to remain hidden from view and so they were required to move through the thick swamp vegetation during the day. In doing so, both Junjudee and Dooligahl are forced into contact with dogs, something that Junjudee

---

[9] Invariably, Fatfoot would stop for lengthy periods behind large trees, or thick bushes. Indicators of this behaviour were flattened grass around the base of the tree and occasionally, a treebite at a height of about six feet on the associated trunk. During the night, red eyes might be seen appearing from behind the trunk depending upon the alignment circumstances. The longest period of time spent doing this was ninety minutes after being provocatively observed from our verandah.

prefer to avoid or minimise. This interpretation appears to be consistent with the early morning encounter with Junjudee by Lynn Pendlebury while walking her dogs on the road. The dogs took after a Junjudee and engaged with it over a long distance as they all ran into the valley below. Dog encounters with Dooligahl and Quinkan, on the other hand, historically suggest that dogs are highly intimidated by these larger 'marsupial hominoids'. Many testimonies by pig hunters tell of encounters with the larger bipeds where their hunting dogs cower, run away, return to their cages on the back of the ute, or simply hide. Ian's anti-personnel dog Toby is another immediate example where a highly capable and trained dog refused to attack, or even corral Fatfoot. Similarly, Ian's cousin Debbie, who lives on a farm near Cowra, NSW, said that her working dogs would sometimes show extreme fear at night and hide under her house. She highlighted this comment by saying that this was extremely uncharacteristic behaviour because her working dogs were a very tough pack that normally showed no fear of anything! At nighttime, it seems that Junjudee are less concerned with being seen and heard, so they openly move about more freely, under the relative cover of darkness.

It is very difficult to say when it happened, but the valley became quiet again for very extended periods after 2006. Like four years earlier, we thought that Fatfoot had died or had moved on. The slow and stealthy movement of Fatfoot through the valley and the distinctive barking response of the dogs were absent. The nightly 'G'day' foot thump at around 2 A.M., followed by me turning the outside driveway lights on and off three times in reply, had ended.[10] As for the Junjudee, we didn't know enough about

---

[10] As can be seen from the Log earlier, Fatfoot was a chronic foot thumper. This behaviour was very annoying during the night, so something had

them at this time to be definitive, but they seemed to be itinerant. To test this notion, I drove several kilometres to higher ground to our west and from an easterly lookout listened for the sound of dogs barking. When overseeing a vista of many square kilometres, a few dogs will certainly be making a commotion somewhere. However, within a very short time, a pack of neighbourhood dogs were identified in a more distant valley that was making a conscientious effort to alert their owners and actively defend their territory. At this time, I was still in possession of my much loved parabolic dish microphone, which allowed me to more clearly identify the distant bark of individual dogs and follow the quick procession of subsequent canine responses along the valley. It seemed quite probable that the Junjudee had moved to another valley, further east of our location.

THE NEXT TRACEABLE activity pattern of Junjudee was located further to our west about three years later. During mid-2009, the Octopus was receiving reports from a family building a house on a remote acreage, who were experiencing problems with an apparent 'mischief' of Junjudee. When I visited them, the two owner-builders, Frank and Dale, were teachers with whom Sandy and I shared many professional contacts, school histories and a strong interest in owner-building. They were constructing a brilliantly designed house made from concrete, that was being built into the side of a hill by the pair. The top concrete storey of the house accessed the road outside on

> to be done about it. The plan was to acknowledge one foot thump by turning the outside driveway lights on and off three times and then ignoring all other thumps. Also, we made the habit not to go outside after a foot thump because the original purpose of thumping was to entice us to come outside and play.

a continuous level, which left no visible, raised building profile that could be seen from the street. The other main living areas were arranged on floors directly beneath the parking area, with the second storey remaining a shell or incomplete stage until the rest of the building could be finished. Rising through the core of the building was a concrete stairwell.

They first became aware of a problem when they started to hear the sound of, what seemed to be several children running around on the concrete floor above them during the middle of the night. When they investigated they couldn't find anyone. Understandably concerned, Frank barricaded the stairwell to prevent whoever was entering the building from above, from running around their upper storey and possibly visiting the other lower levels that the family were occupying.

Their son Rohan would get up early in the morning, typically in the dark or at sunrise, to prepare himself so that his mother Dale could drive them both to school. The truth was revealed when Rohan went outside to the clothesline to collect his frozen-stiff shirt and pants. As he entered the backyard, he saw several small monkey-like animals running around near the clothesline and taking refuge in the scrub after being discovered.

To help the family confirm the identify of the animal, I lent them a game camera that Rohan and I set up on their property boundary where some tracks led to their house. Soon afterwards we recorded a video image of some reflective eyes with the sound of accompanying footfalls, that appeared to be low down on the forest floor. The video reminded me of the eyes seen in the stormwater drain that Gordon and Lynn had captured at their house about sixteen kilometres away a decade earlier. At that time, we thought that it might have been another feral animal because we were not aware of how low to the ground these animals

11: Junjudee 469

can travel when trying to be stealthy. I have observed a Junjudee during the day moving like a commando crawling along the ground, at a height of about eight inches. It was obvious that the animal had detected and recognised the illuminators of the low-lying camera and had not ventured into the open further along the track.

When this video was taken, our knowledge of Junjudee was very limited and consequently, our perspective remained narrow. Over time the possibility that something other than a dog or cat could be responsible became more feasible. We already knew that many Junjudee frequented

Fig. 11.2 Screen shot of probable Junjudee reflective eye (two eyes alternately seen on video), hiding low in the bush on the left of the tree, close to the ground. (Photo: Neil Frost)

the bush in this area at nighttime. We also knew that Junjudee's eyes have a tapetum lucidum. Later on, we found that Junjudee have the habit of crawling very low down to avoid attention. This was a very significant consideration. Also, small dogs and cats are known to be Junjudee prey.[11]

Using the specifications for the camera, the field of view (FoV) was 42°. The maximum detection range is 38 feet (11.6 m). The actual distance to the eyes was measured at 36 feet (11 m), and the distance across the field of view (L) is tan(21) x 36 x 2 = 27.64 feet (8.4 m). Since the eye separation in this instance was 22 pixels and the total width of the image was 2560 pixels, this gives the eye separation a ratio of 1:116 of the field of view. The eye separation, therefore, was 27.64 ÷ 116, or 0.238 feet, which is equal to 72.6 mm. My eye separation is 67 mm.

Unfortunately, shortly after this photograph was taken, the game camera disappeared. The family members and I undertook a detailed search, with family members continuing for weeks afterwards, but the camera could not be found. It had not been secured to anything because it was camouflaged at a low height in some grass in a remote location on private property. As we confirmed a few years later, Junjudee are very mischievous and like to steal objects, particularly anything that smells appealing, such as pheromone packets, or is shiny and reflective, like batteries and much more. Also, Junjudee like to rearrange, damage and destroy surveillance-related objects, such as motion sensors, flood lights, solar panels, cabling and of course, game cameras, which they must find annoying.

---

[11] 2017 was a peak year for missing dogs and cats in our valley. Neighbours Sandy and Michael visited to find out if I had seen their cat recently. Their black-and-white cat was last seen by me in our swamp sunning itself on a vegetation raft at midday while I thought to myself, "This cat is not going to last very long!" It was never seen again. Around this time, numerous 'missing dog' posters were seen on power poles along our streets, typically advertising small, white and fluffy dogs.

11: Junjudee

By late 2010, a large number of Octopus participants, a half dozen researchers, two film producers, a Director of Photography and his crew were working on a feature documentary, with the working title, *The Hairy Man*. As part of the film, a group of these people hiked into a remote Blue Mountains valley and camped there, with the hope of recording some activity. Despite having made a campfire and cooked some aromatic food, nothing happened until about 4 a.m. The only activity that was recorded was the sound of wood knocking coming from the bush to the north of the campsite.

A few days after this, the film crew were undertaking night filming in a neighbouring valley. From experience, unless you are lucky, it can typically take several days or nights to be noticed by any of these 'marsupial hominoids' and for them to respond in some way. A long and tiring night was about to get much worse, as a thunderstorm was starting to roll in. The Director of Photography and I were leaning over a railing away from the others, discussing our plans for retreat ahead of the advancing lightning. During a quiet period in our conversation, I could hear movement in front of us and to our right. A second or two later, the noise became the sound of running and other distracting noises, coming from the dark, about ten metres away. Not wishing to take instant ownership of the incident by imposing my preconceived opinions, I said nothing to Miklos and waited for any response from him. These sounds seemed overly deliberate to me and I am sure that Ian would have agreed. They were the same attention-seeking noises that Fatfoot would similarly employ on occasions, to entice us to play. Fortunately, after a few seconds of processing time, Miklos excitedly asked, "Did you hear that!" There was something in front of us—from the frequency and amplitude of the multiple sounds made, I immediately thought that it was Junjudee. I replied to

Miklos, "I am so glad that you heard that because I wasn't going to say anything unless you spoke first!"

With the storm bearing down on us, we implemented a plan to leave a two-person camera and sound team hidden in the bush, while the rest of us made an overly conspicuous departure back to the cars. The aim was to apply reverse psychology in an attempt to stimulate an anxiety-driven response, by pretending that we were either disinterested or sensorily obtuse. The hope was that it might elicit a similar response to what happened on a previous occasion with Dr. John and Fred, where one of us would conspicuously depart the encounter site, allowing sufficient time for that person to disappear a long way ahead on the track. The next person would depart, usually after five minutes, rendezvousing with the previous person a long way down the track and out of sight. As happened on that occasion, the anxiety that was slowly building within these prankster animals, reached its maximum shortly after I departed, causing one of them to demonstratively give away its location to Dr. John for a second time. Dr. John was very excited about how we had manipulated the circumstances and what we had experienced in response. We called this well-tried technique the 'Bunny Method', because it reminded me of my two uncles who trained greyhounds[12] for dog racing during and after the Second World War. Ian and I had used this technique with Fatfoot and found that generating frustration can be useful, as long as it is conducted with some awareness of the powerful response that

---

[12] I remember that my Uncle Stan would 'fix' dog races. He had an excellent greyhound, Rocket Jet, that won every race that it entered. Uncle Stan took the greyhound down to the Parramatta River and towed the dog behind his outboard dingy, forcing it to swim for a very long time before the race later that night. Stan put a lot of money on the opposition dogs. Rocket Jet won the race. It was a very memorable evening for Uncle Stan and my parents, who had backed the 'sure thing' alternatives.

11: Junjudee

these 'marsupial hominoids' are capable of. Unfortunately, it started to rain and the away team had to abandon their plan and make a lengthy run for the cars in order to protect their equipment as the storm worsened.

The next morning, we returned to the site from the previous night. There was nothing found of any obvious significance, so I returned later in the day towards the sunset with a small Sony VOR digital recorder. I found a downhill track that was very well-worn and near to the encounter site that the Junjudee had most probably used the night before. I spent some time carefully choosing a suitable location, installing and camouflaging the recorder. The location that I chose was a hollow tree stump, about a metre from the ground alongside the steep track. Being a metre off the ground meant that any small animals would not disturb the recording with their scurrying. Anything larger should be clearly recorded by the external stereo microphones, especially if the animal was about a metre or more tall.

The following morning I retrieved the recorder. Clearly and crisply recorded was the sound of a bipedal animal walking and then picking up speed, as it accelerated down the steep track, past the microphones, before stopping. As it stood still, in very close proximity to the microphones, the leaf litter underfoot was variously compressed and expanded, as the biped shifted its weight about. This was followed by an extremely unusual vocalisation, a 1.8-second high-pitched and energetic burst. After a short pause, in the background can be heard a similar, one-second reply (http://tinyurl.com/junjudee). Clearly, the two bipeds were communicating. I had never heard something in the bush with such a high-frequency 'call' before, which was no guarantee of anything, so I asked as many people as I thought might offer some hope in its identification. First on the list was my good friend Gary

Opit, an Australian naturalist who hosts a radio talkback show on the ABC's 2NR. Gary has a very broad knowledge of Australian biota and has had extensive field experience as well. He was unable to identify the animal, even though he was aware of Junjudee and the other 'hominoids'. Similarly, I gave the recording to a university biology professor who was equally discombobulated but regardless, he sought the opinion of others in the field, without success. In any case, interpreting sound recordings is highly subjective. It is like asking an economist for an opinion and getting half a dozen alternative explanations in reply.[13] After all, the only person to fully appreciate the value of a discrepant audio file is the individual who took the first aberrant steps to record it!

BY 2011, THE JUNJUDEE held a major presence in the valley even though there were occasional signs that Fatfoot was still around. It almost seemed that the two 'marsupial hominoid' types had made a time-sharing agreement that permitted the dual occupancy of the valley system, with Junjudee being mainly active during the day and Fatfoot maintaining a reduced dominance during the night. However, there seemed to be occasional breakdowns with this armistice, as when Fatfoot deliberately called for my attention at 11 A.M. by repeatedly foot thumping from the bush to the north, while I was working in the garden.[14] Similarly, Junjudee were occasionally heard moving about during the night. We weren't totally sure, but there seemed

---

[13] After Anthropology, Economics was my second major at university. This was the standard self-deprecating joke used in the profession. Similarly, Anthropology was defined by many as, "The study of stories that native people tell anthropologists."

[14] After determining the origin of the foot thumping, I went for the obligatory walk to investigate the area, not expecting to find anything.

11: Junjudee

to be some sort of territorial battle underway during this time.

To increase our opportunities at photographing Junjudee, we bought another game camera and a bulk quantity of 'D' cell batteries in order to boost the photographic performance. The plan had been to use a quantity of block human and gorilla pheromones as bait[15] and set up the camera nearby. Sandy wouldn't allow me to keep the pheromones inside the house, so I placed it on the verandah table, together with eight of the 'D' cells. The next morning the pheromone block and eight batteries were missing. After a short search, we started to find a trail of bright silver batteries leading across the garden and into the neighbouring bush to the north.

After almost two decades, we were very accustomed to Fatfoot's *modus operandi*. In addition, Fatfoot had some very specific habits that she had developed during our time spent together. The most conspicuous of these was her foot thumping behaviour. As can be seen from the Log, Fatfoot was a prolific foot thumper. Her main objective in doing this was to gain our attention and hope that we would come outside to play. When we started to become resistant to this, Fatfoot's response would be to increase the frequency and volume of thumping—sometimes, through frustration, she would do other things, like throw objects at the house. Whenever someone turned on a light inside the house, Fatfoot would respond after about five seconds, usually by foot thumping and moving closer to the house. Since we couldn't allow this annoying nightly behaviour to continue, we devised a plan that would acknowledge her attention-seeking behaviour, whilst also attempting to limit it.

---

[15] Andrew, another researcher had given me a human-gorilla pheromone block that is commonly used by North American researchers.

We can't remember when we first started conditioning Fatfoot, but it was probably during the latter part of 1993 or maybe early 1994. Whenever we heard foot thumping, or when we had decided that we had had enough, I would turn the outside driveway light on and off three times. Whatever Fatfoot did after this, we would completely ignore! By following this procedure, the projected message was, "We have heard your foot thumping!" By not responding after this, particularly by not opening any external doors, or shining torches about in the bush, the subsequently implied message was, "No one is coming outside to play!" During the following decades, however, both parties seem to have allowed the meaning of the message to evolve. I tend to regard the new meaning to be "Hello", which is also my intended reply using the driveway light signal, particularly after a long absence by Fatfoot. Since Fatfoot occasionally spent long periods away from the valley, it usually took a number of progressively louder foot thumps to reestablish communication, after which Sandy and I would usually affirm, "Fatfoot is back!"

However, we were not alone in this 'marsupial hominoid' recognition process. Apart from us, the other main identifiers were the dogs, which proclaimed their verdicts through their highly individualised barking behaviours. The barking of the dogs was significantly different for each type. With Fatfoot, the dogs mostly barked during the evening with the canid relay making slow progress along the valley. With the Junjudee, the barking tended to be more frenzied, moving rapidly across a broader property front. During the day the dog barking also made rapid progress through the valley but occurred at random times throughout the day. Sometimes we heard the relatively undisguised movement of multiple Junjudee through the bush, with the dogs responding in the appropriate direction towards the 'mischief'.

11: Junjudee

There were also other currently known behaviours that allowed us to reliably differentiate between our two local 'marsupial hominoids', such as how each species responds to light intensity, communication by foot thumping or wood knocking and others. Additionally, there were physical trait differences, such as height, body mass, number of toes, foot size and more. However, as our personal experience broadened to include the newest 'marsupial hominoid' arrival, more differences would be discovered.

OVER THE DECADES, the Octopus received occasional reports from outside areas containing elements we didn't know how to respond to. During the early 1990s, a few active areas mentioned sticks being found that seemed to have been pushed into the ground. Like the 'wood knocking' phenomenon, 'stick planting' was another element that we did not have any direct experience with. Our awareness of these behaviours changed after Junjudee passed through our valley.

The common opinion, or the conscience of inexpungible scepticism, was that these embedded sticks were the natural result of small branches randomly falling from trees and impaling the soil on impact. However, on inspecting the sticks and establishing that they deeply penetrated the compacted soil, it was intuitively obvious that this was the consequence of being forcefully pushed, rather than the aftermath of a dart-like free fall under gravity. This is not to say that on a few occasions a stick may not temporarily attain a sustainable vertical landing—but no significant soil penetration would be achieved and it would be inherently unstable. From the few examples that we had, the minimum soil penetration was forty millimetres, ranging up to seventy-five millimetres. Replacing the stick in the original hole to the same minimum depth required several kilograms of 'weight' (approximately

Fig. 11.3 'Stick plant' found on the morning after the grass was mowed, opposite the neighbour's playground equipment on the other side of the hedge, 23 September 2011. (Photo: Neil Frost)

20 N)[16] and when creating a new hole alongside, the required force was about double that (approximately 40 N).

Each stick was about a metre in length, with the longest being one and a half metres. The sticks were the same as any overhead stick that might have recently fallen from a hardwood tree, being typically Blue Mountains Ash *(Eucalyptus oreades)*, or Tallowwood *(Eucalyptus microcorys)*. The butt end of the sticks had typical diameters ranging from 8 mm to 14 mm, which tapered down along their length. Since these sticks were dead when they were shed, they were air dried with a probable moisture content of about 15% to 20%. These dried hardwood sticks, with a butt diameter of 10 mm and a length of 1 metre, had a typical mass of about 10 grams/m or 0.01 kg/m.

Assuming that (1) a stick begins its free fall under gravity from its maximum attainable height of 40 metres; (2) that there is no friction from air drag slowing down its acceleration; and (3) that the branch falls vertically like a dart: the kinetic energy of the stick at the point of ground impact is calculated as $E=mGh$.

Substituting into the equation:

$E = 0.01$ kg $\times 10 \times 40$; $E = 4$ N

Therefore; a stick falling from a height of 40 metres in a vacuum would only produce 4 N of force, or 10% of the 40 Newtons of kinetic energy required to achieve the required soil penetration. Alternatively, the stick would need to be dropped from a height of 400 metres with the same perfect preconditions, in order to acquire enough kinetic energy for soil penetration to this minimum depth. Therefore, it can be confidently concluded that these sticks have been pushed into the ground.

---

[16] The force required to push a stick into damp compacted soil is calculated using Newton's second law of motion, $F=ma$. Where m = 4 and G = 10 ms^{-2} (rounded up, actual is 9.81 ms^{-2}) and E = 40 Newtons.

Fig. 11.4 Eight mm-diameter 'stick plant', 49 mm to 61 mm ground penetration. (Photo: Neil Frost)

11: Junjudee

The photograph in Fig. 11.4 shows an 8 mm diameter stick after its removal from wet soil. My thumb and forefinger were tightly grasped around the neck of the stick at or slightly above ground level, but lower than the height of the cut grass, acting as a depth gauge. After removal, the hand and stick were photographed together, without alteration and the stick's diameter was measured.

> The length of my thumbnail was 17 mm, or the equivalent of 300 pixels from a screen image.
> The total length of the buried stick butt, to soil height, was measured at 1075 pixels. 1075 ÷ 300 = 3.58 x 17 = 60.9 mm.
> The partial length, up to the wet line on the stick was measured at 865 pixels. 865 ÷ 300 = 2.88 x 17 = 49 mm.
> The diameter of the stick was confirmed at 139 pixels. 139 ÷ 300 = 0.46 x 17 = 7.9 mm

Conclusion: The 8 mm diameter stick was pushed to a total depth of 61 mm. However, the top 12 mm of the soil profile was relatively soft humus. Ignoring any resistance provided by the humus layer, gives a discounted and highly reputable soil penetration depth in firm soil of 49 mm and physical confirmation that this stick had been pushed into the ground.

DURING THE DECADES that Fatfoot had a solitary presence in the area, we never experienced 'wood knocking', or found evidence of 'stick planting'. When Fatfoot wanted to communicate with us, she mostly foot thumped—during the earlier years, she occasionally left territorial markers, like broken trees, as reminders of her powerful presence. Also, Fatfoot never caused any significant problems. She would have taken food to be sure, but she certainly wasn't malicious in any way that required us to take remedial action.

From spring 2011, we started to cultivate a large vegetable garden on a cleared section of our land, using the tractor and the rotary hoe attachment. We knew that doing this would generate a number of additional problems, as food production creates fresh opportunities for existing and newly-attracted animals. Our first task was fencing and keeping wallabies out of a garden is always difficult.

In order to assist with managing the plot, we installed a number of devices. We placed four motion-sensing, solar-powered, PIR lights along the fence perimeter and in the fruit orchard. Each unit had two independently targetable lights. One light would be directed towards the crops and the other was used to deter the wallabies by pointing it into the adjacent bush. The orchard lights were shone along the rows of trees and vines. For passive protection, a large plastic bird of prey was placed within the fenced crop area.

From the beginning of the growing season we noticed an increasing number of anomalies. After several nights, we observed from the house that some of the lights had ceased to function, despite the obvious presence of the 'wallabies'. At first we found that the solar panels were facing different directions, compared to how they had been originally installed. The batteries were undercharged because they were receiving insufficient voltage from the poorly aligned panels. Since most of the PIR lights were attached to tree trunks, we thought that possums must have been pushing past the solar panels, altering the orientation. In response, the solar panels were relocated away from the trunk where possible or the adjustment-locking rings were tightened very firmly and monitored each morning. Incredibly, after several more nights the light failures recommenced, again because of battery power loss and despite our close monitoring of the panel orientation. We then discovered that the power cables from the panels to the main unit had been either completely or partially unplugged at one end, the other, or at both. After inspecting the solar lights in the orchard, we discovered that each of the four lights was now facing down towards the base of their supporting posts, which greatly restricted their range and effectiveness. As these incidents now seemed very deliberate, we placed VOR audio recorders in concealed locations next to the main lights. On replaying the recordings, we could hear disturbances in the adjacent bush, walking and manual interference with the lights. Perhaps more interesting, the recordings also detected that the plastic owl amongst the crops was being tested for signs of life! On one recording, a hollow sound could be heard, consistent with the plastic owl being repeatedly struck with a small rock, followed by the muted sound of each rock hitting and then rebounding off the ground. Several small rocks were found at the base of the post which was supporting the owl. The time interval between the two recorded sounds (0.5 seconds) was the same as the fall time under gravity from the plastic owl that was standing four feet or 1.2 metres above the ground.[17]

Finally, after securely fixing the PIR lights in the orchard to their required positions, a few mornings later we found that a small stick had been inserted through the front panel of one of the PIR sensors and moved about, completely destroying the underlying fresnel lens, preventing that light's activation. The other PIR sensor remained undamaged. Although we could not be certain, either the Junjudee were more intelligent than we thought or they had been observing our repair attempts during the

---

[17] $(2 \times 1.2 \text{ m}) \div (9.8 \text{ m/s}^2) = 2.4 \div 9.8 \text{ s}^2 = 0.245 \text{ s}^2) = 0.5 \text{ seconds}$

day, which might have given them some insight into the critical operation of the lights. Regardless, their guerrilla insurgency was truly amazing!

Around this time, other strange things were happening. Sticks began to appear in grassed sections of our land that were standing vertically and unsupported. It is difficult to say when this first started, because we didn't pay a great deal of attention to them, prior to mid-2011. It would seem that the vegetable garden was encouraging regular visits by the Junjudee and while they were here, they took a look around.

As shown in the photograph of 7 April 2012, a stick can be seen planted in our grassed area immediately adjacent to our neighbour's playground. When the stick was removed, a replacement would appear after a few nights. Our neighbour's daughter Rosemary would play in this area most afternoons. There were swings of various types, a slippery dip and a sand pit. Also at this time, a baby boy named Oliver was born.

Across our valley, another baby could be heard crying most evenings, shortly after sundown. Following these cries, the dogs would start barking in relay up the valley, which would terminate in the swamp directly opposite us. In response to the crying from across the valley, I walked down to the edge of the swamp and sat quietly in the dark to listen. It was very disturbing listening to the baby cry and also hearing multiple movements in the vegetation near the rear boundary of the house. I repeated this surveillance on a number of occasions over the course of a week, until the circumstances changed. These incidents made me reflect upon the advice and warnings given to us by Aboriginal Elder Merve, back in 1993.

About five months after these incidents, our valley was burnt as part of a wider bush fire hazard reduction. The burn off was welcome because it was long overdue, as we

Fig. 11.5 A 'stick plant' opposite neighbour's play equipment, 7 April 2012. (Photo: Neil Frost)

had fuel load estimates that placed us in the 'catastrophic' fire category. At the same time we knew that this reduction in vegetation cover would most likely affect Fatfoot's behaviour, because she disliked any loss in vegetation cover and this was going to make it extremely difficult for her to move about undetected.

As we predicted, visits by Fatfoot to the valley, day and night, ceased because of the lack of cover. The dogs stopped barking in their predictable manner, there was no foot thumping, nor any sound of heavy walking and branches breaking. The Junjudee, after a short pause, seemed to continue on during the night, perhaps moving a little faster than what they did before, but they were missing during the day.

Both of our children had finished school three or four years earlier and were completing their university studies. To support themselves, Avril and Drew had completed higher certificates in the Responsible Service of Alcohol and Responsible Conduct of Gambling, resulting in both working as bar attendants and gambling cashiers at a distant sporting club. As young casual employees, they typically received the undesirable late night shifts which finished during the early morning hours and then additionally, required a long drive home. They would typically arrive home after 2 A.M., so we would leave the outside driveway light on for security and to prevent any sudden nighttime surprises, as was its original installation purpose. Meanwhile, as parents, Sandy and I would instantly wake when the front door knob was turned, signalling their safe homeward arrival. On a few occasions, Sandy and I woke to the noise of the door knob being turned multiple times, followed by no accompanying sound of the front door being unlocked and opened. On inspection, I found that no one had arrived home! Apart from incidents that occurred at the driveway, of which Avril had the most number, this unnerving experience resulted in the purchase of a PIR sensor light bulb, which was installed at the front door and left permanently on. For both of them, having this automatically-activated light dispelled a lot of apprehensions. Although Fatfoot had a history of opening a door and entering a house, as happened to Wilhelmina after she had lost her child's yellow jumpsuit in 1993, it would have been very unlikely for Fatfoot to move past the driveway light blocking the main entrance because of the high sensitivity of Dooligahl to bright light. However, as we discovered during our earlier encounter with the three wood-knocking Junjudee, and those at the Pendlebury property, they were not intimidated or seemingly affected by the flash of the camera, although they prefer not to be surrounded by white or infrared light.

Following the bush fire hazard reduction burn, with the absence of the grassland, there were no wallabies present in the valley for a long time afterwards. We missed seeing the arrival of mating couples to our grassed backyard and watching the progress of joeys as they first began to peek out of their mother's marsupium until they finally became independent in their fight to survive. Over the following year, the green slowly began to dominate, with grass and many weeds beginning to establish themselves. It was interesting to watch the new ecological succession of the valley flora after the dice were cast. The old species of fern, in particular, were replaced by different varieties that are smaller and less dense.

Unfortunately, during this period of denudation, the Great Western Highway authorities and our Local Council were undergoing the final stages of determining how to dispose of the very large quantity of sheet water flow from the highway's hard surfaces. Regrettably, this greatly enhanced volume of mostly unchecked water was channelled down our valley system, which after one heavy storm, stripped our 'hanging swamp' away from the bedrock, sharing the debris across the many rivulets on their way to the Grose River below.[18] We estimated that from our property alone, about ten thousand tonnes of humus and detritus-laden soil was washed away.

The expert wisdom that was provided to us recommended that the many listed noxious weeds that were reproductively very-well established, should remain in place as they offered a stabilising influence over the unprotected soil. However, my elderly foresight predicted that a time of blind reckoning would descend upon us, where the governing authority would demand the removal of such weeds. This would be very costly and labour intensive. Consequently, through early intervention and a combination of manual and chemical weeding, we removed all noxious plants. Instead, we planted many indigenous trees along the newly meandering water course, even though we were advised that this was a waste of time and effort. The other neighbours complied with the authority's demand for inaction. Specifically, this has resulted in very large and dense blackberry patches being established in other parts of the swamp, which have proved to be highly attractive to both Junjudee and Dooligahl when in season.

WITH THE ARRIVAL of several new neighbours to our north in 2016, we wondered how long it would take before they had an encounter that would require some therapy. It seemed that the Junjudee's activity was picking up in intensity and some evenings they could be heard rapidly moving through the bush as a 'mischief'.

---

[18] Interestingly, our council has reclassified our wetland. It is no longer an environmentally-prised 'hanging swamp', but a 'natural watercourse'. Solves a lot of problems!

11: Junjudee

I received a phone call from Mike Williams asking me if I had the time to assist a young cadet journalist with an assignment. Mike said that the assessment task required the cadet to interview someone who held peripheral views on an extreme topic, like UAPs, black panther sightings, the Loch Ness monster and similar areas of controversy, with the aim of writing a newspaper article based upon the interview. As a former teacher, I felt an irrational urge to assist. More importantly, most of us regularly complained about our perception and treatment by the media, particularly when the instrumental track 'Materia Primoris' from the parent album *The Truth and the Light: Music from the X-Files,* is played in the background during an interview, signalling to a naïve and inherently sceptical public that you are barking mad, just in case the tone of the interview or the raising of eyebrows was insufficient to convey this perceptual bias! This was my opportunity to reeducate a journalist before narrow-minded peer pressure could turn him to the dark side.

The cub journalist had to travel to our home from suburban Sydney by train, so I arranged to pick him up from the station. He opened the passenger door, slammed it closed and said "Hi". For a potential journalist, he wasn't very talkative during the return drive. To start the interview, I sat him down at our kitchen table. He took out an audio recorder, pressed record and placed it on the table. I looked at the recorder. Then I looked at him. Then I looked at the recorder again, before having to explain the polite and ethical procedures involving recorder and photographic etiquette.

After two hours, I think that I had given a detailed and rational account of our research and tried to emphasise the importance of empathy when dealing with witnesses. I asked for a copy of the article, but I am still waiting. We walked outside in the dark to where the ute was parked on the driveway and fortuitously, a Junjudee raced across in front of the vehicle and ran off into the bush. After providing a sufficient pause for processing, I asked the cub if he saw what the animal was. He replied, "It was a kangaroo!"[19] I drove him to the station, he said "Thanks" and he slammed the door!

With the increased Junjudee activity in the valley it was only a matter of time before something happened. However, Fatfoot was still around and we imagined that she would have been very keen to take a close look at the new inhabitants, as well.

Around this time I bought a 12-inch (305 mm objective) Newtonian telescope on a Dobsonian mount to enhance my casual sky viewing. It was an interim purchase until I designed and built my retirement observatory.[20] It was a large reflecting telescope with 1149 times the light-collecting power of a healthy human eye with a fully-dilated pupil of 9 mm. The mirror was figured to F4 which, as in a camera lens, is fast and allows for the viewing of faint astronomical objects across a wide field of view. Most importantly, the simple altazimuth design makes the telescope highly portable and does not require polar alignment, allowing the telescope to be quickly set up anywhere, despite the considerable weight of the optical tube assembly.

---

[19] Contrary to common belief, kangaroos like the eastern grey are not prevalent throughout the entire Blue Mountains, being mostly found in the large grassed valleys, like the Megalong to the west. Other kangaroos, like the red, western grey, antilopine wallaroo and the black wallaroo are found elsewhere in Australia. The smaller rock and swamp wallabies, plus the slightly larger eastern wallaroo are more suited to the environment and steep terrain of the Blue Mountains.

[20] I also intend to install a repurposed 800 mm satellite dish as a partially-fixed parabolic microphone, capable of listening to movement a long way down the valley. I completed the 'Dooligahl Observatory' in 2017.

11: Junjudee

Usually I would set up the telescope on the path at the front entrance to our house. This position was advantageous because the hard surface made the telescope stable and it was close to an electrical outlet. It was very easy to place the telescope onto the verandah at the conclusion of observation, without requiring a lengthy pack-up time.

After spending many evenings observing, it became apparent that there was another advantage. Around the front entrance were a number of topiary bushes that provided thick and high cover. Particularly during the later part of the evening, it became possible for me to sit or stand at the eyepiece for hours, scanning the firmament but also paying attention to the increasing activity that was happening in the bush nearby. This was much more interesting compared to the usual alternative of just standing around and waiting for the activity to happen. On several occasions after midnight, as the last of the lights in the valley were going out, less inhibited movement was increasingly heard on a neighbour's land. During one incident, it seemed certain that a number of Junjudee were looking through the neighbour's windows on the southern side of the house. Unlike Fatfoot, Junjudee seem to be risk takers and are less concerned with being discovered.

Fortunately, one of the new neighbours reported an incident that occurred nearby, which was very interesting. He said that another new resident had heard a suspicious noise one evening at the back door. On opening the door, the person was confronted by a very large figure that was a few inches taller than the witness, standing on the covered wooden deck, a short distance from the verandah steps. When attempting to confirm the details of the encounter a few weeks later, the witness revised the story by saying that the large figure was a mature male wallaroo.

Confirming the details of some witness encounters is occasionally problematic if left for too long. Typically many cognitive issues will surface within the intervening period, as the witness begins to reflect on the matter and contemplates how the unorthodox information will be perceived by others. Also, details can become lost or clouded if they are not recorded immediately. A good witness is a recent witness! From the second-hand information that I was initially given, the very large figure was several inches taller than the observer. With the witness being 6'1", this would place the intruder's height at about 6'3". Furthermore, the internal floor of this house at the rear entrance was four inches higher than the verandah, giving a newly revised height of 6'7". As I originally suspected, the very large figure was undoubtedly Fatfoot.[21]

To further complicate matters, the revised report said that the figure was a large male wallaroo. Wallaroos are occasionally found in our valley, with the smaller swamp wallabies being the more dominant species and rock wallabies not having been seen in this area. The eastern wallaroos are the largest macropod species to visit our valley. However, females are short with a standing height between 2½ and 3 feet and carry a weight up to 25 kg. As could be expected, male eastern wallaroos are muscular and bigger, typically weighting up to 50 kg, with a standing height of between 4½ to 5 feet. Additionally, macropods have great difficulty negotiating narrow stair treads with their long macropodoid feet, whereas Fatfoot's shorter feet were able to more easily traverse a lengthy flight of a dozen steps when she entered and exited Wilhelmina's house in 1993.

---

[21] Witnesses tend to become highly defensive after a close proximity encounter at their home. Some time afterwards, three sections of low fencing were poorly erected across the property in a futile attempt to exclude these 'marsupial hominoids', which usually step over such obstacles. The most extreme situation involved Murdock, who erected an eight-foot-tall fence around his property's perimeter. Other than constructing physical barriers, some install motion-sensing lights.

11: Junjudee

During this very active encounter period, Sandy experienced a number of her own situations. After spending some time listening to the Junjudee moving about in the bush to her north, she saw a number of figures in the faint moonlight approaching our house. She thought that they must have been aware of her presence because they took up positions behind some of the larger trees, about twenty metres away. Sandy returned to the house and quickly explained the situation to me. We both went outside and Sandy pointed out the three dominant trees that the Junjudee were last seen hiding behind. After we moved down to the driveway, several Junjudee simultaneously broke cover and ran back towards the garden and the denser bush lying behind it. It was very difficult to see the bipeds in the faint moonlight, but the task was made much more difficult because our line of sight was interrupted by the large trees that they had been hiding behind, a situation that the Junjudee consciously seemed to be taking advantage of. In order to gain a better view, we moved a few metres down our land and could occasionally glimpse something moving behind a tree. After the majority of Junjudee had broken cover from behind the stringybark, one delayed its escape but maintained the same line of escape that the others had taken. It was running away across a relatively open area of grass and vacated garden for a distance of about twenty-five metres until it obtained cover again.

This Junjudee appeared to be black. Its height would have been half of mine, so about three feet. From the silhouette, it was impossible to determine any significant body detail, other than it was lightly built or gracile and had no obvious tail. The limb proportions were too vague under the conditions to accurately determine. However, a very noticeable trait was the presence of a waddling gait. From behind, the upper body moved through an easily detectable but small arc of several degrees. There are many possible reasons for this gait but the most likely relates to hip configuration.

On Saturday, 8 April 2017, around 10 A.M., I was working at the front of our house and needed to get some potting supplies from the old greenhouse at the back of our property. I could hear that the weekly Saturday morning soccer practice was in progress, further down the neighbour's property. Next door, young Oliver, who was also celebrating his recent birthday, could be seen in his full English uniform practising soccer manoeuvres with his Pome[22] father in their backyard, prior to departing for the weekly soccer game. We had purchased a new greenhouse to replace the old hail-damaged one, which we had previously moved near to the wire boundary fence, out of the way for the time being. I took the long way around our house and was walking up the eastern side of the old, semitransparent greenhouse, approaching the corner of the building nearest to the wire fence, when an explosive movement took place on my right side. The six-strand wire fence directly in front of me was making a very loud acoustic resonance, consistent with something having heavily struck or being forcefully passed between the wires. The resonance very rapidly radiated away from the source, moving leftward towards the house. The sine waves that moved along the wires were very small, but visible for a few seconds, having been restricted or choked by the small holes in the star pickets. The rightward movement of the resonate sound waves could not be seen because of the obstruction made by the forward section of the greenhouse, but its progress down the fence line could be heard for a

---

[22] 'Pome' is Australian slang for an English person. As Australia was a penal colony until 1868, new convict arrivals were known as 'Prisoners of Mother England' or POME, which distinguished them from free settlers and earlier convicts who had served their time.

more lengthy period. The origin of the disturbance had been the opposite corner of the greenhouse, 1.8 metres to my right. Following this commotion was the loud sound of running through the bush, diagonally away from me towards the 'natural watercourse'.

The soccer practice continued, with the players remaining oblivious to the commotion thirty metres away. On this rare occasion I was carrying my mobile phone with me, so I went through the lengthy log-in procedure, opened the camera application, selected video and started recording.[23] Of course, by now there was nothing around to warrant videoing, but it seemed that I should go for a long bush walk while recording the surveillance attempt, just in case I flushed something out. After about thirty minutes spent searching a small section of our bush with an approximate area of two to three thousand square metres, I returned home. Particularly as it was a daytime encounter, I was undoubtedly being watched. I wasn't disappointed, because from experience these native predators are very good at maintaining their concealment. No wonder that they cannot be found at nighttime, with some attributing them with supernatural abilities!

Also around this time, neighbours further to our south were undertaking extensive renovations and building extensions, some of which were close to the swamp. The work was very noisy and involved the use of heavy excavation equipment in the placement of many large rock boulders. The sound from the equipment would have certainly

---

[23] I timed this startup procedure, from opening the phone cover to the start of recording, at thirteen seconds. The majority of encounters would have been completed well within this short time. On a few occasions, involving the potential use of a charged and readily accessible FLIR, I made the decision to observe rather than try to record a partial image because it seemed to be a better use of the available encounter time.

radiated a long distance into the valleys below and beyond. As the building work took place over many months, we often wondered how Fatfoot would have reacted to the construction, if she was currently around. We hadn't heard foot thumping for some time. With the Junjudee being present, we weren't certain of what to expect.

During late April and early May, I had been extending our southern fence in an attempt at controlling the grazing habits of the wallabies, but in a way that would not completely exclude them from our land, by leaving the bottom section open. I had extended the length of the wire fence by about a hundred metres, to within twenty metres of where the hanging swamp used to be. On 5 May 2017, I had mostly completed the work. All that remained was for me to install a strainer rail onto the terminal post and then restrain the six fence wires.

I started to work around 9 A.M. by placing all the anticipated tools and materials into the ride-on trailer and driving them down to the end of the fence. The end post was made from a mature Turpentine *(Syncarpia glomulifera)* log that was left oversized so that it had more mass, making it suitable as a terminal fence post. Hardwood turpentine can grow to heights of 58 metres, is extremely durable and was typically used in the construction of wharves because marine organisms could not eat the timber. Similarly, chains and other saw blades give off sparks during cutting, which happened on this occasion because of the timber's high silica content. Being in a bushfire-prone area, this timber was also chosen because of its high ignition threshold. Using the tractor auger, a four-foot-deep hole was drilled because the soil was mainly swamp peat that compacts and supports rammed soil very poorly.

While I was working on the strainer, I could hear the builders about seventy metres away. In the distance, the dogs were barking to the north so I became more vigilant.

11: Junjudee

During the next hour, I did some measurements, dug a hole for the strainer footing, notched out the post and cut the rail to length with the chainsaw. I stopped to listen for a few minutes and noticed that the dogs were now barking diagonally across the valley, with the woofing front moving south. At this stage, I began to pay closer attention to the surrounding bush. I continued to work, drilling and countersinking a hole for the coach screw that was going to secure the top of the rail to the notched post. In reality, I was working slowly, being more concerned with what was happening around me.

The barking of the dogs had ceased. My best guess was that, whatever it was, it had moved away from the fence line where the dogs were being constrained, a behaviour that Fatfoot would frequently do when she wanted to approach our house more directly and with stealth. This acutely reminded me of hunting Fatfoot with Ian, twenty-four years earlier, with the difference being that Ian was not around anymore and it was daylight.

When being stalked, it was very necessary to maintain a long memory of events, the sounds and particularly their position, always keeping a lookout for extended patterns of activity or behaviour. I started to hear frequent twig snaps and other subtle noises that were spaced out, both in time and location. The pattern seemed to move slowly past where I was working, so I continued to pretend that I was fully occupied with my work. I took an extraordinarily long time tightening the coach screw that was securing the rail to the post, doing this actively, but very quietly. Over the following fifteen minutes, I became aware of movement coming from the south—the opposite direction to where the movement had approached from. I thought that perhaps this animal had had a look at the builders first, probably something that it had regularly done over the previous months and had then decided to come back and see what I was up to.

Fig. 11.6 Terminal fence post, 5 May 2017. (Photo: Neil Frost)
A sample audio file from April 2017 demonstrating the movement of Junjudee through the valley, tracked by barking dogs, can be heard at http://tinyurl.com/swamptrack.

As the minutes went by, a regular and consistent pattern of movement was being established that was focused on my location. I had no clear idea of what to expect. The top contender was a Junjudee but not Fatfoot, as the behaviour was different. It wouldn't be a swamp wallaby, because they immediately hop away from any potential predator and humans. I even considered the possibility of a black panther stalking me because I had seen one during the day, a few months after the hazard reduction burn, bounding through the swamp, although they tend to be extremely cautious with people also. The Octopus reported on panther sightings, which we passed on to Mike Williams and Rebecca Lang. John Appleton had photographed a suspected panther kill when bushwalking at Springwood, although the predator could have been a Hairy Man.

Like a countdown, the movement was becoming closer and I was confident that I had accurately determined the position from the last sound heard. I was facing the source, but keeping my head down, pretending to be fully absorbed with my work. On hearing a twig snap, I was confident that I could expect to see something, so I lifted my head and looked directly at a small face looking at me from very low to the ground and from within a clump of grass. Using my ears to accurately target the animal's location had been the best sensory method because I don't think that I would have directly looked that low down on my first visual attempt.[24]

I had been preparing for this moment for decades! I was well aware that encounters are typically fleeting and it was essential to make the most of this brief opportunity.

---

[24] Aboriginal Elder Merve told me to look up into the tree tops when searching for these hominoids. The pair of 'blink comparator' photographs (automatic film camera photographs #3 and #4) taught me to look very low to the ground as well.

Having interviewed many hundreds of witnesses who had exceptionally good encounters, the initial hurdle was typically overcoming the shock from unexpectedly stressful circumstances. Clearly, by being unprepared and trying to quickly adjust to the demands of the situation, observational and processing time is usually lost. For me, however, there was no shock and I felt extremely confident and in control. In fact, I was so prepared and relaxed about the situation that I spoke to the Junjudee, saying, "I can see you!" This was a cheeky response that I had not anticipated and probably shouldn't have been said, because it may have shortened the encounter.

Another important aspect of encounter preparation is predetermining where to concentrate your attention after a sudden opportunity arrives. For me, I already had a separate bucket list for each species. A common point of clarification needed, across all of the 'marsupial hominoids' was to confirm the presence of a marsupium and/or the absence of breasts. The facial appearance was important too, particularly the eyes and the shape of the pupil. Limb proportions and the number of fingers for each species were also on the priority list.

After the face appeared from underneath the sword grass and I had given my salutation, I only had an estimated five seconds to absorb as much detail as possible before it began to move away. I took accurate notice of certain features of the face and for a few seconds afterwards, details of the left foot and buttock as the Junjudee turned to move away. The Junjudee quietly crawled away beneath the vegetation, did not rise up in height and disappeared by keeping a very low profile for the entire distance of its hidden departure.[25]

---

[25] As determined previously, it takes about thirteen seconds to initialise the video application on a mobile phone and probably half that time if a

11: Junjudee

As it seemed pointless in pursuing a Wildman that is very fast and can move very easily through thick vegetation, I held the image of its face in my memory and returned immediately to the house. I found a few sheets of paper and immediately began to sketch what I had just seen. After that, I wrote down in point form and random order, anything that I could remember about the encounter.

Following is this random, descriptive list:

> For about 15 minutes prior to seeing the animal, I became increasingly aware of something stalking my position. Twig snaps and movement towards me. Actually reminded me of hunting Fatfoot with Ian. You need to maintain a long memory of sounds and their locations over time, especially during the night.
>
> After it became apparent that something was approaching me, I pretended to continue to work on the post, whilst listening intently to the approaching source of sound.
>
> I heard a gentle twig snap and knew its direction.
>
> I looked directly at the face, without needing to look around. This lasted for five seconds, during which time I said, "I can see you!"
>
> The face was looking at me from underneath a clump of grass.

basic digital camera is used instead. However, the total available time of this encounter was about seven seconds. In a best case scenario, with the video recorder already in operation, it would take a number of seconds to frame the subject and zoom in enough to provide a worthwhile image, which is just barely feasible.

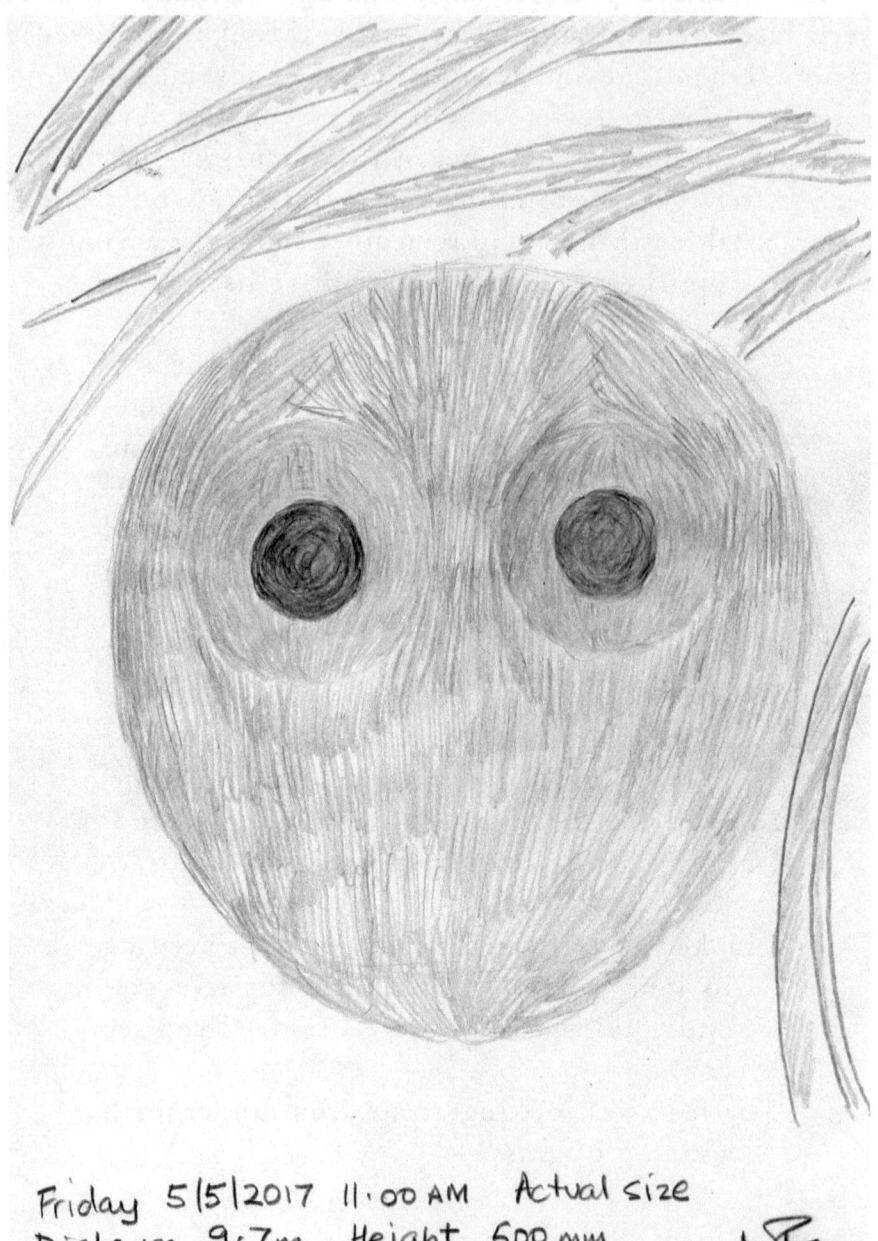

Fig. 11.7 Drawing of face looking up from under a clump of sword grass. (Drawing: Neil Frost; view full color at http://tinyurl.com/junjudee-sketch)

- When I first saw it, I thought that it was a swamp wallaby.
- The face was sharply defined, having fur on the face (and buttocks) that seemed very short and 'groomed' like a horse's coat.
- It looked eerily like a barn owl, in terms of its clearly demarcated, facial regions.
- The eyes were jet black.
- Each eye was surrounded by a wide, sharp ring of brownish-yellow fur.
- I don't remember seeing any other facial features. (I think that I simply ran out of observational time.)
- The face was expressionless.
- When it turned right, away from me, it maintained its head height.
- It turned on the spot.
- After it turned away from me, I saw its backside and left heel.
- The fur was black and short.
- It had no tail.
- Its left buttock was firmly sitting onto its left heel and then lifted up slightly.
- As it began to turn away, it moved like a Cossack dancer, in that it moved with its buttock mostly resting on the heel and the head still inclined horizontally.
- I observed the turn and movement away for about one and a half seconds. A very sharp turn. I could hardly hear any sound of movement away.
- As it moved away, it did not rise up in height, keeping a very low profile.
- There was no animal visible after this, despite the low vegetation height.

There was no hopping heard.
I could have easily observed the animal if I had moved along the fence line towards the swamp.
Realised that this animal was a Junjudee.

After writing this list I returned to the site to take some more refined measurements and see if there was anything that had been overlooked. The time of the encounter was 11 A.M. The weather was fine and mild. The distance from the post to where the face was observed was 9.7 metres. The height of the top of the head from the ground was 600 mm. The size of the face was about the same size as mine. The breadth of the Junjudee's head was calculated to be 164 mm, which was determined from the 210 mm wide, A4, scale drawing. The head width of the Junjudee compares to my actual breadth of 160 mm. Having such a large head is compatible with the animal's high problem-solving intelligence. By using the sketch as a relative guide for scale, the diameter of the drawn central black section of the eye, or potentially the iris and pupil, was calculated to be 25 mm. No sclera was visible. The brownish-yellow annulus around the outside part of the eye was similarly calculated to have a width of 13 mm. The face was ovoid or egg-shaped, with the narrow end being where the jaw should be located. The head had no sagittal ridge or any obvious robust features like teeth, including prominent canines, or muscular jaw, which made the face look delicate. The overall body morphology was gracile. Based upon the small size of the leg seen, I estimate that the overall size of the biped was similar to a ten-year-old child and therefore, possibly weighing around 30 to 35 kg. The left leg and foot looked gracile, like a child, despite being covered all over with short, well-groomed fur. When the animal moved off, it travelled close to the ground,

Fig. 11.8 View from the end of the fence, next to the post, looking south. Circle identifies where Junjudee was laying under sword grass—distance 9.7 m. Line shows heavily worn track leading to the post. (Photo: Neil Frost)

which reminded me of a 'commando crawl', however the observed movement seemed awkward. Having fur means that the animal is a mammal. In appearance, the hair of the animal reminded me of wallaby fur, but slightly finer like a horse's coat.

This was the first daytime visual sighting of a Junjudee in our valley. Another visual sighting in a higher valley system was made by Rohan, who simply described the Junjudee as "black" and "monkey-like". Lynn and Zoe Pendlebury from lower in the mountains, described them as being like an "overgrown monkey" that was "screaming as it ran away on two legs". It was "black and about four to five feet tall. As it ran, its back was perfectly straight and it may have had a tail, although I was not very certain because the dawn light was so poor." However, these sightings were not the first reports in the Blue Mountains.

There was also a newspaper article from *The Illustrated Sydney News* that related an encounter by the wife of the caretaker for Sir Henry Parkes,[26] at Faulconbridge in 1886. According to the newspaper article:

"It seems that she was in the act of gathering a few sticks when a commotion amongst the fowls attracted her attention, and on looking up before her stood a Thing about seven feet high. The black hair growing on its head trailed weirdly to the ground, and its eyeballs were surrounded with a yellow rim."[27]

This historical encounter from Faulconbridge occurred 131 years before our 2017 sighting and was separated by a distance of only fifteen kilometres. Unfortunately, as

---

[26] Sir Henry Parkes (1815-1896) was an Australian colonial politician and the longest serving non-consecutive Premier of the Colony of New South Wales.

[27] The legend of Lindon: The grisly details. 1889. *The Illustrated Sydney News* (Thursday, 3 October) p. 13.

was typical of sensationalist newspaper reporters at that time, some 'grisly details' about the 'Thing' seem to have been wildly exaggerated for greater impact and higher paper sales. However, the unimaginable description of the animal's eyes may have survived interference: "its eyeballs were surrounded with a yellow rim."

We continued to have multiple encounters with Junjudee during this time, mainly because of the provocative and mischievous nature of these animals. As a consequence, we set up a number of game cameras and digital VOR recorders at various locations throughout the valley. Many of the new neighbours were getting an introduction to cryptids and some needed to get some urgent advice.

In our backyard, we had plenty of pasture after the hazard reduction burn and autumn rain. We had two wallaby joeys about 400 mm tall, that hopped around our backyard most mornings and afternoons. On 16 May 2017, eleven days after my encounter with the Junjudee while fencing, the game camera had some relative success. I had been moving cameras around in the swamp area for many days and was getting nothing except videos of wallabies and their joeys tearing through the undergrowth. The audio recorders were obtaining plenty of interesting nighttime recordings, where bipeds could be heard moving throughout the area, with some pushing over small trees and making many strange, previously unheard noises. After reviewing the videos from that day, we could see one of our small joeys racing out of the swamp shortly after sunset and briefly pausing in the cleared area and looking towards the camera (Fig. 11.9), before hopping towards the camera and disappearing in front of the newly installed terminal post to the right of the frame. After this, the flyer or mother came out of the swamp from the same track at the left of the frame and also briefly surveyed the area from the same position, before passing through the bush in the opposite direction to her joey—away from the camera. A short time afterwards, the very distinctive sound of a bipedal running could be loudly heard, which passed around the post and behind the camera, before audibly fading into the distance along the line of the newly installed wire fence (http://tinyurl.com/joeyhunt). It was obvious that our two young joeys were the target prey being pushed by the Junjudee.

Soon after this, I sent several of the recent audio recordings to Russell (aka, Rusty), a researcher who had had a great deal of success with his highly innovative work. The audio recording of the Junjudee walking used in his analysis was taken from the video of the joey, on 16 May 2017.

Russell had pioneered a number of remote sensing video and audio systems, that were designed to be extremely passive and operate for lengthy periods of time in remote forests. Russell was also a very talented audio engineer

Fig. 11.9 Young joey briefly pausing on the edge of the swamp (left), before hopping towards the camera near the post and continuing along the newly established fence line, being pushed by a Junjudee. The presence of the Junjudee was revealed by the audio. (Photo: Neil Frost)

11: Junjudee

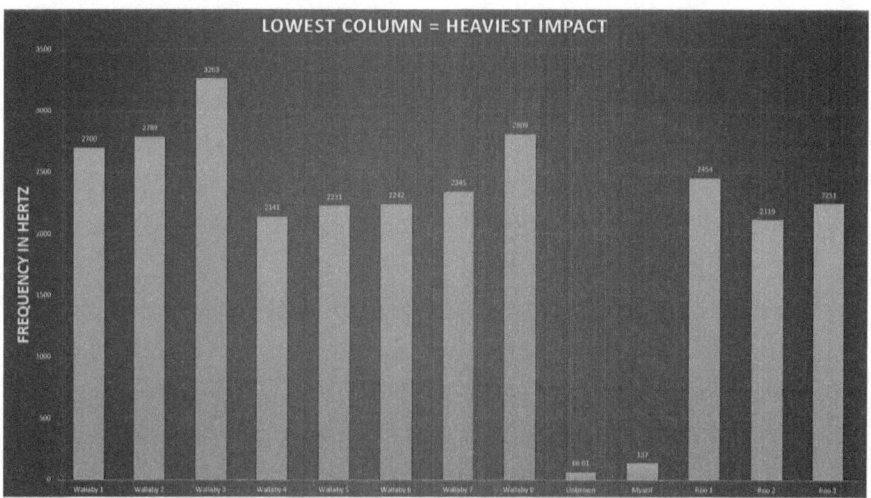

Fig. 11.10 Chart #1: showing relative impact of four bipeds (from left to right): wallaby (8); Junjudee (1); human (1); and kangaroo (3). (Chart: Rusty)

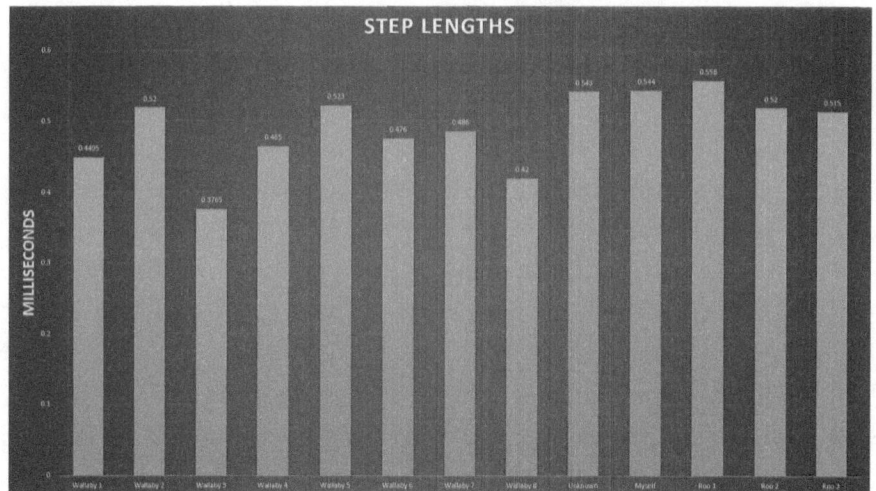

Fig. 11.11 Chart #2: showing relative step length of four bipeds (from left to right): wallaby (8); Junjudee (1); human (1); and kangaroo (3). (Chart: Rusty)

who was able to analyse the sound from our recording and determine the relative impact of the feet and their relative step lengths, across four different bipeds. As shown on both charts: the first eight readings were from wallabies; followed by the single 'unknown' or Junjudee sample obtained from the video recording; then a single recording from Russell ('myself'); and finally, three readings obtained from kangaroos.

I have no understanding of the acoustic theory and mathematics behind Russell's calculations, but Chart #1 clearly shows that Junjudee and Russell are both relatively heavy walkers. Wallabies and kangaroos are both lighter on their feet, which is not surprising considering the high-impact injury risk inherent in a hopping gait. Male red kangaroos (boomers) are the largest and heaviest hopping biped known, with a typical body mass between fifty and one hundred kilograms. From Chart #2, it can be seen that the average step of wallabies is less than the other three bipeds, which could be attributable to their smaller stature. Junjudee, humans and kangaroos all have similar step lengths, which is an interesting result.

After thinking about the recent changes to the fence line and the survival prospects of the swamp wallabies and joeys, it now seemed that our partially confined property was facilitating the herding and capture of these prey animals. After the video, we didn't see either of the joeys again!

Junjudee mostly hunt as a 'mischief' using the pushing method. The size of these predators is reflected in the largest prey that they take, with the two biggest 'marsupial hominoids' mainly relying upon mature swamp wallabies and Junjudee hunting swamp wallaby joeys and domestic animals. In our valley, 2017 was the worst year for small dog and cat losses, with many missing pet posters appearing on power poles on a very regular basis, with some pet owners also conducting door-to-door investigations. During earlier times, the Pendleburys lived in a less densely populated rural community and like residents elsewhere, their well-known Australian film critic neighbours Bill and Joan found their small pet dog had been mutilated. It had severe head, eye and rear-end wounds and had been partially disembowelled.

The following day, Sandy went for her usual morning walk. She was particularly interested in looking for the joeys and any signs of Junjudee activity. As she was walking along the edge of the swamp, near to where the joey had been videoed, she came across an extremely powerful and overwhelming stench. She said that it was like rotting meat and very repulsive. This was new! It was another alleged Junjudee characteristic that had been occasionally reported by Octopus encounter witnesses, something that we had never experienced before with Fatfoot.

After we mentioned this olfactory assault with neighbours, we were surprised to hear that Doreen had recently experienced something very similar on three separate occasions. According to Doreen, all three experiences had happened during the previous few months when she had been reclining on the lounge room couch reading a book, in the hope that she might become tired enough to sleep. All incidents had taken place after midnight in the lounge room, on the southern side of the house where the bedrooms of the children were also located at that time. The intense smell had penetrated through the partially open window next to where she had been reading. She heard and experienced nothing else during these three malodorous encounters.

Coincidentally, the neighbours had been completing renovations on their house, while the Junjudee activity was taking place around it. Rosemary was about to enter high school and like most children of that age, she wanted her own room. The old main bedroom on the northern side of the building became Rosemary's new room and her parents moved into the newly constructed master bedroom at the rear of the house. Furthermore, finding a stick planted in our lawn was not very common, so finding two was a bit of a surprise! This latest stick planting had occurred immediately after Rosemary had moved into her new room. If someone was prone to look for patterns with hidden meanings, then it might be suggested that these two sticks were aligned and possibly used to point at the newly occupied bedroom belonging to a person of interest.

Many years before this spate of stick planting in our neighbourhood, we would often mentor selected people with a strong interest in researching these 'marsupial hominoids'. This had always been done on a one-to-one basis with the aim of sharing and preserving knowledge and experience, which had taken our community so long to acquire and was in danger of being lost unless it was passed on![28] The conspicuous loss of traditional knowledge within the Aboriginal community had been the motivation behind our actions. However, it seemed that this low-profile activity was going largely unnoticed by a few, which led to calls for the holding of larger-scale tutorials. Consequently, a substantial group of interested people were invited to camp overnight in the hope they would get something out of it. As part of the experience, we organised an excursion into a nearby valley which seemed to be active with Junjudee at the time. The group reported seeing and hearing some strange things, including wood knocking. The following morning, a stick was found planted in the freshly mowed grass on the opposite side of the house to the campers. Interestingly, the stick was impaled at the

---

[28] This motivation is also the prime reason for publishing this book.

11: Junjudee

Fig. 11.12 Dual 'stick plants' on our land, opposite Rosemary's newly occupied bedroom on the northern side of their house, 24 May 2017. As the second stick may be difficult to resolve, it is located midway between the forward stick and the clothesline post. (Photo: Neil Frost)

entrance to the track where we had encountered the three stick-wielding Junjudee on 25 September 2004. I am certain that every member of the group benefited from the broad experience.

A VERY IMPORTANT aim of this research was educating the community. Before we became Yowie-aware, the perception held by the general Blue Mountains population was that anyone who claimed to have had an encounter with one of these three 'marsupial hominoid' types was strange, or a nutter.[29] At the very best, encounter reports were politely ignored and at worst, vulnerable victims were bullied, as was the case with high school student Troy in 1986. The other component of this aim was to provide support to encounter victims, who were seeking answers and solutions to the immediate problems and trauma that they were facing.

Regular community contact was an important part of the educative process. Once the Octopus had become established in our local community and was autonomously networking, encounter reports were increasingly received, at a rate of about three and a half per week at the height of notification.[30] Using a conservative estimate of 1.7 reports per week over the 22 years from 1993 that the Octopus was mostly active, gives a total number of reported incidents equal to 1950. As the total population of the Blue Mountains was 79,195 in 2020, this gives a cumulative

---

[29] Our current estimate of encounter witnesses in the Blue Mountains as a percentage of the local population is about five percent.

[30] This high weekly reporting frequency had been formally regarded as impossible, as very few encounters were previously received over a year using orthodox methods. This proved that the encounter activity rate had always been high, except that earlier witnesses had been keeping "as tightlipped about it as an Aldebaran Shellmouth". *Star Trek*—Dr. McCoy, 'Amok Time' [S02E01] 1967.

sighting total, as a percentage of the population, equal to 2.46%, or one person in forty. The actual figure should be considered as being higher than this, as many witnesses remain 'tightlipped'.

In late 2017, I was asked to speak about our research at the Twenty Mile Hollow Cafe at Woodford. This was the first of three two-hour talks that were aimed at increasing local awareness. The initial talk highlighted the large number of case study examples found in the immediate area, starting with an early encounter report directly across the road from the cafe and then radiating concentrically outwards with other reports. The audience was then invited to search my personal database of encounters by nominating areas within the mountains that interested them. Just over a year later, I gave a second talk which followed a similar pathway to that detailed in this book. The third talk, in mid-2021, dealt with some of the audio and video evidence plus some infrared imagery. Members of the audience were invited to speak about their own experiences and questions were taken.

A significant aspect of these talks was how many people had had an encounter or knew of someone who had. Of the three 'marsupial hominoid' types, the encounters that most easily stood out were Junjudee. This was probably because Dooligahl and Quinkan tend to remain mostly hidden during their activities, with the main indicators tending to be the loud sound of walking, branches breaking, or the sighting of reflective red eyes. Junjudee encounters tend to be more provocative and conspicuous.

One of the more interesting Junjudee encounters wasn't discussed at the cafe but on my way to it. I had gone to pick up my new spectacles, just in time for the talk that evening. I had the intermediate reading glasses made to an unusual focal length that was more suited to reading notes from a rostrum or a suitably-distanced computer screen.

These 'intermediate' spectacles are somewhere between reading and distance glasses in focal length. As reading glasses usually have a focal distance of 400 mm, I find that this is too close for rostrum reading and prefer the focal distance to be 700 mm. This only requires an adjustment to the lens prescription. When I was picking up my spectacles, the sales assistant noted the unusual focal length and asked me if I was an optometrist. I told her that I wasn't and after some further discussion, I mentioned that I ground my own telescope mirrors when I was young. As it happened, Gayna was also an amateur astronomer but no longer observes because she had a terrifying experience recently.

Gayna said that she enjoyed observing but there was nowhere clear enough in her backyard to see the night sky so she would take the telescope more than a hundred metres into the bush and set it up during the late afternoon. She would usually get up in the early morning and walk to her observing site. On the last occasion, she had arrived at the clearing when she heard walking at multiple locations around her. She became increasingly afraid as the walking by many individuals continued to circle her and she could also hear movement in the treetops around the edge of the clearing. Gayna said that she didn't want to appear afraid, so she slowly walked back to her house. Like most people, she had no idea of what she had been dealing with, so I spent some time trying to relieve her apprehension.

Apart from a few encounter reports from the Western Plains of NSW and from country Victoria that mention seeing monkey-like animals climbing trees, Gayna's encounter clearly involved Junjudee and was the first to mention movement "through" the trees. This was confirmation of Merve's early advice that told us that "they live in the trees".[31]

In a similar way, an audience member recounted her experience while camping at Wentworth Falls. She said that during the night a group of monkey-like animals had been moving about in the treetops above the campsite, making unusual noises. Unfortunately, the woman left the cafe before I could speak to her in detail. Chris, the cafe owner was unable to contact her on my behalf, saying that all he knew was that she was a nurse.

These talks were also an opportunity to meet with some of the interested outsiders who were present. I met three Scandinavian women who were holidaying in Australia and happened to come across an internet advertisement for the cafe talk, making the effort to attend by traveling from Sydney by train. Also in attendance was Rob, one of the film producers of *Myth,* a dramatised account of a Yowie encounter in the Blue Mountains that I had previously consulted for. Similarly, I met with Attila Kaldy, the producer of *Track,* a Netflix movie about encounters in the mountains and provided a short opinion piece.

JUNJUDEE, OR BROWN JACKS, are the smallest of the three species of Australian 'marsupial hominoids'. Mature individuals are typically three feet or slightly less than one metre in height, with some estimates being five feet or about one and a half metres. Their skin is black. The body is covered all over with fine dense hair, which is black and about ten millimetres in length. The hair appears to be well-groomed and is similar in appearance to possum fur. The head is large for the body size, but slightly smaller

---

[31] When Merve told me that our hominoids "live in the trees", I didn't seek immediate clarification. From experience gained in interviewing tribal Elders in Papua New Guinea and other Melanesian islands in the late 1970s, I was more familiar with 'spirits' inhabiting objects in nature and the local people, rather than taking the meaning literally.

than a human skull. The face has an ovoid shape that has a vertical, long axis of symmetry, with the apex or apical point located at the jaw. The eyes have a wide interpupillary distance, with a photographed individual having an IPD of 72 mm, giving an excellent depth of stereoscopic vision. No sclera was visible and the pupil could not be differentiated from the iris, which suggests that it is a predator, however, photographs of eye shine indicate that the pupil dilation is very large. Junjudee are less affected by bright light when it is used by people as a deterrent, compared to the other 'marsupial hominoids'. Since Junjudee are successful diurnal and nocturnal predators, it would make considerable sense for them to have vertical slit pupils, like Fatfoot, which give greater control over a wide range of light conditions and provide a number of superior options in determining the distance to prey through stereopsis, motion parallax and defocus blur. Interestingly, our Blue Mountains Junjudee have a wide annulus of brownish-yellow fur surrounding each eye that has not been reported elsewhere in Australia, suggesting that this population might be a subspecies. No other details have been closely observed on the head, which surprisingly, includes the ears. Similarly, teeth have not been observed, however, some outside sources have commented on the presence of canines.

The appendages of Junjudee are gracile and do not appear to be out of proportion, unlike the robust and longer limb lengths found on Dooligahl and Quinkan. The arms and legs have proportionately small hands and feet that are prehensile, because of their observed arboreal habit. Each foot and hand have five digits with black claws, that indicate that, apart from other factors, they are not related to the great apes, which have pink, flat nails. Consequently, Junjudee are not relic hominins that have been affected by insular dwarfism, such as *Homo floresiensis*.

In form, Junjudee are very similar in appearance to monkeys, but have no tail. They have an erect posture with a straight back. When walking on the ground, one individual was observed having a waddling gait that moved through a detectable arc of several degrees. This gait is similar to an echidna and some reptiles. After the swamp encounter where the face was seen, the initial departure could be described as being like a 'commando crawl', with the abdomen and limbs keeping a flat profile, that didn't break the low vegetation cover. By itself, this crawling ability would explain why Junjudee appear to 'vanish' after being observed, particularly during the night.

Even with the absence of a tail, Junjudee are also well adapted tree dwellers. As advantageously placed encounter witness accounts could not be fully verified through follow up, it was not possible to confirm the exact motion of Junjudee through the trees, but several less suitably placed observers described their movement as 'rapid' or 'fast' and that many Junjudee were involved. More commonly, many more witnesses have observed the easy ascent and decent of trees by an individual.

Junjudee are extremely adaptive mammals. Their current evolutionary position could possibly be regarded as transitional, in the sense that they are physically suited to both terrestrial and arboreal environments, which is a very rare ability that must assist greatly with their survivability under harsh Australian conditions. This makes additional sense when their locomotion is considered. Junjudee can be considered to be facultative quadrupeds, meaning that they are primarily adapted for walking on two legs, but can also walk on four. Bipedalism is normally associated with terrestrial locomotion, whereas the efficient movement within trees is generally regarded as requiring four legs, or five limbs when a tailed possum or monkey is considered. To complicate matters, Junjudee have been observed crawling on all fours, 'commando'-, or 'lizard'-style and on one occasion, by rolling like a ball down hill.

Considering all of the physical evidence suggests that Junjudee have travelled a different evolutionary path than the other 'marsupial hominoids'.

JUNJUDEE ARE HIGHLY social animals. They live and move about as a group, or 'mischief', which can contain a large number of individuals. They possess many of the essential attributes that define and make them a coherent social group. Their most important quality is their intelligence. They are capable of solving complex problems—for example, the ability to disable motion-sensing equipment using a variety of methods, that would otherwise, interfere with their covert activities. Most probably because of their intelligence, Aboriginal Elders have described the nature of Junjudee as being 'mischievous' because of their rascally behaviour and tendency to play tricks on people.

Unlike the sedentary habits of Fatfoot, our local Junjudee appeared to be itinerant. At the beginning of our study, Junjudee were not immediately conspicuous across the broader Blue Mountains landscape, with their population seemingly restricted to the lower foothills. Other reports received by the Octopus tended to be outside our local area, involving solitary sightings of monkey-like figures climbing on branches, or descending from trees, in addition to a few group encounters, with a number of reports coming from rural NSW and Victoria. Over the years, reported sightings of Junjudee seemed to migrate up the mountain until they started becoming interspersed with the activities of Fatfoot. This caused some initial confusion with the separation of behaviours and our general ability to differentiate between the two Wildmen, which tended to obscure the disappearance or presumed expulsion of Fatfoot from the local area. Clearly, there was a protracted turf war underway, with many conflicting issues and interests needing to be resolved between the two genera.

An unusual behaviour of Junjudee is their tendency to be cathemeral or the flexibility to be active at any time during the day or night. As a survival strategy, being cathemeral allows these bipeds to adapt to any environmental opportunity. It also allows Junjudee the relative freedom to observe humans and their activities at any time, more so than Dooligahl and Quinkan in particular, which have greater difficulty moving about and remaining hidden during the day. Like the other megafaunal 'marsupial hominoids', Junjudee tend to be fascinated by humans at work but have a more worrying preoccupation with children.

To successfully coordinate their social agenda within the 'mischief', Junjudee use several methods of communication. 'Wood knocking' is used as an auditory signal at night, or when individuals are widely separated, with the intention of broadcasting and confirming an individual's relative location within the group. High-frequency utterances from individuals may also serve a similar purpose. 'Stick planting' is a visual form of communication that is not well understood either, but seems to be used as a means of identifying things of shared interest to other group members, for example, where children play. It might also be used as a marker that displays the recent presence of an individual at a location, or the flagging of a position that has earned some recognition, for example, where an encounter with people has occurred. Two additional and clearly understood methods of communication are 'tree ripping' and 'tree clawing', where Junjudee show their displeasure with human activity by conspicuously damaging the bark of trees at significant locations. Most importantly, Junjudee communicate via a spoken 'language'. After a

Fig. 11.13 'Stick plants' continue to the present, December 2023. (Photo: Neil Frost)

Fig. 11.14 'Stick plant,' December 2023. (Photo: Neil Frost)

shared encounter with Junjudee while filming a documentary, an mp3 audio recording, named "Animal X", was made of a 'conversation' between two individuals that clearly records them having a high-frequency vocal exchange.

Junjudee are thieves and saboteurs. They are attracted to bright, shiny objects like batteries, or other attractive things, including fragrant items. They will steal these objects, but some may be recovered by searching the nearby bush. They will destroy anything that annoys or hinders them, like motion-sensing security lighting, or fencing.

As suggested above, Junjudee are manually dextrous and are competent tool users. At a simple level, they use available sticks to wood knock, stick plant and beat the bush to drive prey or attempt to intimidate people. Consequently, for these purposes, it is assumed they have an opposable thumb, however, from tracking evidence they surprisingly do not have an opposable hallux to aid with their tree-dwelling activities. Further evidence of their fine motor skills includes the ability to: manipulate door knobs in an attempt to enter a building, unplug DC power connectors and reposition flood lights and movable solar panels. Rock throwing is used to show disapproval, intimidate and in one instance, help to determine if a plastic owl was alive.

Even though Junjudee are not robust like the other 'marsupial hominoids' and do not possess the same dental armoury to bite through hardwood timber to extract tree larvae, like birds and other animals they are capable of achieving a similar outcome by biting through and snapping the branches of smaller wattles. As omnivores, known sources of food are fruit and vegetables stolen from gardens and swamp wallaby joeys, herded into and corralled by local fencing, which would supplement native sources of fruit and other animals. Judging from the large number of community posters that appear during periods of high

Junjudee activity and after discussions with a few resident owners, small dogs and cats are also on the menu.

When confronted, or nearby, Junjudee sometimes give off an extremely powerful stench. Doreen mentioned smelling this repulsive odour on three early morning occasions, at the same location while reading a book on the lounge. Sandy has smelt a very powerful smell, similar to rotting meat on several occasions, usually on the edge of the swamp while walking the barking dog. There are no malodorous reports associated with Fatfoot.

Fig. 12.1 Common larval galleries and bore holes in the ring-barked sapwood of a large Blackwood (*Acacia melanoxylon*) with a Janka rating of 5.2 kN. (Photo: Neil Frost)

12: Alibi

ABOUT SIXTY-FIVE THOUSAND years ago, the first Australians to eat wood grubs were most probably inspired to do so after watching parrots easily harvesting them from the hardwood of eucalypts and acacias. Parrots evolved in Australia and consequently, it is not surprising that they are very well adapted to the unique requirements of this challenging task. Visually and acoustically, the most conspicuous predators of witchetty grubs throughout the Blue Mountains are yellow-tailed black cockatoos—it would seem very difficult to imagine how this predation could have gone unnoticed and not resulted in a human analogue of this practice. Even with the late arrival of European settlers to Australia, these agriculturalists from Britain could appreciate the importance of supplementing their meagre rations with the unfamiliar grubs when times were hard or when scavenging off the land whilst travelling. It could be assumed that the early settlers were equally capable of learning from the wildlife, although this unpalatable alternative to beef and mutton was most probably introduced to them by the Aborigines. Similarly, any other witchetty-aware[1] predator could be expected to have learnt

---

[1] Many animals observe and learn from each other within their environment. Feeding the local carnivorous birds to entertain our children marked the start of our intense interaction with Fatfoot in early 1993.

Fig. 12.2 Bullseye borer gallery in the trunk of a Yellow Bloodwood (*Corymbia eximia*) with a Janka rating of 9 kN. (Photo: Neil Frost)

from observing the Parrots, just as Fatfoot learnt from us when feeding the carnivorous birds or our dog Bess.

IN THE PAST, the indigenous vegetation would have been different, but most native Australian trees are hardwoods that flower and produce enclosed seed pods. Australian hardwoods are evergreen, unlike most northern hemisphere species, like elm, maple and oak which are deciduous. Most Australian trees tend to be slow growing and are consequently very dense and hard.[2] An excellent example of Australian hardwood is Turpentine *(Syncarpia glomulifera)* which has a high silica content. As a consequence of its extreme hardness, which blunts the file-like teeth of marine borers, the timber is highly desirable for maritime applications. This hard, silica-laden timber causes saw blades to give off sparks and quickly blunt and burn. It has a Janka Rating[3] of 12 kN, compared to a rating of 6 kN for American oak and 6.1 kN for new English oak. Several other Australian hardwoods have higher Janka Ratings—ironbark is 14 kN and grey box is 15 kN, although the hardest known timber is either Lignum vitae at 19.5 kN from the Caribbean and the northern coast of South

---

[2] A small to medium-sized turpentine (12 kN) that I cut up after it was severely storm damaged had a trunk diameter of 250 mm. It required three new chains to efficiently complete the felling and docking. Cutting produced many sparks and quickly removed the edge from the new chain cutters. Counting the growth rings showed that the tree was more than 300 years old, which is around the time that Macassan Trepangers were trading with Aborigines along the Northern Australian coastline around 1700. This would suggest that the much larger turpentine at the edge of the swamp and others cut down by ignorant local residents were probably about 1000 years old.

[3] The Janka Rating measures the timber hardness of flooring and was developed by the U.S. Department of Agriculture. A 11.28 mm (7/16″) steel ball is pressed to a depth of 5.6 mm into a sample and the required force is measured.

America, or Australian Buloke *(Allocasuarina luehmannii)* with a contested rating of 22.5 kN.

WITCHETTY OR WITJUTI is an Aboriginal name for a specific larvae. It is commonly used to inaccurately describe a larger group of wood-boring larvae that are part of the family Cerambycidae, which contains more than 1200 species of longicorn (or longhorn) beetles. The eggs are typically laid on a eucalypt or acacia where the female beetle has detected a chemical signature identifying the tree as being damaged, stressed or generally having a lowered immunity for some reason.[4] In addition, other researchers suggest that with Australian plantation hardwoods, for example, increasing larval infestations are influenced by poor commercial timber practices involving single-species planting, overcrowding, poor weed control, soil infertility and other factors. On hatching during late spring or early summer, the larvae burrow beneath the outer bark and begin to inwardly devour various concentric levels of the tree, starting from the phloem, cambium, sapwood and ultimately, the heartwood, depending on the larvae species and its stage of development. Ultimately, these infestations result in eucalypt dieback, involving some of the tree's branches or the entire tree.

In its defence, a healthy eucalypt will excrete a dark red resinous substance called 'kino' to drown and immobilise the growing larvae. The kino also plugs the hole to seal the wound. A healthy acacia will similarly produce gum. It seems that the local Yellow Bloodwood *(Corymbia eximia)*

---

[4] We have found healthy eucalypts with no apparent immunity problems that have borer infestations. As suggested by the Department of Environment and Conservation Western Australia, this may be due to water stress, but it is too difficult for us to determine decisively. [Information Sheet 37 / 2010 Science Division]

Fig. 12.3 Female (grey eye-rings, bone beak) yellow-tailed black cockatoo eating seeds from *Banksia integrifolia* pods. (Photo: Neil Frost)

Fig. 12.4 One of many species of wood-boring larvae, or what is commonly called a witchetty grub. (Photo: Neil Frost)

and Red Bloodwood *(Corymbia gummifera)*,[5] like most acacias, have relatively poor timber compared to many other native hardwood trees and are more prone to damage. Our yellow and red bloodwoods have Janka Ratings of 9 kN and 10.9 kN respectively. Bloodwood trees are appropriately named, as the species exudes large amounts of dark red kino.[6] Larvae continue to chew within the tree for up to several years, depending upon the host, the temperature, humidity[7] and larval species. On occasions, the noise from their chewing can sometimes be heard by humans and undoubtedly by cockatoos and many other predators as well. From our repeated observations, it seems that by placing their beaks against the trunk or branch, the cockatoos could either feel the larval presence or could more easily hear their chewing, like when

---

[5] Bloodwood grows on poor sandy soils of the Hawkesbury Sandstone region west of Sydney. The timber, though tough, is extremely poor and has extensive gum lines (particularly in the red bloodwood) in comparison to other Australian hardwoods. The wood is very susceptible to termite and insect larvae infestations that limit their lifespan, while still having Janka Ratings from 9.0 kN to 10.9 kN.

[6] Yellow-bellied Gliders *(Petaurus australis)* and Sugar Gliders *(Petaurus breviceps)* are gliding marsupials that are convergent with primates like the Masoala Fork-marked Lemur *(Phaner furcifer)* of Madagascar. Gliders typically wound bloodwood and eucalypts in order to eat the weeping flow of kino and gum. Our large family of sugar gliders live in a hollow tree near the house. Bigfoot researchers have mentioned that Sasquatch wound and lick the sap from maple trees, as do bears, deer and doubtless other animals. It would be fairly safe to say that any food has more than one obvious predator.

[7] Over the past thirty years, there has been a very noticeable increase in longicorn beetle destruction and consequential growth in larval populations in the Blue Mountains, due to upwardly migrating temperature gradients (isotherms), shifting wind patterns and the more frequent extremes of drought and flood. Perhaps the most affected regions in Australia are the alpine ecosystems of northern Victoria and southern NSW with eucalypt dieback found in snow gums. These species have nowhere higher to go!

Fig. 12.5 One of many species of longicorn beetle. (Photo: Neil Frost)

using a soundboard to enhance the audio quality of a musical instrument. This more precise method of detection would probably be utilised by other predators. Mature larvae are holometabolous, which means that they typically bore into the hydrocarbon-rich sapwood to pupate before becoming adults. On exiting, they leave a borehole to the outside together with a noticeable saw dust deposit. This type of damage typically results in the death of the tree or the affected branch.

Longicorn beetles are most nutritious when they are in their final larval stage. Of the many Australian species a typical variety is *Phoracantha semipunctata* also known as the Common Eucalypt Longicorn. In addition to this widespread species, in the Blue Mountains, a more dominant

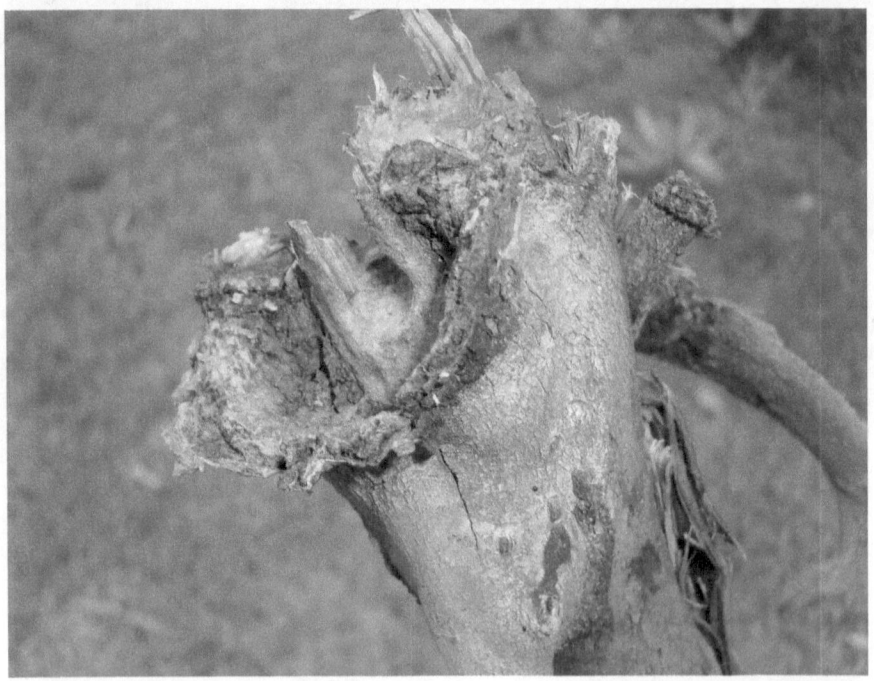

Fig. 12.6 One of a number of branches from the upper 50 m canopy of a Flooded Gum (*Eucalyptus grandis*), weakened by a bullseye borer gallery. (Photo: Neil Frost)

species is *Phoracantha acanthocera* or the Bullseye Borer. Both produce large, 20 to 40 mm long, cream grubs,[8] with rusty-coloured heads and prominent mandibles. Having a powerful bite force[9] and a suitably hard cusp edge to initiate wood fractures, allows the grub to easily find a passage through hardwood and a researcher's finger if incautious. For this reason, it is not advisable to eat the head, as the mandibles continue to function long after being decapitated! Nutritionally, the larvae are very high in protein and fat with substantial amounts of vitamins and minerals, with a fresh almond taste.

PARROTS OR *PSITTACIFORMES* evolved in the Gondwanan Rainforests of Australia. There are 56 species of Australian parrots, including budgerigars, galahs, rosellas, lorikeets and the ancient species of palm cockatoo. Parrots are highly intelligent birds. Most use tools, have language, problem solve and engage in play. As a result of their intelligent versatility and ability to fly, they have spread and adapted to other parts of the world. They have four grasping toes, with the outer two toes on each leg being the longest. Most parrots are left-footed and use their prehensile foot like a hand to skilfully manipulate objects. Both the lower and upper beaks can move independently of each other, which further increases their oral dexterity. As eaters of large seeds, yellow-tailed black cockatoos need to be able to manipulate the difficult-to-control cones and pods found in their ecological environment, through the skilful use of their dual leveraged beak and a dextrous left foot, cognitively coordinated with the aid of their left eye. The beak is curved to allow for the easy penetration of tough

---

[8] Some species' larvae can grow to 180 mm in length.
[9] The bite force of a hardwood-boring grub might be worth measuring.

Fig. 12.7 A very small part of a larger flock of approximately three hundred yellow-tailed black cockatoos undertaking early migration to the Blue Mountains. (Photo: Neil Frost)

fruit skins and nut casings by concentrating the shear bite force at the tip. The large muscular tongue is used to manipulate seeds and other food inside the mouth, being particularly involved in the de-husking of seeds. The bite force of some parrots is very considerable with typical estimates ranging from 300 to 500 psi,[10] or capable of very easily removing a human finger.

In the Blue Mountains, the principal parrot predator of wood-boring larvae in hardwood eucalypts and acacias is the Eastern Yellow-tailed Black Cockatoo *(Calyptorhynchus*

---

[10] In having a bite force of about 1300 psi, a gorilla is perhaps the closest analogue to a Dooligahl, in having a similarly-sized mandible musculature and a sagittal ridge. By being larger, a Quinkan could be expected to have a greater bite force. Being gracile, a Junjudee does not have any of the exaggerated physical traits needed to produce a hardwood treebite.

funereus).[11] Other non-parrot predators, including the carnivorous Grey Butcherbird *(Cracticus torquatus)*, have also been observed feeding off larvae beneath the bark, by using the downward hook at the tip of the beak to skewer and extract the grub from its chamber. Another parrot, the Sulphur-crested Cockatoo *(Cacatua galerita)* will also eat larvae from within the smaller, softer and more easily assessable acacia branches and stems, by crushing the foliage with its powerful beak.[12] Like other parrots, the normal food of these large birds is seed, typically obtained from native sources such as *Banksia, Casuarina, Hakea, Acacia* and other native seed pods. With the widespread cultivation of radiata pine plantations, the unfamiliar seed from these difficult-to-harvest, non-indigenous cones has now become an equally important and localised food source. Normally, during early to mid-spring the yellow-tailed black cockatoos can be seen migrating to the forested regions of the Blue Mountains from the broader seed-producing plantation areas, to take advantage of the increasingly more abundant animal protein and fat, which are the essential elements required for successful egg fertility for this species.

Extracting the seed from *Banksia* pods and radiata pine cones is a very difficult and highly skilled task that these large parrots are skilfully capable of. The dietary rewards obtained from extracting pod and cone seeds greatly

---

[11] Yellow-tailed black cockatoos are diurnal, 550–650 mm in length and 750–900 grams in weight. A high protein and fat-supplemented diet from larvae is essential for this species of large cockatoo, to prevent egg infertility.

[12] Sulphur-crested cockatoos have been observed by us eating the insects and grubs removed from inside acacia wood branches. Although these smaller cockatoos have a reduced ability to rip open hardwood trees for larvae, it is not their habit.

Fig. 12.8 Male King Parrot (*Alisterus scapularis*) de-husking sunflower seeds with its beak and tongue. (Photo: Neil Frost)

exceed the physical effort, particularly through the consumption of high-energy vegetable fats. Nutritionally, seeds are high in protein and minerals. They are a significant energy source derived from their fat and complex carbohydrates, although seeds tend to be deficient in some vitamins—mainly A, C, and E, with vitamins D3 and B12 being mostly missing. Consequently, seed-, nut- and grain-eating birds need to supplement their diet from other sources, with many parrots obtaining these missing nutrients by eating insects—in the case of these large cockatoos, from larvae. It seems that global warming is responsible for multiple changes to this ecosystem. Migration timing seems to be moving back towards winter. There is an increasing incidence of eucalypt dieback in the Blue Mountains and elsewhere, in part caused by, or related to, the increasing larval populations brought about by environmental changes. As we have found in the Blue Mountains, their increasing larval numbers are not limited to the trunks of trees. Many borer infestations occur in the crowns of stressed eucalypts, as found within branches from our Flooded Gum *(Eucalyptus grandis)* and other species.

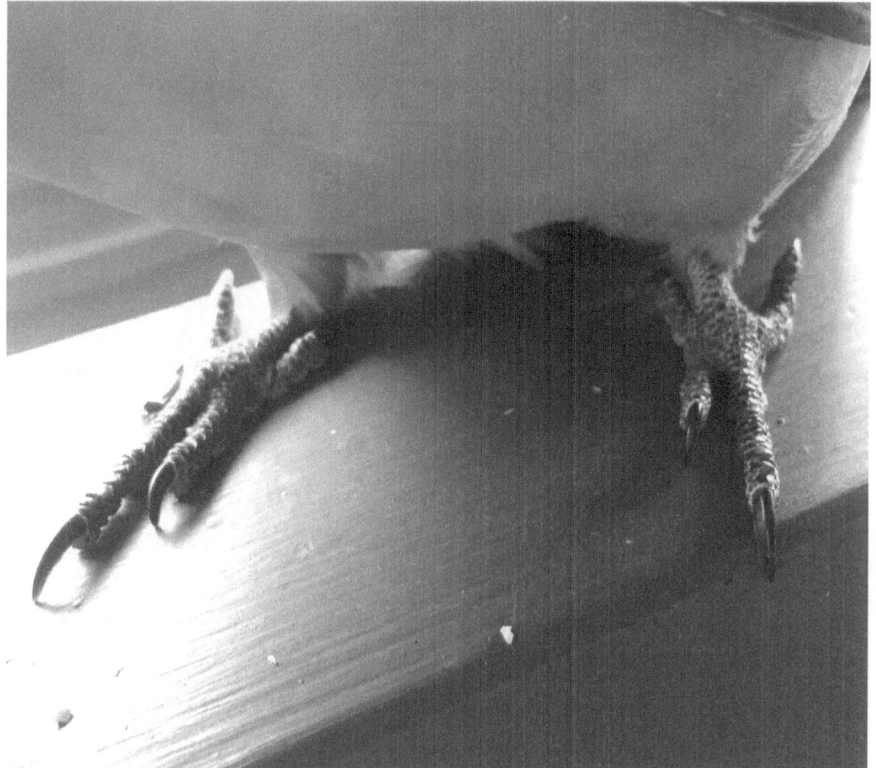

Fig. 12.9 Male king parrot feet, outer toes longest. Left foot and eye laterality. (Photo: Neil Frost)

As Ian Price would often remind us: "Animals prefer to conserve their effort. If they can do something that requires less exertion for the same reward, they will do it instead." This observation was derived from Ian's extensive biological knowledge and experience, gained from his very wide reading of scientific journals and books, his empirical and theoretical studies working as a biological assistant at Macquarie University and the CSIRO, together with his expertise gained as a herpetologist studying reptiles and amphibians. He appropriately applied this rule to our 'marsupial hominoids', across many issues that we would regularly discuss when sitting around our verandah table, whilst having a smoke and a cup of coffee. For example, Ian accurately predicted that Fatfoot and the others could mostly be expected to travel through the landscape by maintaining their altitude, following the isoheight or natural contour lines. As he simply explained, changing altitude requires more effort and extra food. This regular pattern of movement became increasingly obvious, almost from the beginning when we would, almost exclusively, pursue Fatfoot along the interconnecting valley floors. Similarly, considerable energy savings can be made by taking shortcuts, rather than going the long way around, even though the opportunity cost would be a greater risk of being seen. This was proven in multiple Octopus reports, where shortcuts were made over and under bridges that saved effort whilst also increasing exposure risk and generating encounter hotspots. However, he also predicted that there could be exceptions, where it would be easier and more energy efficient to move across the isoheights, where the vegetation is too difficult to move through. This was also shown to be correct in 1997 when Ian and Cheryl found tracking evidence of Fatfoot using the cleared easements beneath high-voltage power transmission lines to connect between tracks at different altitudes to our north.

Similarly, any extra food-gathering effort needs to be compensated with a higher yield to be worthwhile. Not surprisingly, with an overabundance of easily accessible larval infestations occurring in the less mature and softer branches of the tree canopy (Fig. 12.6), where a complex filigree of perching opportunities also abound, the yellow-tailed black cockatoos are not required to expend large amounts of effort in establishing a vertical perch and labouring to extract the embedded larvae from the seasoned, 7.5 kN hardwood trunk. This learnt strategy by the Blue Mountains cockatoos is a more energy-conserving extraction method that does not require any futile effort to establish and maintain an artificial grasp on the tree trunk, or the convoluted manufacture of a bark ledger.[13] Perching on the canopy branches enables a more natural resting position, provides more than enough larvae within easy reach and requires less energy to harvest.

Cutting larvae from the monumental trunks of hardwood trees requires much more effort than from the smaller and less substantial branches higher up. It also requires a different knowledge and skill set, if one is available. Yellow-tailed black cockatoos use their intelligence to learn the required identification and extraction techniques from their parents.[14] Rather than being instinctive, the bird's

---

[13] We have never seen a large yellow-tailed black cockatoo attached perpendicular to a tree trunk in the Blue Mountains and for good reason! Assuming that the cockatoo has ignored the other, more accessible and easily-extracted larval sources nearby and the bird's claws were capable of maintaining a secure attachment long enough to accomplish the lengthy task, there should be obvious and considerable collateral damage on the trunk, of which there is none!

[14] Young cockatoos that have been raised in captivity do not know how to locate and extract larvae. See: McInnes, R. S., and P. B. Carne. 1978. Predation of cossid moth larvae by yellow-tailed black cockatoos causing losses in plantations of *Eucalyptus grandis* in north coastal New South Wales. *Australian Wildlife Research* 5(1): 101-21.

behaviours are learned and are very unique to a particular region. They are specific to the dominant host species of *Eucalyptus* and specific to the local bird population. On the North Coast region of NSW around Coffs Harbour, the birds determine the location of the larvae chamber by looking for external borer holes and swelling under the bark. The parrot then cuts at the top of a vertical section of bark that is parallel to the timber grain and pushes it down to make a temporary perch. This is repeated until the accumulated bark ledgers can freely assist in supporting the bird's weight. The cockatoo then "chops out the wood to expose the larval gallery, hooks in her beak and pulls out the grub".[15] This explanation does not seem convincing considering the large size of the cockatoo and the initial starting platform that is required to be built before the supporting ledger itself can be constructed. When compared to the confirmed treebite made by an observed Dooligahl after sunset (Fig. 12.21), there are certain similarities between the photograph and the description of a 'bark ledger.' The timing of the Dooligahl treebite is important because these cockatoos would already be resting in their trees. It would seem, therefore, that the author is describing damage made by a Dooligahl rather than a cockatoo.

Like other parrots, yellow-tailed black cockatoos are very intelligent, adaptive and problem-solving birds. In the Blue Mountains, the circumstances and the inherited culture of larval extraction for this local population of cockatoos are different from the North Coast birds and most probably, from cockatoos in other parts of Australia as well. Here, the dominant tree species that host the larvae are bloodwoods rather than flooded gums. Having

---

[15] Ibid.

followed these very shy and cautious cockatoos around during the day with binoculars on many occasions and for long periods, there seem to be some very significant differences between the bird populations from these two regions. With bloodwood being the dominantly affected tree species in the mountains, compared to the smooth cream bark of flooded gums of the Coffs Harbour region, identifying a larval vent tends to be less obvious because of the bloodwood's course fibrous bark. However, with advanced infestations, there is considerable swelling under the bark, particularly where the bark has fallen away, as well as a very prominent red kino discharge. On most occasions, the upper or lower branches of the tree or a close neighbouring eucalypt, or bush are used to perch on, with the bird typically gaining hold with the aid of its clawed feet on one tree, the other, or both and sometimes assisted by the beak. As many of these bloodwood trees have multiple infestations across a broad range of heights,[16] it was very apparent that the cockatoos can afford to be selective in their choice of predation site, ignoring the ones with more difficult or impossible perching access that are lower down, in favour of easier and safer locations higher up in the canopy. Not surprisingly, these cockatoos have never been observed clinging to the side of a bloodwood trunk—unlike *Eucalyptus grandis* trees, the bark and cambium are too brittle[17] to allow the construction of a 'bark ledger' for the bird to perch on. However, according to the North

---

[16] Bullseye borer infestations in bloodwoods within the Blue Mountains extend from the lower trunk up into the taller reaches of the tree many metres above and are mostly located at a shallow wood depth compared to other larvae species.

[17] The bark of a Yellow Bloodwood *(Corymbia eximia)* is insubstantial, fibrous and crumbly. From an engineering perspective, its bark is incapable of providing an adequate bond with its own cambium layer. It is benevolently described by botanists as: rough, tessellated, flaky, scaly.

Coast study, "For some unexplained reason, less than 1% of the many thousands of galleries the team located were more than 2 metres above the ground. The average height was half a metre."[18, 19] This vertical distribution of borers is in sharp contrast to the flooded gum that we planted on our land in 1983, where larvae are only present in the top 20 or so metres of the 50-metre tall canopy and no borer infestations, at all, are present in the trunk.[20] Such an inverted discrepancy in the vertical distribution of the damage could be explained by different borer species being present at the two locations.

Since the bark of bloodwood trees is fibrous, brittle and insecure, for these very large birds, attempting to attach vertically and securely to the trunk with their claws would be suicidal and when this contact does occur on smaller horizontal branches, these large birds leave very obvious signs of their attachment through the easy and conspicuous removal of the fibrous bark. In addition to using obvious visual clues, like swollen bark and kino discharge, the Blue Mountains cockatoos listen for chewing and feel the transmitted sound vibrations of larvae movement by placing their beaks against the tree. I would imagine that any of the Wildmen would similarly place an ear against the trunk to achieve the same result. When excavating the larval chamber, the beak is used to cut across the wood fibres at one end, then the splinter is worked free and torn

---

[18] McInnes and Carne, *op. cit.*

[19] 'Marsupial hominoid' treebites found in bloodwoods in the Blue Mountains are typically below 1.8 metres.

[20] Steve, a person with forestry experience from the North Coast of NSW, commented on the absence of borers on the main trunk of our flooded gum, saying that the tree would be very valuable as plywood veneer if it was harvested.

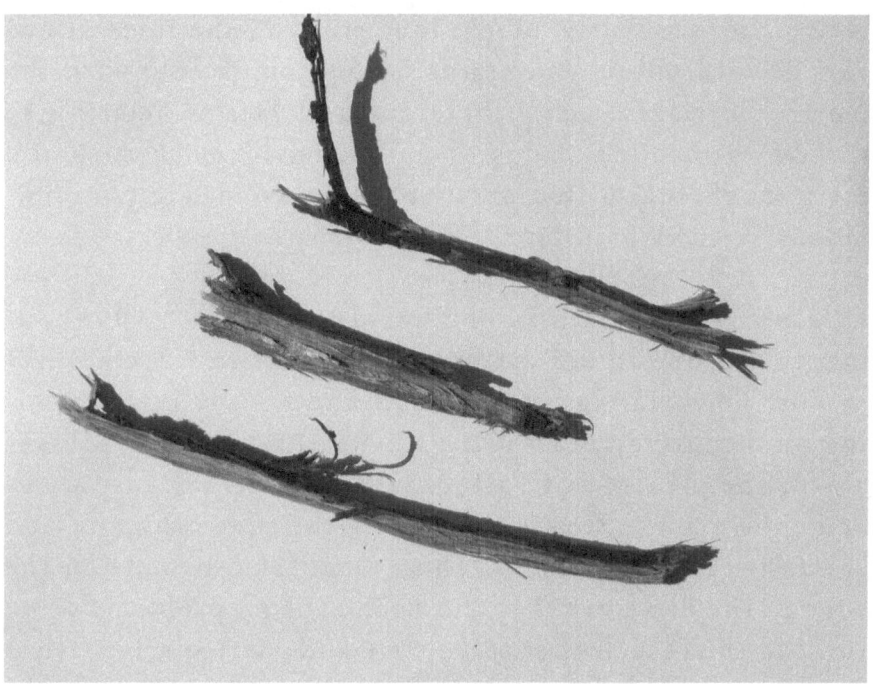

Fig. 12.10 Sample timber bite spoil from yellow-tailed black cockatoo treebite, showing random length timber splinters, with one cut end and an opposing ripped end. (Photo: Neil Frost)

away at the other end using the beak and left foot, producing a timber shard that is cut on one end, ripped on the other and of a random length (Fig. 12.10). The result is similar to the indiscriminate wood fragments removed from timber railings and gable trim by other destructive cockatoos.[21] With subsequent cutting and tearing actions, the same result is achieved, but each timber splinter is of a different random length depending upon the range of variance for this bird's ripping procedure. Once exposed, the grub is hooked with the beak and pulled out of the

---

[21] In Australia, the sulphur-crested cockatoo is the main destroyer of household timber. In New Zealand, kea have a reputation for being thieves, criminals and vandals, because they take objects and destroy cars in an attempt to get inside.

cavity. Consequently, our unique study of the infestations found in the Blue Mountains is not comparable with the North Coast research, where cultural factors relating to the two bird populations of yellow-tailed black cockatoo are undoubtedly different. Also, bloodwoods are not commercially cultivated on the North Coast.

During the afternoon of Thursday, 21 March 1996, our family was due to depart Australia for a two-week holiday in Fiji. The evening before, at around 8 p.m., I was standing on our driveway enjoying a beer and listening, as usual, for the presence of Fatfoot. I heard nothing suspicious and continued to slowly walk along the edge of the bush. I came alongside a group of three small bloodwoods on the edge of the driveway that had been causing problems when turning the vehicles for some time and particularly, earlier that day. After briefly thinking about the obstructions, I decided to cut all of them down on our return from holidays. I went to bed.

Typically, having had trouble sleeping the night before our holiday departure, Sandy got out of bed at 4 a.m., about three hours before sunrise. After making a cup of coffee, she sat outside at the verandah table. Over the following hours, her attention was increasingly drawn to a small section of a tree on the edge of the driveway that seemed lighter in colour compared to the surrounding bush. As dawn approached, the contrast between this and the nearby bush had become so noticeable that she believed that the tree had been severely damaged in some way. As this aberration was something new to the driveway, Sandy went inside and woke me, so that we could jointly inspect it.

When viewing the tree before dawn at a distance of fifteen metres from the verandah, the exposed white heartwood indicated that the tree had sustained a substantial injury of some kind. After inspecting the tree more closely, it could be seen that a section of timber had been removed. More than this, it was also immediately obvious that the tree had been bitten, rather than randomly chopped at by a cockatoo, or mechanically damaged in some manner. Deeply cut through the heartwood were two parallel tears and other evidence of unusual marks in the wood. As a result of this, a large cavity within the timber was exposed where a larva had been burrowing. The intended purpose of removing the timber in this unusual manner had been to extract larvae for food, just as cockatoos do. We were awe-struck. The immediate impression was that the damage was caused by large canine teeth and that a great bite force had been required to achieve this key type of destruction in hardwood. Since I knew that these trees were undamaged earlier in the night, the damage had been caused within an eight-hour evening window and the damage had occurred at least three hours before sunrise—yellow-tailed black cockatoos could not have been responsible for the damage. Some other culprit must have been responsible and the immediate suspect was Fatfoot.

The median elevation of the tree damage from the ground was about 1.84 metres (6 feet). No known extant ground-based native animal can attain this height, except swamp wallabies, which do not inflict this type of severe injury. After many decades spent observing wallabies, we have not seen them biting trees, or even licking the kino, though they do peel citrus fruit, particularly orange and mandarin, eating the rind and discarding the pulpy interior. Sugar gliders have no effective restriction on the height that they can reach to tap a tree, but they do not inflict severe damage with this particularly deep and striated pattern. Introduced pest animals, such as goats, like to damage the lower trunks of bloodwoods, eucalypts and other trees with their horns and incisors to a height

Fig. 12.11 Fatfoot treebite from the edge of our driveway, showing larval chamber with bore hole. Nighttime provenance, between 8 P.M. and 4 A.M. Showing multiple, stacked, parallel, 60 mm canine tears. (Photo: Neil Frost)

approaching six feet so that they can lick the kino. Such behaviour is typically the result of the goats being neglected and malnourished, which results in the ringbarking and death of the tree. Fortunately, at that time there were no domestic or feral goats in the valley. Feral pigs have been seen biting at the bases of eucalypts to obtain the larvae and also lick the kino from the damaged sapwood.[22] Similarly, cattle have been known on occasion to bite the bark on trees and rub up against the trunks. We have no cattle in our valley. Regardless, as herbivores they are not known to seek this source of food and their broad chisel-shaped, lower incisors would be incapable of producing this characteristic damage.

Clearly, as with most food niches, there is more than one larvae predator. Without a doubt, it seemed to us that an unacknowledged 'marsupial hominoid' was responsible. Apart from possessing the height and the robust cranial musculature required to provide the bite force needed to extract the larvae, the damage occurred between 8 P.M. and 4 A.M., the typical time range where Fatfoot was most active. Also, close to the base of the three trees there was a considerable amount of flattened native grass that was identical in appearance to trampled areas that Fatfoot was previously known to have produced and which had been earlier studied in detail, along with nests or beds. Obviously, a cockatoo would not cause grass flattening during its aerial predation.

As time was limited before our departure, I photographed the bite and made several measurements. Perhaps the most important measurement was the distance between the parallel tears. Since it was apparent that these were

---

[22] This goat behaviour was observed by us, involving some irresponsible landowners. This feral pig behaviour was observed by local professional shooter Steve Crofts in other areas outside of the Blue Mountains.

Fig. 12.12 Another top view of Fatfoot's treebite showing multiple, stacked, parallel, 60 mm canine tears and the larvae chamber. (Photo: Neil Frost)

tooth marks, I referred to them as the 'canine separation' which was measured at 60 mm (2⅜ inches). The diameter of the small bloodwood tree was 100 mm (4 inches). Background information from Aboriginal Elders told stories and produced art that portrayed Dooligahl and Quinkan as having prominent canine teeth that visibly extended over the lips.

At the base of the tree where the native grass had been flattened, were a large number of bark and timber fragments. Looking at these it could be seen that both ends were cut—a consistent fit for length was possible by placing them between the canine tears. Measuring the timber splinters confirmed that they were each under 60 mm long.[23] All fragments were bagged and labelled. The remaining task was to preserve the treebite. As the time for our international departure approached, I decided to roughly cut the tree below the bite with the chainsaw. With the tree safely on the ground, I used a hand saw to neatly trim and remove the unnecessary upper and lower sections. The tree sample was bagged and together with the fragments it was placed in a postal box for immediate mailing.

Going overseas for a few weeks meant that we were impatient for an opinion on the unconfirmed predator, so I sent the sample to Steve Rushton, a good friend, sponsor and fellow researcher. Believing that Steve would be equally excited by the find, I knew that he would relentlessly seek alternative bird opinions from many sources in

---

[23] Much later, in January 2002, when we had a brief period of scientific support, these sample fragments were considered for DNA analysis by members of the CSIRO but it was thought that too much contamination would be present for a reliable result. Two decades on, I would imagine that it would be much easier to obtain cheek cells for analysis if the support was still available.

our absence. On our return from holiday,[24] I immediately telephoned Steve. He told me that biologists from the University of Queensland in Brisbane believed that the damage was caused by black cockatoos. Needless to say, Steve attempted to suggest that perhaps another perpetrator was responsible for the different pattern of damage. In discussing their conclusion, Steve and I agreed that they were conveniently oblivious to the orderly arrangement present in this sample, compared to the commonly found randomised bird damage seen in trees. We thought that, at least, they should have commented on the unusual aspects of the physical evidence and were perhaps being too readily dismissive and narrow-minded. It seemed that the university scientists were simply being academically safe, after suspecting that Steve was approaching the subject out of left field. By not taking any intellectual risk, no scientific progress was made, but at least their collective reputations were kept intact and their current fields of study remained unchallenged and inquiry free. We thought that the nocturnal provenance of the find should have been regarded as highly significant and worth further consideration. In conclusion, we felt that on our part, we had been extremely naïve in expecting an open discussion. After all, these animals don't exist! Such conservative attitudes are a common obstruction that prevents the revision of established paradigms—they were to become a portend of what to expect in the future.

---

[24] On the evening that we returned home after spending two weeks in Fiji, from the car we all saw the red eyes of Fatfoot follow us down our driveway. I think that she missed us! Coincidently, a number of us wondered why Fatfoot had chosen one of these three bloodwoods from among the many alternatives. Apart from being infected, it seemed reasonable to assume that Fatfoot had been observing my keen interest in these specific trees and wanted me to be aware that she was watching my activities.

12: Alibi

Subsequently, when presenting future evidence for comment, it became advisable to introduce the samples naïvely and without any suggestion or reference to what we already knew. To do otherwise would inevitably invite spontaneous ridicule, during which it was usually possible to observe 'the shutters coming down'. I called this wary approach "avoiding the 'Y' word"—where 'Y' is a reference to anything Yowie. Regardless, even after exercising extreme caution and restraint when dealing with academics, scientists and government authorities, their preferred method of disposing of the variant evidence, was to make it forcibly fit into the established paradigm, no matter how badly. However, when the researcher's frustration and intolerance inevitably reach their elastic limit, the developing situation demands that these set ways and blind conclusions should be challenged, whereupon it is usually interesting to notice the appearance of an all-knowing smile or authoritative tone of voice, either of which is guaranteed to raise the Dalai Lama's blood pressure. Sometimes the respondent might even hint at or use the 'Y word' themselves, which creates a rare opportunity for banter!

There are many aspects of the new evidence that should, at least, raise the critical attention of open-minded people. Some details of the evidence are less obvious than others because it requires a background knowledge of certain subjects that few understandably have, but enough information should remain that could be classified as common sense. For example, the treebite occurred at night, one day after the autumnal equinox, between the hours of 8 P.M. and 4 A.M., when the assumed co-predator is perched in a tree asleep. By itself, this should be enough evidence to suggest that the bird has an alibi and that some other assailant did the job.

On the first weekend opportunity after returning from our holidays, we spoke to Ian. Shortly after showing Ian the tree sample and without being prompted, he said, "Something with canines has extracted the larvae and would that something happen to be our hairy beastie?" After a brief discussion, Ian agreed with us that the damage had not been done by a yellow-tailed black cockatoo, commonly observed throughout the mountains. He similarly commented on the uniformity and neatness of this treebite compared to the messy, random damage inflicted by cockatoos in general, across a wide range of wooden targets, including gable barge boards. He went on to add that some of these wood borers can be very large and would make a substantial contribution to an animal's

Fig. 12.13 In this photograph of an *Acacia* branch, an incomplete 60 mm canine bite by Fatfoot can be seen. The left canine has completely cut through the timber shard, with the right canine leaving a compression mark in the wood without completely removing the splinter. (Photo: Neil Frost)

12: Alibi

Fig. 12.14 One of the largest of the thirty Quinkan treebites found along the edge of the road (linear distribution) and nearby that day, close to Jerry and Sues' house. About six feet from the ground, the lower canine separation was 55 mm with a 6 mm rounded canine tip ripped through the bark, being measured by Jerry. Upper canine separation of 82 mm not seen in this photograph. (Photo: Sue O'Connor)

diet with their high protein content and various fatty acids. When we told Ian that Sandy had found the damage during the night and that we were certain of its nocturnal provenance, Ian felt reassured that our conclusions were correct. However, Ian thought that we should visit his boss, who was a professional shooter and wildlife expert and show him the sample. Ian and I got into the ute and we drove to Steve's house.

When I asked Steve Crofts to comment on the treebite, he said that the damage was "extremely impressive". He said, "Everyone knows that black cockatoos eat the grubs and that it was clear that the responsible animal, in this case, had large canine teeth and the required power to easily bite through the hardwood timber. This damage is different!" The purpose in doing this, he said, "was to extract the wood grub from within the heartwood which is a rich source of protein that Aborigines frequently eat". He said that he "had tried this bush tucker on a few occasions". To achieve this, he used an axe to deeply cut away the timber on all sides as "the grub would maintain a tenacious hold." Looking closely at the sample, Steve said that "whatever did this was quite lazy",[25] as the bite was simply not deep enough to allow easy extraction. He thought that "the animal would have required a long claw or perhaps a stick to dig the grub out of the deep cavity".

THE TREEBITE WAS sustained at a median height, or midpoint of the damage, 1.84 metres or six feet above the ground. The tree was a yellow bloodwood with characteristically light cream- to yellow-coloured heartwood, with rough fibrous bark. There were no immediate branches capable of supporting a bird above the damaged area within a metre or more and none at all below it. There was no bird-related collateral damage in the bark below the treebite, suggestive of any attempt to latch onto the tree or make a temporary perch.[26] The fibrous and brittle nature of the bloodwood bark would not have permitted the construction of a supportive perch, or ledger, as is the supposed behaviour of North Coast cockatoos, because the tree material is clearly unsuitable. However, there was some bark removed around the edges of the bite where

---

[25] Steve's comment was in agreement with Ian's belief, that all animals conserve energy by doing the bare minimum amount of work.

[26] Including supposedly microscopic claw marks that were the fabricated explanation proposed as a method of bird attachment to the tree on the ABC TV science programme *Catalyst* in 2002.

the 'canines' had passed through and dislodged peripheral fragments.

Around the base of the tree had been some undisturbed native grasses that occupied the confined space between the three immature bloodwoods near to the northern side of our driveway. After the incident, these grasses had been flattened and compressed in the same manner that we had become accustomed to seeing, after witnessing regular encounters around our property. On the flattened grass and around the base of the bitten tree were the timber shards that had fallen to the ground. On inspection, the length of the timber splinters fitted neatly and fully within the bite, with many consistent lengths of slightly less than 60 mm.

From the beginning, I would have thought that the patterned and ordered cuts contained within the tree damage would have at least raised a few eyebrows. The precise nature of the damage seemed manufactured, rather than natural. Apart from the severity of the bite, there were two parallel tears repeatedly gouged through sequential layers of hardwood, which I thought should have screamed of something unusual taking place.[27] If people were already familiar with cockatoo damage when examining this sample, particularly from the notorious sulphur-crested variety, they noticed the difference immediately. This is because cockatoo damage is typically very messy and destructive, as these birds are not afraid to use their bills for fun through random and meaningless vandalism or possibly as a legitimate means of keeping their bill sharp and to size. They do not produce regularly arranged cuts

---

[27] After this treebite sample was passed around the 300+ audience of the Myths and Monsters Conference, Sydney, 20 October 2001, from a show of hands, the majority of people recognised the likelihood of canines being responsible for the damage.

with parallel striations of uniform spacing throughout the bite and more significantly when compared across multiple treebite examples. The nature of the damage seemed to be obvious, as with timber that has been cut using an axe, compared to timber cut by a saw. Frequently felt was the overwhelming urge to point out the difference in the destructive style of cockatoos, versus whatever did this. For those who were not familiar with the randomness of

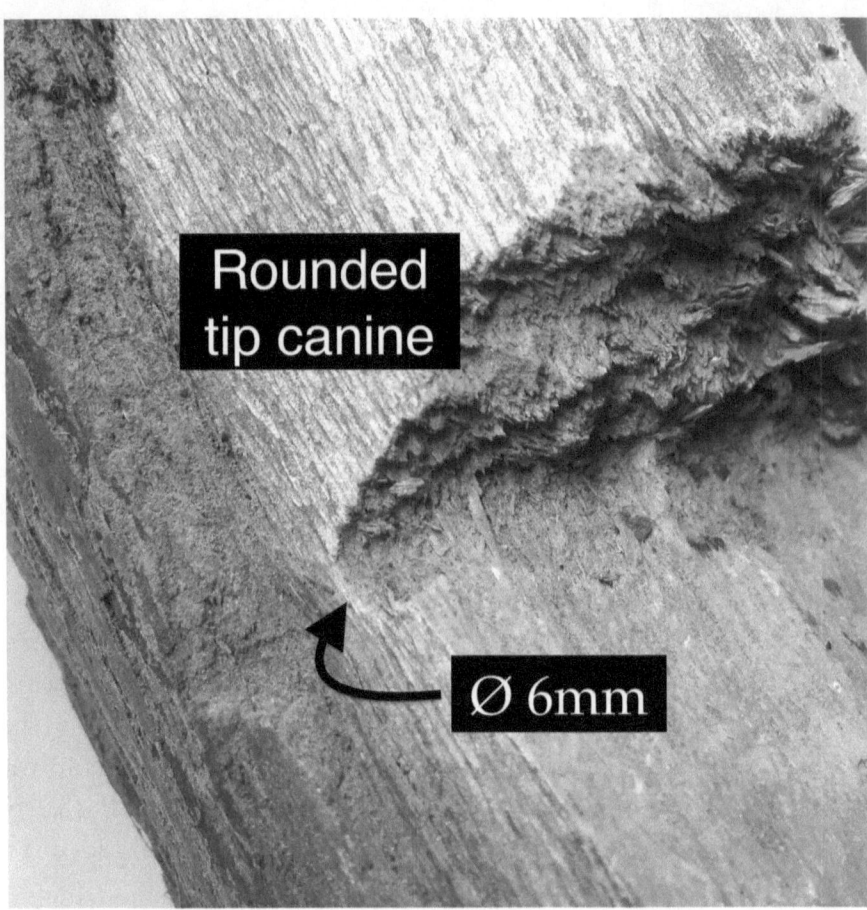

Fig. 12.15 Round-tipped canine impression, 6 mm Ø, in yellow bloodwood with a Janka hardness of 9 kN. (Photo: Neil Frost)

cockatoo destruction, this orderly damage did not seem questionable, like those who have never cut timber.

Instead, we began to realise that most casual observers were regarding the parallel tears as a chance alignment of two separate cuts, on a sample considered in isolation. This simple explanation might have seemed easier to accept, compared to the alternative possibility, that the precise tears were the repeated result of two fixed canine

Fig. 12.16 A 6 mm trench can be seen at the basal level of the 60 mm wide treebite, with numerous layered cuts being visible on the opposing side. (Photo: Neil Frost)

teeth belonging to a powerful animal with a sagittal ridge and secretly lurking in the Australian bush. I think that the subconscious thought was too horrendous for most people to openly acknowledge.

Proving that the continuous rips were a matched pair was not particularly difficult. Firstly, it should be noted that the two tears are not only parallel, but they both start and finish in the timber adjacent to each other at right angles. Clearly, this indicates that the two cutting edges were at a fixed distance apart as they were dragged concurrently through the timber, entering and exiting the timber, usually at the same time. If the two tears had each been separate incidents, then the likelihood of this dual alignment being both parallel and equally terminated, as well as being replicated throughout multiple layers of timber, would be highly improbable. Secondly, as the wood had not been totally removed in a single attempt, as clearly indicated by the multiple wood fragments, several layers of timber had to be progressively detached with each overlapping bite. Measuring the remaining whole fragments revealed a consistent length of slightly less than 60 mm (2⅜ inches), which was marginally within the separation of the two canine rips. This consistency of splinter length across the majority collection of fragments, further confirms that canines, or some other fixed, gauged pair of ripping knives, were most likely responsible for the regularly produced damage. This conclusion is further supported by identical fibrous cut marks being present at both ends of most fragments. If a bird had raked the timber with its solitary beak, then the damage at one end would most likely be dissimilar to the opposing fragment end. This is because observations of cockatoo behaviour reveal their habitual tendency to cut through a section of timber and then lift the freed end with their beak and, aided by the left foot, twisting and levering the piece until it is torn away.

12: Alibi

By later attempting to assemble some of the willing timber fragments within the excavated cavity of the tree section, it was possible to crudely reconstruct a few of the many bite layers. This was a lengthy and very incomplete process which was not particularly satisfactory because the originally hydrated timber fragments had substantially warped and shrunk as they had dried during their long return postal journey from Queensland. However, the exercise did reveal, as best that it could, that the splinters were repetitiously made as a series of diagonal cuts, that were later cross-cut to finish excavating the larval cavity. After multiple passes had been made, the angle of attack was adjusted and mirrored in a series of slanting rips at

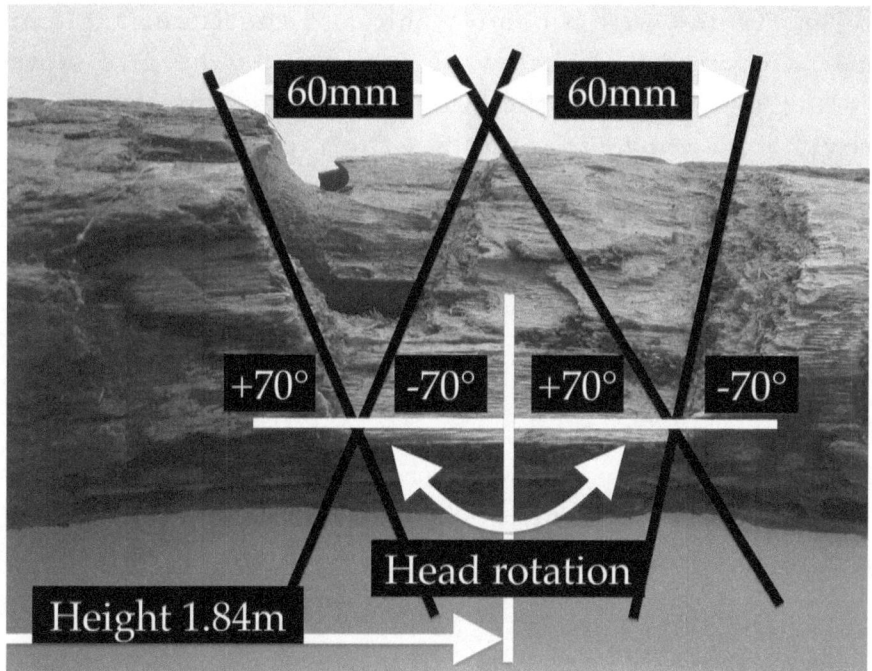

Fig. 12.17 The central axis of the treebite showing the height from the ground at 1.84 m; head rotational axis; lower and upper paired canine tracks with a common central origin. Note the perspective distortion present in this photograph. (Photo: Neil Frost)

an opposing angle. Like an axeman who is methodically chopping a log, cross-cutting in this manner allows the timber spoil to be more easily removed with less effort. Each opposing pair of cuts traversed the timber at an angle of 70° to the outside edge of the tree. For Fatfoot, changing this ripping angle would simply require a tilt of the head, as the common point of neck rotation would easily allow.

Alternatively, for a bird, there were no underlying or overlying supporting branches on the bloodwood, no constructed ledger or any visible evidence of it being attached by its claws, despite the claims of a forestry expert. If the cockatoo had been able to attach itself horizontally and the bird was attempting to rip this tree in the same uniform manner, with a comfortable and consistent angle of attack coming from below, it would either need to move to the other side of the larvae infestation or reposition itself from above to complete the cross-cutting task. A vertically-attached cockatoo would be unable to replicate these cutting actions. In either case, it must be remembered that the immature bloodwood was only 100 mm (4 inches) in diameter which would make the exceedingly cramped working conditions impossible for the bird to accommodate.

Clearly, the canines that undertook the majority of cutting for this excavation were the upper set,[28] because subsequent treebites confirmed that Fatfoot's upper canine separation was 60 mm and the lower was 55 mm. The

---

[28] The lower canine damage shown in Fig. 12.14 were from a Quinkan with a lower separation of 55 mm. The upper canines with a separation of 82 mm are not clearly shown in this photograph because the timber substrate that they were cutting through were less substantial and fell away to the ground as spoil. This photograph provides evidence of a large gape angle.

opposing presence of the lower canines is not readily evident in this tree section, except for a possible canine compression point on the rival edge of the bark, which indicates that Fatfoot's jaw articulation or gape angle, similar to the thylacine,[29] is particularly large. Being able to apply an opposing force using the lower canines is obviously an essential requirement for Newton's Third Law of Motion. In comparison, the gape angle of my mouth would not be able to circumvent the tree's diameter, forcing me to nibble away at the timber in a similar way to the action of a bird.

The impressions made by the upper canine set can be seen at several locations on the timber. The clearest and best preserved canine impression appears at the beginning of a rip (Fig. 12.15), where the maxilla has made its deepest and final pass of the timber. At the start of its run, the canine made a very clear circular compression mark of 6 mm in diameter in the hardwood timber.

By using some basic domestic equipment it was possible to simulate the compressive damage made by Fatfoot's bite. A pair of 8 mm, fully-threaded, metal bolts were used to represent Fatfoot's upper canines. The ends of both bolts were rotary ground down with a 3 mm radius (6 mm round, overall cross-sectional profile), that were rigidly set 60 mm apart using 30 mm square section steel and some hexagonal nuts. The metal jig was fixed beneath the end of a hydraulic cylinder. Bathroom scales with a maximum capacity of 120 kg were used to measure the downward force in kilograms. With the original treebite

---

[29] A thylacine jaw can open to an angle of eighty degrees (see Fig. 5.11), although it is doubtful that a Dooligahl or Quinkan could achieve this maximum angle and still maintain a powerful bite force. Based simply upon engineering vector principles of force and mechanical advantage, bite force decreases with an increasing gape angle.

Fig. 12.18 Bite force test using a hydraulic press, bathroom scales, simulated metal canines (60 mm separation and 6 mm Ø) and the original timber sample. The actual canine compression mark used for depth comparison and the trailing tear through the timber can be seen to the right of the metal jig.

placed between the jig and the scales and secured with a clamp, a downward force was progressively applied to the sample. As the hydraulic pressure increased, a subtle noise could be heard as the timber fibres compressed, until the depth of the impression matched, but did not exceed the canine depression visible alongside.

The 120 kg load is equivalent to the mass (m) multiplied by the gravitational constant (g), 9.81 ms². This gives a value for the bite force of m x g of 1177.2 Newtons (N). Since the total load was equally spread across the tips of two metal canines, each with a diameter of 6 mm, this gives a total compressive area of 2 x 28.3 mm² or 56.5 mm². Dividing 1177.2 N by 56.5 mm² gives an International System (SI) force of 20.8 N per square millimetre or 20.8 MPa. Converting this to a more commonly understood but non-scientific standard gives a bite force of 3017 psi at the canines.

High bite forces are typically associated with carnivores that take large prey. Such an impressive bite force is consistent with the large heads and correspondingly massive musculature and sagittal ridges as seen on Dooligahl and Quinkan but not Junjudee. It also means that the skulls of these animals must be robust or rigid without flexure, in order to support the muscular strain. Being a very high bite force, it is generally equivalent to jaw pressures found in carnivorous ambush predators like the Nile crocodile and American alligator that seize their prey by mouth, although with these 'marsupial hominoids', having arms and hands would be more effective in holding and dispatching prey. Understandably, there is a degree of variability in quoted bite force measurements, that can be influenced by many intervening circumstances. However, the animal with the greatest quoted bite force is reportedly the saltwater crocodile, at about 7700 psi. The bite force of large herbivores like gorillas and hippopotamus is estimated at about 1300 psi and 1800 psi. Interestingly, large parrots like macaws have lesser force estimates of 400 psi, with sizeable cockatoos being frequently quoted at about 350 psi. The lower bite force of parrots is able to be supported by their keratin-based beaks, which must rely upon their shape for strength. Such beak pressures, combined with their oral and manual dexterities, are sufficient for their granivorous needs and other purposes.

In mammals the outer layer of teeth, including canines, is made from enamel, an inorganic substance. Although enamel is very hard, it does wear away and is not replaced. This is in strong contrast to the curved and pointed keratin beak of all cockatoos, that leave sharp but not rounded tear marks in timber. Like human fingernails, parrot beaks are constantly growing—keeping them sharp and narrow-tipped is a normal consequence of their abrasive feeding activities.

On further examination of the timber section it was possible to reasonably determine that Fatfoot had been standing on the eastern side when she bit the tree.[30] This is because the two sets of diagonal bites intersected at a 'V' on the eastern side, where it could be imagined that the head would have been as it easily pivoted when changing the angle of attack from -70° to +70°. As we had determined the height of Fatfoot over the years to between 2.08 m (6 foot 10 inches) and 2.13 m (7 foot), the 'V' intersection measurement of 1.84 m, plus the height of the head, would seem to be reasonably consistent. As the bite rips extend fully across to the far side of the 100 mm diameter tree, this might fairly suggest a comfortable bite depth and articulation that has not reached its full

---

[30] The damaged section of the tree had been facing south towards the house.

limit.[31] By comparison, my maximum and comfortable bite depth measured to the canines is only 26 mm (1 inch). This was calculated by forcibly placing a length of dressed softwood across my mouth and pushing it as far back as it would go, biting down hard on it and measuring the depth of the tooth impressions made from the edge of the timber. My jaw articulation, or gape angle, is only 25 degrees and I would clearly be incapable of producing such damage.

From its easterly position, Fatfoot made two angles of attack on the bloodwood with its canines. During the first series of bites, the head was tilted down at an angle of 20° below the horizontal and the canines were continuously ripped up through the timber, over the top of the central larval chamber. The partial remains of this series of rips can still be seen in the timber for the left canine only, at the bottom of the treebite. The number of passes made by the teeth is uncertain, although gauging from the number of fragments and striations still present in the timber, about eight traverses seem to have been made. After opening up the tree and partly exposing the larval chamber, the position of the head was rotated upwards from its centralised location, with the jaw now biting down through the timber, also at an angle of 20° above the horizontal, creating the characteristic 'V' at the convergence of the canine rips, close to the mouth. This later series of bites partially over-cut the previous set, removing earlier evidence

---

[31] The thylacine or Tasmanian tiger *(Thylacinus cynocephalus)* is a convergent 'marsupial dog' having a jaw articulation, or gape angle of 80 degrees and a powerful bite force. The Tasmanian devil *(Sarcophilus harrisii)* is a convergent 'marsupial hyena' that also has a jaw articulation of 80 degrees and has the most powerful bite force relative to the body size of any mammal at 1200 psi. It is claimed to be the largest extant Australian marsupial carnivore, although, clearly it is not!

of the damage made by the right canine. Clearly, this rotation of the head would have allowed the right canine to asymmetrically concentrate the bite force on the remaining and less supported ledges, while the left canine repeatedly bumped along the exposed base, creating a 6 mm trench. This final canine pass would have given greater access to the insect larvae. As with the first series of bites, about eight passes seem to have been made during this second attempt, with the final canine pass being pressed deeply into the timber heartwood. Since both canines have left well-defined compression marks in the timber along their journey, the thickness of the rounded end of these canine teeth was measured at 6 mm or ≈ ¼ inch. Such blunt compression marks would not be expected if the damage had been made by the pointed beak of a large cockatoo.

In combining the evidence so far, other conclusions can be drawn about the probable characteristics of the jaw and head. Firstly, the maxilla and mandible are obviously large. The upper canine separation is 60 mm (2⅜ inches). Another name for these tearing fangs is the 'eye teeth' because they initially descend from below the eyes in the maxilla. Consequently, Fatfoot's upper canine separation of 60 mm compares favourably with her measured inter-pupillary distance (IPD) of 66 mm, which was determined from a photograph. However, this 60 mm canine separation compares with only 28 mm (1⅛ inches) for me, bearing in mind that I have the same IPD, but a narrow palate and associated crowded dentition. When I was young, the dentist had to remove several of my upper and lower teeth to allow enough room for the remaining teeth in the jaw to develop without excessive crowding. My anthropometric variance at the bottom of the normal range is most probably the result of my northern European genetics, where long and narrow faces are an adaptation to cold climates because the restrictions produced cause increased nasal turbulence, which assists with warming the intake of air. Interestingly, the photograph of Fatfoot's face suggests that her nose is similarly leptorrhine. Susan's encounter testimony stated that the "nose appeared normal but narrow" and that the "head and face looked like a normal human, except that the head might have been longer".

Expressed as a factor, Fatfoot's canines are spaced 2.14 times further apart compared to mine. Dental research would further suggest that my crowded teeth and poor jaw development are also the result of eating soft foods, unlike Fatfoot who has an obvious disposition for tougher foods, particularly involving Australian hardwoods. Consequently, the density, shape and robustness of our mandibles are influenced by the forces that are applied to the jaw bone by the attached muscles. For Fatfoot, having a sagittal ridge would allow addition muscle attachments that would further increase the bite force and directly affect the mandible's structure and robustness. Confirmation of Fatfoot's harsh dietary habits are also evident by the extensive bluntness of the canines (6 mm or ¼ inch at the tip). As for the condition of the other teeth, in many early human societies, dental health and hygiene played a major role in determining longevity, where tooth infections from abscess caused many early deaths. These infections were typically the result of having an abrasive diet where, for example, the grit from grinding wheels contaminated the flour, resulting in the abrasive wear and the penetration of tooth enamel. Similarly, in comparing bite depth, Fatfoot's demonstrated reach of 100 mm (4 inches) is, at a minimum, 4 times greater than my 26 mm (1 inch) extension. In addition to other potential factors, such a deep bite reach would indicate that Fatfoot, most probably, has a large gape angle. In considering this, these simple comparisons would indicate a jaw and head that is physically much larger than mine and further reinforces the non-human origin of these bipeds. Consequently, when these characteristics are viewed from within a marsupial context, the morphological evidence conforms closely with the known traits of other, lesser Australian carnivores, such as the Tasmanian tiger and Tasmanian devil.

In having a larger head, this finding was very reassuring since many eyewitnesses, including Ian and Sandy, had reported that Fatfoot's head was visibly at least 50% larger than a human's, though a significant proportion of this was probably hair. Most sightings occurred in silhouette when the backlight conditions were favourable and the witness was able to view most of the body. This allowed some sense of proportion and scale to be determined. On a few occasions, the use of night vision binoculars permitted similar observations, although determining scale was always difficult under those circumstances.

Finally, during an unpalatable and very silly attempt at simulating a treebite, I was able to easily remove the outer layers of bark and some sapwood from a bloodwood eucalypt with my teeth. The following attempts to bite into the sap and heartwood were clearly pointless, because my jaw's size and gape angle were grossly insufficient. However, what became obvious was that my incisors were mostly in contact with the timber and not my canines. Therefore, it was apparent that Fatfoot's canines must be highly prominent for them to be able to do the work required. This confirms Aboriginal knowledge and artwork depicting protruding canines on Dooligahl and Quinkan.

SOMETIME AFTER FINDING the treebite on the edge of the driveway in 1996, we started to find other damaged trees in the bush around the house that were very similar. We were mostly seeing the bites in yellow and red bloodwoods,[32]

---

[32] We would very occasionally find a bitten *Angophora costata*.

typically at heights between four to seven feet, but mostly around six feet. They had the typical V-shaped tear marks that were found in the original sample and were located in positions around our property that also happened to provide a good view of the house. We already knew that the house was being watched, but finding treebites, like areas of flattened grass, was further physical confirmation of the association and helped to identify these viewing locations more precisely. We were also aware of a viewing cycle, where Fatfoot would usually start foot thumping from the northwest as we prepared the evening meal, followed by thumps from the northeast when we had moved to the lounge room and finally, from the southern and southwestern sides as we prepared for bed. Over the seasons, these locations tended to be where the new damage would appear. It also meant that while Fatfoot was observing us, she was having a larval snack if available. This was also a period of local bandicoot extinction as we started to notice a rapid decline in the population and we didn't think that cats and foxes were necessarily to blame!

As spring approached each year, the yellow-tailed black cockatoos would begin to migrate to the mountains and we would see the large birds loudly calling as they flew in large, loose formations overhead. On arrival, they would move through the treetops in the valleys, looking for and extracting the grubs. It also meant that the Hairy Man treebites would start to appear as well. As a result of the seasonally-timed supply of larvae, both major predators and a few lesser ones would also begin their synchronous predation. Anyone who stayed in the Blue Mountains was aware of the yellow-tailed black cockatoos during this time, mainly because of their raucous calls and slow, labouring wing beats. However, few were aware of the similar larvae-harvesting activities of the Hairy Man and how to differentiate their damage from the big birds!

By the spring and summer of 1997, we started to find more of the characteristically different treebites in other areas away from the house, with some being found many kilometres into the surrounding bush. Since these characteristic treebites were associated with the Hairy Man, we were on alert to find intermediate examples that might indicate a route through the valleys, by using a simple process of connecting the bites. Up until then, we had mostly been using a network of cotton threads strung through the valleys in order to detect the regular paths that were being taken. This bite identification and tracking procedure was greatly assisted by Ian as he drove his truck around the local area delivering building supplies, interviewing locals and looking for damaged trees and other evidence on the side of the road. On a number of occasions, particularly on weekends, Ian and Cheryl would take long walks in areas where Ian had found roadside evidence during the week or had been told about recent activity by local clients. Mostly, we were finding treebites with the 60 mm and 55 mm canine separations, but occasionally we were finding differently spaced samples and trees that had been very extensively damaged so we started to consider alternative explanations.

Over the next few years, we noticed that there was a strong linear association with the treebites where damaged trees would be found along the sides of bush tracks. There were also fields of bites in certain places where bloodwood trees more densely dominated the forest, with this association being more areal. We also became very proficient at identifying previous season treebites that had become callused over. As most of the areas that we were investigating were Crown Land, I would quickly drop a tree with the chainsaw, dock it to a manageable size that would fit the car boot and drive it home for a treebite autopsy and in the hope of not drawing too much local attention to this eccentric behaviour. Fortunately, on a number of these tree-napping expeditions I was accompanied by Steve, a NSW police officer, whom I hoped would be able to explain away any difficulties that we might encounter with council rangers or local police. By using a handsaw and a few very sharp chisels, it was possible to expose the old hidden treebite damage by easily removing the vertically callused tissue. On one occasion, Steve videoed the disclosure process so that he could post the video for other researchers to see the very likely association of a particular callus pattern with earlier treebite damage.

By late 1999 it seemed worthwhile to purchase a GPS receiver and make the transition from film to digital photography. Having a GPS handset would allow for the accurate recording of artefact positions as they were found in the bush and more easily assist with relocating them afterwards. It would also allow the position of treebites to be easily and more accurately plotted on a topographic map, making it simpler to visualise any patterns or relationships. Since GPS receivers were much more expensive to buy in Australia at that time, I purchased a Garmin eTrex that was much cheaper from the USA, even in spite of the poor AUD exchange rate. Similarly, I thought that purchasing a digital camera would have a number of advantages over film, by being instantly accessible, not requiring the cost of film, no processing costs and no lengthy delays involving strange looks from photographic assistants. Admittedly, the poor resolution and volatility of digital photographs stored on magnetic media was not fully appreciated at the time. Regardless, I bought an early 640 X 480 or 0.3 Mp digital USB camera. For the first time, it became possible to take extravagant photographs from multiple views of every treebite found with the digital camera at no exposure cost and accurately log its position with the GPS.

In early 2000, we were contacted by Jerry and Sue O'Connor and I visited them. After a lengthy introductory discussion and a driving tour involving some points of interest, we all went for a walk around their local area and apart from many other things, we soon found thirty treebites in their surrounding bloodwood trees. All of the damaged trees were digitally photographed, which wouldn't have occurred if film was still being used unless there was a need to store a higher-resolution image. Fortunately, Sue continued to use film for her records.

There were a number of similarities between our treebites. Apart from being the same tree species, the rips through the timber were very similar, having deep parallel gouges that produced a number of cut splinters of equal length, but the method used to remove the timber shards seemed different. Rather than having a V-shaped pattern resulting from alternately angled rips that were the result of Fatfoot's head rotation, it appeared that the canines had been raked straight across the tree trunk without any obvious angle of attack. Also, the upper and lower canine separations were more obvious because their entry and exit points were clearly visible in the bark of the trees in most cases. The upper and lower canine separations across the thirty samples were consistently 82 mm and 55 mm, which clearly suggested that they had all been made by the same large animal. The rounded width of the canine tip was measured at about 6 mm, with ninety percent of the treebites being at a median height of six feet or slightly more. It immediately seemed that the responsible individual had a more powerful bite force and larger dentition.

With the guidance of Jerry and Sue, I was later directed to an area near the wetland below their house that they had been exploring. There were a large number of bloodwoods that dominated the wetland vegetation and many of them were currently and historically damaged. This was the same location where Brad Crofts had witnessed the Quinkan crossing the road in November the year before and together we had witnessed the biped move through the swamp on its way to, what we would soon find out to be, the O'Connor's backyard.

With the GPS and digital camera, I began the long task of logging the position of each treebite and taking the required photographs. It was obvious that one individual was responsible for the collective damage and that based on the measurements, it was most probably the larger species of three-toed Quinkan that Jerry and Sue had been regularly experiencing. Over the following months, the O'Connors assisted with the work, along with a number of others, including Steve, Sean, Pat, Rob and Scott. Although the task was never fully concluded, more than 200 samples were recorded and plotted from this clumped distribution area alone. It was apparent that this clumping of treebites was simply due to the high density of yellow bloodwood trees that were present in this wetland.

Similarly, there was a field of treebites to the north of our house that Ian had found after driving his truck along a dirt road as far as it could go on several occasions throughout 1997. However, it wasn't until after the bulk of the work was completed at the O'Connor's treebite site towards the start of 2001, that GPS and photographic records of more than a hundred examples were begun at this new clumped distribution location. Most of these treebites were from Fatfoot, with canine separations of 60 mm and 55 mm, although there were a significant number of anomalous samples that didn't seem to fit the definitive standard. Also around this time, treebites were found many kilometres to the east and west of our house. The eastern samples seemed to have more in common with Fatfoot, while the western treebites had more in common with the O'Connors' Quinkan. Clearly, the examination and mapping of the treebites were providing information on a given biped's identity, movement and possibly its territory, but these were still early days.

Apart from treebites in bloodwoods, evidence of a significant amount of insect predation was also occurring amongst *Acacia* bushes. *Acacia*, also commonly known as wattles, are highly susceptible to attacks from many varieties of insect, including wood-boring larvae, which are a common cause of plant mortality. Compared to bloodwood, *Acacia* infestations require much less muscular effort and weaponry to penetrate and efficiently extract food morsels. Consequently, the range of potential predators of *Acacia* infestations broadens beyond cockatoos and the larger 'marsupial hominoids' to include other birds and probably other animals.

Mature acacias tend to be affected by a wide variety of insects that infest all parts of the plant, including the bark, timber and roots. A common method of predation for a bird or other animal is to simply lift or remove the bark. Similar to treebites, some *Acacia* branches are bitten or crushed to obtain access to any embedded larvae. Some of these *Acacia* bites are made by Dooligahl and Quinkan where they leave bite marks with known canine separations, similar to what is found in hardwood treebites. Alternatively, another quick method of extraction by 'marsupial hominoids' is branch breaking, where the application by hand of force at both extremes of the timber, results in it snapping at points made weak by the infestations. This is analogous to the method of preparation for fresh asparagus, where the transition from tender to woody vegetable stem takes place at the breaking point after a moment is simultaneously applied to each end.

It would seem to be very unlikely that Junjudee would not avail themselves of this food, when butcherbirds, yellow-tailed black cockatoos, Dooligahl and Quinkan were.

Fig. 12.19 A quick method of detection and extraction of larvae by 'marsupial hominoids' is branch breaking, where applying force to a length of timber results in it snapping at the points made weak by various infestations. We call this the 'asparagus method'. (Photo: Neil Frost)

Having briefly seen a Junjudee during the day, it would also seem that they do not have the physical hardware to extract wood-boring larvae from hardwood trees. In form, the Junjudee that I saw was gracile and wouldn't have been very tall. When this daytime sighting is combined with an earlier nocturnal encounter where the Junjudee ripped up a turpentine tree, I would estimate their height to be four feet. Also, it had a rounded head that was slightly smaller than mine, with no sagittal ridge and no obvious protruding canines. However, I suspect that they do eat the grubs from *Acacia* and from rotting trees. This is because we found a number of local acacias that had been split down

Fig. 12.20 An *Acacia* that had been freshly pulled in two, to expose the internal larvae. The damage in this particular circumstance was associated with a 'mischief' of Junjudee that we were closely following from behind during the evening. Clearly, a bird would not be present at night to achieve this damage and regardless, two hands would be required to easily accomplish this characteristic task. (Photo: Neil Frost)

the middle, whilst we were following behind a 'mischief' of Junjudee one night (Fig. 12.20). This strongly suggested that something with two hands had grasped each branch of the bush and pulled it in half in order to access the larval cavity. It was apparent that this was not bird-related predation because of the manipulative requirement for the fresh damage and its nocturnal provenance.

WHEN I WAS teaching at a Western Sydney school in early 2001, an office assistant told me that her family regularly went four-wheel driving, mostly through very remote areas. Although I had been at this school for four years by this time, I had only told two other people about my cryptozoological interests, one person as I was leaving, because the school and its population weren't surrounded by a bush environment and were not likely to have had any encounters with the phenomenon. From experience, there was no point in cultivating ridicule when there was clearly nothing positive to gain from the pain. However, in this instance, disclosure potentially seemed to be worthwhile, considering the remote bush travel that was regularly undertaken.

As part of my typical introduction to this very difficult topic, I asked the 'standard question' and Sue replied that they had never seen anything strange whilst driving and exploring. Over the following months, Sue told me about their four-wheel-driving and camping plans for the weekends and school holidays, so I asked her if she could be aware of certain types of vegetation damage. On returning to school after the 2001 winter holidays, I was very surprised and pleased to be told that Sue and her husband had been four-wheel-driving in the Wang Wauk State Forest, NSW, 220 km north of Sydney, where they saw a medium-sized, black-haired, bipedal animal walk across the Stoney Knob fire trail that they were climbing in their 4WD vehicle.

Sue said, "The height of the animal was difficult to accurately estimate because of the poor light and the distance that we had seen it from, but it was slightly smaller than a man.[33] It was black haired and clearly walked upright on two legs. It stopped on the left side of the road. When we arrived at the location where it had been standing, we found that a young tree on the side of the road had been ripped apart. We were both too scared to get out of the 4WD, so I wound down the window and took a photo from the passenger side seat. We could still hear the animal noisily walking through the bush below the edge of the road."

This was an excellent sighting by reliable and experienced bush-savvy witnesses. A biped had been clearly identified walking towards a tree that had been bitten in order to extract the wood larvae. By the time that the 4WD had climbed the steep fire trail in low range, the winter light had dropped sufficiently to require the use of the camera's flash. As with Sandy's nighttime observation, this photograph was further confirmation of the cockatoo's alibi by two other credible witnesses. Unfortunately, Sue was unable to identify the tree species and was only prepared to take one photograph from inside the vehicle. From the photograph, the larval passage can be seen along the entire length of the heartwood. Having described the solitary animal as being "slightly smaller than a man" and considering the extent of the tree damage, it would seem highly probable that they saw a Dooligahl, either a juvenile or female.

---

[33] Sue said that their sighting occurred after sunset at an uphill distance of about a hundred metres.

12: Alibi

Fig. 12.21 A very fresh treebite made by a medium-sized, black-haired, bipedal animal along a steep 4WD section of Stoney Knob fire trail in Wang Wauk State Forest, NSW, shortly after sundown, mid-winter, 17 July 2001. (Photo: Sue)

Sometime in mid-2001, I was contacted by Paul Cropper[34] who invited me to speak at the upcoming *Myths and Monsters* conference, to be held in Sydney on 20th October. This was the first cryptozoological conference held in Australia. Few of us had spoken publicly about our research and fewer had shared their work with others. To a large degree, speaking about Hairy Man research had been avoided because of the predictable reaction from the general public, which at that time was still highly sceptical.[35]

One person at the conference that we knew well was Dr. Helmut Loofs-Wissowa. He had been invited to speak about his cryptozoological work in Southeast Asia over many decades, in a paper that he entitled "Would the real orang-utan please stand up and be counted? In search of unidentified relic hominoids in Southeast Asia." We first met Helmet after being introduced to him by Steve Rushton.

Sandy and I met with Helmut and Ziggy on a number of occasions in Canberra and at our home. We also exchanged letters, telephone conversations and emails regarding our shared research into Australian 'hominoids'. Mainly because of his Asian research, European background and having been a close colleague of Dr. Bernard Heuvelmans,[36] Helmut favoured a classical interpretation for both the Asian and Australian Wildmen, where their origins were either hominin or 'pongid' and certainly

---

[34] Rebecca Lang was also a co-organiser of the conference.

[35] This was in contrast to the developing attitude of Blue Mountains residents over the previous eight years, who seemed to be tempering their scepticism because of the widespread infiltration of the Octopus into the community, which was cultivating a cooperative, supportive and educational network.

[36] Bernard Heuvelmans and Ivan Sanderson are considered to be the founders of cryptozoology.

12: Alibi

placental. Consequently, this was also the path that Sandy, Ian and I initially followed until the model didn't seem to fit our isolated and incompatible Australian context any more. In addition to his broad Asian hominoid research, Helmut's other emphasis was primarily centred on ethical behaviour. In particular, Helmut was worried about the welfare of these sentient beings and wrote letters and submissions to the United Nations in order to have the rights of unrecognised hominoids protected under international law, in particular, through amendments to International Human Rights Law. Helmut was also concerned about the unethical behaviour of a few crypto-researchers that operated outside acceptable practice, who threatened to use a variety of lethal and immoral methods in order to obtain physical evidence, at any cost. Similarly, he also shared our growing concerns regarding the rise of individuals and groups entering cryptozoology with no relevant training or adaptable skills. Entering a new field of study, like cryptozoology, is always going to attract a certain number of non-professionals, simply because there are no barriers to entry for the untrained, compared to established disciplines.[37] Recognised practitioners from related fields, on the other hand, will be more likely to resist the sideways move, seeing little value in risking a professional change that will damage their established reputations. As a new and unrecognised branch of study, many new crypto-researchers did not have any formal academic training from related scientific disciplines and some had no tertiary

---

[37] I describe myself as a 'cryptozoologist' despite having no specific qualifications for this nascent discipline. However, I do have degrees in Anthropology and Economics, with tertiary qualifications in Advanced Technology, including Engineering, as well as Education. Each of these disciplines provides compulsory ethical training for their practitioners.

qualifications at all.[38] Apart from not having any relevant skills and knowledge to draw from, participants were not exposed to any ethical training which is a mandatory requirement of all academic and professional fields of study, that value, guide and determine acceptable standards of behaviour and conduct.

Just as we had Helmut to mentor us and Steve Rushton to sponsor our work during our formative years of research, we similarly made a conscious effort to mentor and support other new researchers in this difficult field of research, although not always successfully. We also provided workshops to deliver specialised training and skills that otherwise may not have been available. Our collective aims were to share and preserve information and not take sole ownership of it.

For us, Helmut provided background information and context, advice, support, contact with international researchers and most important of all, encouragement in the face of very difficult odds. Occasionally, in return, Helmut would ask us to tolerate and cooperate with German film crews and magazine groups on stories about our Australian Wildmen. It seemed to us that German audiences are more obsessed with cryptid stories, compared to other groups.

By lunchtime, even before speaking at the conference, I was approached by two Australian scientists from the Commonwealth Scientific and Industrial Research Organisation (CSIRO) who wanted to talk in private. Both were geneticists that were undertaking a number of active Australian and international research projects. We spoke

---

[38] Having tertiary qualifications is not a prerequisite for success but it helps. Ian Price (aka: "Lizard") is an excellent example of someone who had no tertiary degrees but who was, without doubt, an expert biologist, herpetologist and cryptozoologist.

for longer than the allocated lunchtime break, giving them a narrow and cautious summary of our research work, because experience had taught us that too full a disclosure tends to discount veracity and encourage scepticism. However, this didn't happen on this occasion. They were very interested in a number of discussed topics, particularly the three red hairs that Tom had found on the top strand of a barbed wire fence in the swamp but were lost after a really, really, really bad decision to surrender them for microscopic hair analysis many years before. They said that red hair is an adaptation to life at dark, high latitudes by some mammals, as it allows more and higher frequency light to reach the skin for the production of Vitamin D. I also showed them the treebite sample that I had brought to the conference, although it was clearly too contaminated to provide any useful genetic data from imbedded cheek cells. They told me to find a fresh treebite sample with a similar provenance and call them so that they could immediately collect it themselves, even though it would require a return journey from Sydney. This would be very difficult to fully comply with because they required visual confirmation of the nocturnal activity, as with the situation with Sue and her husband whilst off-roading, or both before-and-after nighttime observations of a tree, as happened with Sandy. After looking closely at the treebite sample and some photographs of others, they were both convinced that the damage seen was made by a terrestrial animal with prominent canine teeth. For the first time since this all began nine years earlier, people were voluntarily queuing up to see our research, rather than us trying to spruik the merits of our work, to small numbers of disinterested individuals, during infrequent windows of opportunity.

Over the following months and years, I kept in contact with the two CSIRO scientists, mainly by email and sometimes by phone. Harry came to our house on several occasions and a group of us would go for long walks through the valleys, looking for other indicators of traceable activity. Harry also invited me to his house in suburban Sydney where I met many of his colleagues working on various projects. At that time I saw some amazing work, including early genetic tracing of the migratory routes of red-haired, tartan-wearing, fort-building Celts throughout northern China and Mongolia.

Kosta was the other genetic scientist, a friend and colleague of Harry. Kosta was a more frequent visitor to our house and he occasionally brought other people with him to speak with us. Over the next few years, Kosta moved away from the CSIRO and worked in private business instead, but we continued to maintain our contact over the next nine years. He and his family moved to a waterfront property near Ku-ring-gai Chase National Park, sometime in early 2011. Shortly after moving there, during the middle of the day, Kosta parked his car on the side of the windy road within the park so that he could more closely inspect some damaged trees. He said that the damage was very similar to our typical treebites, with parallel tears through the timber, at the height of a man. Some of the nearby trees were also bent over and snapped. He said that he had only been standing next to the trees for a moment when a large branch was thrown over his head, which continued on to hit and damage the side of his car. The more troubling aspect was that the branch had been thrown from underneath the rock ledge that he had been standing on. After Kosta phoned me the following day, Guy and I drove to the park to confirm the tree evidence. As further pointed out by Kosta, the coastal geology of the area was mostly well-weathered Hawkesbury River sandstone which was networked with many substantial rock caves capable of comfortably sheltering one or more large animals. While we were there, we told Kosta about the encounter stories that local NSW Rural Fire Service members and the Octopus had told us. During the recent bushfires in the Ku-ring-gai Chase National Park, rocks had been thrown at the firefighters and unfamiliar noises and calls had been heard coming from the bush. We also reminded Kosta that he should ask the 'standard question' to potential witness candidates within his community because undoubtedly, many local residents would have had encounters over the years and despite being initially cautious might only need a nudge to recount them. A few days after this, Kosta phoned me to say that he had questioned the park ranger at the gate entrance that morning, asking if he had ever experienced anything strange when on duty within the park. Kosta said that the ranger told him that he had recently seen two campers running and screaming down the road as they ran towards his ticket gate. They said that they had been harassed during the night from overhead and from on the ground and were eventually chased from their camping site by a large, hairy, bipedal animal early in the morning.[39] The park ranger said that the two campers were extremely terrified.

Looking into the conference audience from the podium gave no indication of the background of the people who were present and what was motivating their interest. Apart from

---

[39] Being intimidated by these 'marsupial hominoids' overnight at a campsite is perhaps the most common or stereotypical scenario received by the Octopus—for example, as happened to our Scout Patrol at Blue Gum Forest during a camp in 1966. Typically, however, the intensity of these encounters tends to diminish as the Sun rises because of the increasing risk of exposure. Reports of overhead activity are very rare. Consequently, it would seem that from encounter experience, a dominant and very aggressive male Dooligahl may have been responsible for this extended dawn confrontation although the overhead activity is more typical of Junjudee.

the two CSIRO scientists and Helmut, I recognised no-one else. I hadn't prepared a speech and Paul Cropper started the open discussion by asking me a few questions. As the talk progressed, the audience members took control of the convention and after speaking about our experiences with Fatfoot, the conversation turned to the conspicuously damaged treebite that was laying on the podium table. As the sample moved around the audience, it was gaining increasing interest, most likely because it was presented as a physical artefact from a probable cryptid and partly because the damage seemed orderly and purposeful, unlike random bird damage. After a lengthy explanation comparing potential predators, the dominant conclusion voiced by many in the audience was that the damage had been made by an animal with protruding canines and not a beak. This vote of confidence was very welcome in contrast to the earlier evaluation of the treebite made by biologists from the University of Queensland, but it was not unexpected because the physical characteristics of the evidence were patently obvious to anyone without a mindset.

Another surprising aspect of the conference was the attention that it generated across a broad range of people and interests. On the day of the event, a Blue Mountains councillor drove to the conference after seeing a promotional article in the daily newspaper. After my talk he wanted to know if I would like to be involved in a NSW Tourism Promotion targeting Japanese groups. I left my options open, but nothing eventuated and thankfully so!

Also at the end of the talk, I was approached by two film producers, one of whom made documentaries. After exchanging contact details I left the conference and drove home feeling unusually optimistic.

During the following week I was first contacted by the documentary film producer and then Mike Williams, who both wanted to meet on Saturday at our house. The other film producer made no further contact. I did not know

Mike at that time. He was the partner of Rebecca Lang, a co-organiser of the conference. I immediately asked Ian Price to be present because of his extensive involvement with the project and his provocative charm. Ian and Cheryl had sold their house in June 2001 and Ian would need time to ride here on his hog from the Gold Coast in Queensland.

On the next Saturday, both interested groups unfortunately arrived within a few minutes of each other which generated a degree of competitive tension between them. Rebecca and Mike were very interested in a free-flowing exchange between Ian and myself, that touched on most aspects of our communal work. This resulted in the six-part YouTube series of interviews that is located on Mike Williams' channel (https://www.youtube.com/@mrnobodyanybody) and was partly the inspiration for a *Fortean Times* magazine article, 'Monster Hunters', and other followups. The documentary film producer, on the other hand, was mainly centred on a few aspects of our unusual research work that were still being narrowed down, which included the treebite samples. As a result of the competitive interviewing environment, Bob and his cameraman were not obtaining the required recordings that they wanted,[40] so it became necessary for me to detach from the broader interview with Ian and provide Bob with what he specifically required.

Over the following weeks, Mike contacted us to organise an opportunity to have a nighttime experience and interview other members of the Octopus community.[41]

---

[40] Bob can be heard in the background of Rebecca and Mikes' third-part interview (3:04 m) trying to ask one of his questions: "Have you ever thought about putting sand down the track?"

[41] Mike Williams would frequently help us with a variety of video recording tasks and Paul Cropper would do follow-up interviews of Octopus witnesses and record them for shared use in his book with Tony Healy and elsewhere.

Mike accompanied us and others, including Sue and Jerry O'Connor, on a number of occasions despite having a hearing problem at that time, that would have made his experience very difficult. Bob contacted me and said that he had prepared a promotional video on our research and was seeking funding.

Several months later, I received a phone call from an Australian Broadcasting Corporation (ABC) producer. She said that they were interested in making a short programme on our research as part of the ABC *Catalyst* science series. She mentioned the treebites and other aspects of our work, so I knew that this contact was one of the contacts from Bob's submissions to be funded by a television network. When queried about the matter, the producer said that they could not afford to fully fund Bob's submission, but could possibly finance the production using existing in-house resources instead. She then asked if I was interested in providing an interview to their science team. I told the producer that I would speak to Bob first and would call her back in a few days. When I spoke to Bob, he already knew that his submission had been rejected, supposedly for financial reasons.[42] He was understandably

---

[42] Particularly over the following decade, as interest in the subject intensified, there would be a number of other television and film requests made. Some of these invitations for involvement were made by U.S. television networks supporting a number of Bigfoot-related programmes. Fortunately, our experience with the *Catalyst* programme had made us aware of the ruthless behaviour that is typical of this industry, but there would be no comparison with the harsh treatment dished out by U.S.-based networks. Consequently, after careful consideration, nearly all filming requests were eventually declined. The exceptional experience was the Australian film *The Hairy Man* by producers Cassandra McGrath and James Anderson which was heavily influenced and supported by most members of the Octopus. Unfortunately, this very promising feature documentary failed because of financial constraints and with Miklos, the Director of Photography, having unexpectedly passed away.

very disappointed but could understand why I might want to proceed with the ABC offer without him.

The next day I phoned the producer back to accept the invitation to be interviewed. With hindsight, this was a big mistake. We should have allowed Bob to produce his documentary and allow him the opportunity to sell it to some other interested party! Bob's treatment of us would have been much better and he would have eventually been paid for his investigative work.

THE FILMING OF the science documentary was not as expected, for a variety of reasons. It was blatantly apparent that the only part of our work that was of interest concerned the treebites. Even the sound technician knew that there was a problem. While the presenter and producer were having a private conversation an inaudible distance away, the sound man, who was listening live to all of our wireless conversations, approached me for our own discrete discussion. He said, apart from other things, "Unfortunately you have chosen the wrong team to represent your case." He said that personally, he found our information to be credible and very interesting and if I ever needed a sound engineer for another project, he would be very pleased to assist. Torsten then gave me his professional card.

Fortunately, I had prearranged for Helmut to be included in the documentary as my academic support, so the next day I telephoned him for some very serious mentoring. I told Helmut that the interview had not gone as anticipated and that it was clear that there was a hidden agenda. True to his gentle nature, Helmut was very reassuring and told me not to worry. He told me that as an anthropologist and French Foreign Legionnaire, he would present supporting evidence of Wild Men that he had obtained whilst working and serving in Laos, Vietnam and other Asian countries over many decades. Helmut also had a strong association with the new field of cryptozoology because of his close friendship with Bernard Heuvelmans and fully understood the difficulties associated with doing research outside of the box. In conclusion, Helmut said that he would ask leading anthropologists and other academics from the Australian National University to provide additional academic support for this highly controversial field of research.

Helmut was interviewed at his home in Canberra the following week, together with other academics from the Australian National University. He presented his work on Wild Men and the team filmed his comments on this 'worldwide phenomenon' that included Australian sightings.

Helmut strongly believed that the credibility of indigenous people was often heavily discounted, in favour of foreign scientists from outside biogeographic regions. He strongly believed that the reported encounters by rural people in many Asian countries were typically being dismissed as superstitious nonsense and the like. Additionally, academics that dared to side with the local inhabitants were being accused of conducting or supporting pseudoscience, a common insult thrown about by sceptics. I remember reading a number of encounter reports that Helmut had given me to study. In one of these reports, Chinese villagers had complained to their local government authority that Yeren, or Chinese Wildmen, were regularly stealing food, animals and other items from their village. Nothing was done about it for many years, despite the frequent complaints made to local authorities, until a minibus load of Chinese engineers saw Yeren crossing the road in the vicinity of the village and reported the incident to the local government. The engineers' report was taken very seriously and was immediately investigated by the Chinese Army, presumably because these highly-educated engineers could tell the difference between a carbon steel, high-tensile 'C' section beam and a Wildman, whereas the villagers were ignorant of the austenitization steel-hardening process and also knew nothing about their local, natural environment! Similar dismissals involving 'superstitious nonsense' with First World witnesses aren't as easily gotten away with, so alternative rationalisations needed to be found that are appropriate for a less spiritually-based and more scientifically-aware audience. Consequently, more sophisticated social psychology arguments are required in First World countries to discredit witnesses. For example, it is frequently argued that all witnesses have a predisposition towards perceiving things a certain way because of their expectations, or perceptual set, also known as pareidolia. Such perceptual aberrations must be present during eye-witness accounts because, after all, the alleged evidence cannot be true. In encounter situations involving more than one witness, the mutual confirmation of evidence can be ridiculed as a conformative side effect of mass hysteria. Such frequently-used arguments by sceptics commonly carry the imprimatur from a supposedly higher intellectual order, by simply appealing to authority, however this form of argument remains fallacious. Additionally, where other corroborative information is supplied in support, the common assault on credibility is usually based upon the supposedly unscientific methodology used, crucial inaccuracies being made during observation, sloppy work or some newly-generated criticism. Many other potential rationalisations can be employed to sweep away the reality of encounter reports, but if all else fails—and quoting Helmut from his documentary interview, "You say you've seen one—you're lying! You must be lying because they don't exist!"

In addition to highlighting the presence of intellectual snobbery in all of its forms, when broadly dealing with witnesses of different cultural or educational backgrounds, Helmut was also concerned with other ethical issues. As a full-time academic, research anthropologist and archaeologist, Helmut taught and practised in the field, which made him well aware of the many ethical issues affecting research participants and how they should properly conduct their work. For example, he believed that in order to protect Wildmen, the current whereabouts of research locations should not be deliberately publicised by others, particularly when the researchers involved had specifically requested that the whereabouts remain secret. He was also highly concerned with the sentient rights of hominoids, worldwide. For this reason, Helmut spent a lot of time petitioning the United Nations, arguing that these Wildmen, whether they were related to humans or not, should have their rights proactively protected, under the Universal Declaration of Human Rights. Of particular concern was Article 3, "Everyone has the right to life, liberty and security of person", since some aspirational groups openly declare that killing a hominoid is acceptable in order to obtain scientific proof. Overall, it would be Helmut's innately strong sense of justice and ethics that would prevail nearly a decade later, when our scientific research into Australian 'hominoids' was brought to media trial.

As promised, Helmut obtained the support of his anthropological colleague, Professor Alan Thorne (1939-2012), an internationally recognised Australian authority, who agreed to be interviewed about Wild Men. When at university studying anthropology I read Alan Thorne's work from Lake Mungo and Kow Swamp, which was beginning to expand our basic knowledge and understanding of early Australians. As an anthropologist, he was frequently controversial. He challenged the validity of the 'Out of Africa' hypothesis and proposed alternative theories explaining the early dispersion and regional evolution of humans. I believe that some of his work may be relevant to the evolution of Australian 'marsupial hominoids'. When being interviewed, Alan Thorne emphasised the importance of thinking outside of the envelope and keeping an open mind. He said during his interview,

"There in the totally searched out Blue Mountains, west of Sydney, there suddenly pops up—the Wollemi Pine. I think that it is a new family of trees. It's a warning that we don't know everything and maybe there are still a lot of surprises out there and if those turn out to include Hairy Men, then so be it!"

A few days after the conclusion of the interviews with Helmut and Alan, I received a highly unexpected telephone call from the producer who wished to have a return visit to the mountains and another opportunity to continue our interview. This was a total surprise and a reversal in attitude considering the concluding remarks made by the producer a few weeks before. Helmut thought that he and Alan had made a significant impact, which I am certain they both did.

On this second rebooting visit,[43] they brought Richard with them, a tree damage expert. To begin with, I reemphasised that more than one predator could be expected to harvest these larvae. As before, I showed the team a number of similarly damaged trees, with parallel tears having

---

[43] Seventeen hours earlier, at this remote interview location, I had an encounter with Fatfoot. I was having difficulty logging a GPS coordinate under the thick bush canopy and was standing silently as she was on her approach to our valley after sundown. I didn't mention anything about the previous night's incident, mainly because I thought that, like all the other encounter reports, they would have thought that I had imagined it! If I had told them, then there was a remote possibility that they might have wanted to hang around until sunset and I didn't believe that they, of all people, deserved to be rewarded with the opportunity of an encounter!

the same spacing and a common height range of around six feet. I pointed out the lack of collateral bark damage on the lower trunk that should have been present if a large cockatoo was desperately clinging on whilst open-cut mining for witchetty grubs. I also pointed out that there was a conspicuous absence of any supporting branches below the Wildman treebite examples. I emphasised the parallel torn nature of the bites with a consistent set of 60 mm and 55 mm separations, also mentioning that several hundred samples could be found in one location, with many more examples with different but consistent parallel tears of 82 mm and 55 mm, that could be seen at the southern field location if we visited it. I emphasised to the expert that this tree damage mostly occurred during the night when the yellow-tailed black cockatoos were asleep and that this occurrence was observed by us one night, across from our house. I drew their attention to the other trees within the field with messy, random damage much higher up in the foliage. After this lengthy monologue, I felt that I had just delivered a soliloquy to a deaf audience!

Following my monologue, the film crew took a wide variety of film cutaways where the presenter and I pointed to particular features on the trees and responded to questions about the damage. When the producer and presenter had another private conversation, I took the opportunity to have a similarly private discussion with the forestry expert. I explained the circumstances that our community had been living through and how we were extremely confident with what we knew. In particular, I told Richard that there was more than one predator of these larvae, that we were able to differentiate between predators and I had been studying, mapping and keeping various records of their areal and linear distributions for a long time.

Despite the seemingly worthwhile discussion with Richard, when the interview filming resumed, the focus was immediately restored by maintaining that a cockatoo was responsible for all of the damage. The filmed example showed a bloodwood tree that had been bitten without any collateral damage to the bark. Despite this, even after our conversation, Richard attempted to defend the indefensible by imagining that a large black cockatoo had left invisible marks on the soft, fibrous outer bark. He said,

"There is damage. You can see tiny little bits of bark that have been removed. These critters, when they grab, have got claws that have to dig in to support a large bird."

Clearly, there was no collateral damage caused by bird claws on this tree! I know this for certain because I was there and examined the bark with Richard in microscopic detail. Even though the video quality is not excellent, there were no visible marks of any size to be seen on the bark of the tree. The only concession that Richard said on camera in response to our private talk was, "The midnight theory is interesting!"

After *Catalyst* was aired on ABC Television on Thursday, 13 June 2002, at 8 P.M., it was clear that this targeting programme could have been worse. Clearly, I could have done a much better job in defending our research and I am certain that Ian would have been a much better, take no prisoners, advocate. I think that the majority of the countervailing success lies with Helmut and Alan, whose academic credibility, reputation and stalwart support played a dominant role in preventing this minor aspect of our work from reflecting poorly upon the whole. Both of them kept the door of reason and open-mindedness ajar.

The presenter of *Catalyst* was awarded the "Australian Sceptic of the Year in 2002 for working to counter pseudoscience".[44]

---

[44] Sceptics regard cryptozoology as a pseudoscience because they claim that "it doesn't follow the scientific method".

IMMEDIATELY AFTER THE broadcast of *Catalyst* had finished, my mother telephoned me. Mum was very interested in all aspects of science and regularly watched science programmes on television. She was extremely angry. We had decided not to tell Mum of our problems with Fatfoot. We didn't want to terrify her because she was a country girl with a very long family history of rural farming and we weren't certain what she may have known. Mum gave me a very lengthy talking to that lasted longer than the television show that we had just finished watching. Her main concern was for her two grandchildren and the potential problems involved with having a Wildman lurking in the bush surrounding our house.

After *Catalyst*, I didn't do much more work in the treebite damage fields. The treebite database, GPS coordinates, photographs, distributional maps[45] and the earliest manuscript of this book,[46] were all lost soon afterwards when the external USB hard drive used to store them crashed and was unrecoverable. I didn't actively search for treebites and record their data for more than a decade afterwards. After moments like these, my father would often say, "Experience is the great teacher."

---

[45] The distribution of hominoid treebites (O) was either linear or clumped. They were linear when they were associated with roads or tracks, and clumped when associated with a particular point of interest, like our house, or when bloodwoods were the dominant tree species within a favourable bush environment. The messy and variable-height bird-related treebites (X), were simply randomly distributed.

[46] The original soft copy manuscript of this book (Version 1), entitled "Encounters with Fatfoot" MS, 1999, was lost except for an earlier hard copy version. This book is Version 3, 2023.

13: Eyes

During my early childhood, our family would travel to Tungamah in Northern Victoria, to spend time with my mother's father and the extended family. During these stays, we would live in the basic accommodation normally provided for the shearers. My elderly grandfather lived here throughout the year and in return, he would cook meals for the shearers and clean up their accommodation while they were working. He would get up every day at four in the morning, usually in freezing temperatures, restart the smouldering open fire in the shearing shed kitchen with mallee roots and light the wood stove. I still have very fervent memories of the very cold, pre-dawn starts, the scent of lanolin and the associated smells of the mallee root fire, the baking damper and the brewed black, billy tea.

The large station grazed sheep for meat and wool and had extensive fields of wheat. To keep my brother and I entertained each day, the farmer would provide a number of different activities for us to participate in. We would muster sheep, help repair and build fences, pull the scarifier repeatedly around a very large paddock with the tractor, ride motorbikes, fish, go yabbying, trap crayfish,[1] search for early settler ruins and Aboriginal sites and travel very long distances to shop or visit people.

At night, we would sometimes eat a large roasted meal with other family and neighbours at the main homestead, sit around a roaring campfire, tell stories and look at the stars, play cards and board games, learn magic card tricks and play cribbage with my grandfather. Sometimes we simply went to bed early, because we were overtired.

On other occasions, my brother and I would accompany the farmer and his sons as they completed their daily farming routine, travelling in the tray of their ute with the working dogs. During the day, kangaroos and sometimes rabbits were shot as opportunities arose, mainly because the farmer and his sons always carried rifles in their utes to accomplish a variety of lethal tasks. The kangaroos were shot because they were considered pests that ate the limited grass and drank the scarce water. After being shot, the carcasses were thrown into the back of the ute and we were given a hammer in case we had to finish the job ourselves. The bodies would be taken back to the yard to feed the working dogs.[2] Rabbits were shot for similar reasons, but taken back to the shearing shed and used to feed us. My

---

[1] 'Yabbying' is the act of catching small, freshwater crustaceans or yabbies, from country dams typically using an old nylon stocking with a suitably smelly bait inside that has been cast into the muddy water on the end of a set line and then slowly retrieved with the aid of a trailing net. The stocking prevents the loss of the bait and helps to snag the spiky claws. Crayfish are large freshwater crustaceans or lobsters. We would catch crayfish from a boat by lowering a partially open wire basket with an attached sheep's head inside, into a river and quickly raising the trap later in the day.

[2] Out of necessity, grazing is a brutal occupation. As a child, I remember sitting alongside two ewes that had been left outside the working dogs' yard earlier in the day. Both had lambing sickness but I spent many

grandfather liked cooking rabbits, as well as wild ducks that he shot during the day. Some of the wild ducks would prove to be very tough and grandfather would say that he included a stone in the baking dish, which would become soft when the birds were done. The meal usually included lead pellets that required careful chewing.

Throughout the region, it was also common practice to shoot wedge-tailed eagles when they were on the ground, usually when they were feeding off road-kill, because they often took lambs to feed their chicks. These large birds with wingspans as large as nine feet, would be tied to the wire fences with their wings outstretched as a warning to others to stay away from this territory. When we drove around the district we would see many shot eagles displayed in this way and often we would see the eagles flying very high overhead looking for food.

Sometimes we would go 'roo shooting at night. This culling of the kangaroos was also a social occasion, with other local males competitively testing their shooting abilities. Shooting at night with a spotlight made the task much easier, with the red reflective eyes of the macropods brightly illuminating the central cranial target.

When targeted by the spotlight at a distance, the kangaroos would become blinded and dazed. This made the shot easier to make. Also, the scattered light from the light beam would usually be sufficient to highlight other nearby targets. If any of the macropods in the mob looked slightly away, then the reflected red glow from the tapetum lucidum would shift to yellow.

hours feeding them fodder and water. When the farmer returned, he told me that I was wasting my time. He took out a knife and slit both their throats. I remember that the blood from the sick animals was purple rather than red.

Although Australia's extant kangaroos are herbivores, they have excellent night vision that helps them avoid nocturnal predators. In the past there were a number of carnivorous kangaroos, such as *Balbaroo fangaroo*, which had teeth similar to a wolf, were about the size of a human and galloped rather than hopped. There were also other carnivorous megafauna such as *Thylacoleo carnifex*, the marsupial lion, a convergent placental analogue. It was a two-metre long, 160 kg, cat-like predator with a very powerful bite force, shearing teeth and a semi-opposable thumb with a retractable, hooked claw.

In 1966, at the age of twelve, I became very interested in astronomy. Living in suburban Sydney was not an ideal observing location with many of the faint celestial objects being too difficult to see because of light pollution. Encouraging my interest, my father bought me a 60 mm, f11.7 Ascotron refractor telescope, the best that he could afford. The quality of the two-element, achromatic (achro) objective, was average for that time. As a doublet, it could only correct for two (red and blue) of the three (red, green and blue) incoming spectral colours of light, that human eyes are primarily sensitive to. This resulted in chromatic aberration or false colours being observed and other refractive defects, such as spherical aberration. Consequently, when looking through the eyepiece, bright objects tended to have a coloured 'greenish' halo of uncorrected or unfocused light. Other false colours might appear, depending upon several factors, including their distance from the central optical axis. As the human eye is trichromatic (LMS or RGB wavelengths), better quality and more expensive refracting telescopes usually have three elements that correct each of the RGB frequencies used in human vision and are known as apochromatic (apo) triplet objectives.

Alternatively, higher quality but more cheaply produced two-element objectives that use specially formulated extra-low dispersion or ED glass, mostly correct these three essential colours and are known as semi-apo doublet objectives. Compound objectives with four or more correcting elements are referred to as superapochromatic (sapo). Objectives with four or more elements are very expensive and are mostly used in cameras because they are of a smaller and therefore, more affordable diameter compared to telescope objectives. They correct or focus a minimum of four colours to a common focal point, reducing chromatic aberration or colour blur, plus other potential optical defects.

When I transitioned to high school in 1967, I became friendly with another student whose father was an industrial chemist. Having a science background, Shane's father researched the options and bought his son a 60 mm, f15 Unitron refractor. This was a very expensive telescope. Even though it also had a two-element objective, it's superior optics and longer focal length reduced the significance of chromatic and most other aberrations. It also came in a fitted box made of timber, rather than cardboard. After several shared viewing nights, I realised the greater importance of optical quality in astronomy but knew that my family could not afford a better telescope. It was around this time that I promised myself to buy a high-quality telescope and build an observatory to house it when I was older.

At school over the following year, the librarian noted my interest in telescopes and astronomy. I was always looking at telescope catalogues and reading anything related to astronomy. He would regularly purchase books on astronomy and optics and I would similarly check the new book stand, although he would usually keep the new purchases aside for me. Afterwards, he would quiz me on these purchases and occasionally bought some very expensive follow-up books. I particularly remember a book on telescope optics that included detailed instructions and side notes on how to grind telescope mirrors and how to detect and correct optical errors, or aberrations at an atomic level. With this book, it was possible to produce spherical, and eventually parabolic mirrors, with an accuracy of ⅛ the wavelength of light. Included with this book was a photolithographic copy of Ronchi rulings, used to create light interference patterns across the surface of pyrex mirrors, using very basic, homemade optical testing equipment. For example, by making a sealed lightbox with a very small hole at one end and a low-wattage incandescent light bulb at the other, the escaping light cone would

Fig. 13.1 The 'Dooligahl Observatory' (Photo: Neil Frost)

13: Eyes

be further constrained by a tight slit formed by two new, traditional razor blades. The resulting narrow line of light that was similar to a primitive laser was bounced off the optical surface, back towards the observer. At the radius of curvature, the intercepted light beam would be passed through the Ronchi rulings, which produced an interference pattern that showed incredibly fine surface detail, highlighting any hidden hills or valleys present in the glass surface. These irregularities might only be a few atoms in height. In order to achieve this high order of precision, it was essential that the basic equipment was set up correctly and accurately aligned. The light source and the observer's eye needed to be on the same plane, close together and in this case, at the radius of curvature. If they were not, then the reflected interference pattern of the fully illuminated mirror disc would not be visible.

Grinding a telescope mirror had the advantage of producing an objective that was much larger in diameter than could be affordably purchased commercially and the process was simple. This was the main appeal for my father. Alternatively, grinding a lens was technically complex and effectively impossible for an amateur to achieve. Additionally, with a parabolic mirror, distant light of all frequencies is reflected at the interface by the same angle as its incidence and therefore, achieves a common point of focus. This is because the light never physically enters the optical medium causing the reflection and does not react with it. On the other hand, light entering a lens is refracted at the change of medium density, by an amount determined by the difference in the refractive indices of the media and the light's frequency. Therefore, as with a prism, the various frequencies of light cause them to deviate or refract at different angles and will not form a common point of focus unless the lens is made from multiple compound elements with appropriate refractive densities and ground in such a manner as to achieve a common focus for the targeted light frequencies.

To begin with, Dad took me to Astro-Optical Supplies at Crows Nest in Sydney and we purchased the smallest and cheapest, four-inch (100 mm) grinding kit. This option was recommended by the owner, most likely as an inexpensive precaution based on his experience—in case I was not up to the task. However, in the shop were these beautiful, Japanese Unitron refractor telescopes, like Shane's but much larger. They not only had superb optics with a high f-stop that reduced the effect of optical aberrations, but more importantly, they had presence. The Unitron Model 132, 4-inch (102 mm) f14.7 equatorial refractors were so large that they would not fit easily into a room of standard ceiling height. It wouldn't matter if you didn't use them because they looked so good!

As it turned out, the four-inch grinding kit was good practice! On seeking further advice from the shop owner, he recommended that my Pyrex mirror and grinding tool should both be ground back to a blank! In other words, recycled! After gaining a trade-in on the failed mirror, Dad bought me a ten-inch (250 mm) pair of blanks and a fresh supply of grinding powders and pitch. As failure was no longer an option, any optical imperfection that appeared after interference testing was ground out of existence shortly afterwards by a new round of figuring. Each time that a correction was made, the focal length of the mirror became shorter. Consequently, the f-ratio of the mirror, the focal length divided by the objective diameter, became progressively shorter and faster, during my repeated attempts to achieve perfection. In the final reincarnation, the ten-inch mirror was parabolised at slightly more than f4, having had an initial target of f8. These f-stop settings are also found on a camera lens, where the variable diaphragm allows control of the incoming light for exposure purposes and the manipulation of the depth of focus.

With astronomical mirrors or lenses, the f-ratio is fixed. In having a low f-ratio, the f4 mirror has a wider field of view and is optically faster (produces a brighter image), than a similarly sized objective with a higher f-ratio. The optical characteristics produced by the mirror are similar to a variable diaphragm camera lens when it is set to f4. For the 60 mm Ascotron, f11.7 refractor telescope that my father bought me, having a high f-ratio and therefore, a long focal length, had several advantages. Although the image was duller and the field of view narrower, typical of higher f-ratio telescopes, the long focal length f11.7 refractor minimised the problems of chromatic and spherical aberrations normally associated with a shorter focal length refractive lens.

With the human eye, determining the obtainable range of variable f-stops is not straightforward. Having a typical focal length of about 22 mm and dividing this by the diameter of a fully dilated and healthy entrance pupil, gives a maximum relative aperture quotient of f/2.4 and progressively higher f-stops as the pupil dilation decreases in brighter light conditions or as the person becomes older. However, many other factors such as 'accommodation' or the ability of the eye to maintain focus on an object as its distance varies, having binocular vision and the post-processing of the image by the visual cortex affect the relative aperture of a lens.

Abandoning the four-inch mirror for the ten-inch had obvious advantages. In astronomy, a major consideration is light-gathering power. As a mirror or lens increases in diameter and area, more light is available to increase the brightness of celestial objects, improving contrast and allowing fainter bodies to be seen, as well as increasing resolution. A ten-inch objective is 6.3 times the area of a four-inch mirror, or in other words, it collects 6.3 times more light. Another advantage of a larger objective is greater resolution, or the ability to see close adjacent objects, separately. In other words, it allows finer detail to be seen.

When completed, an astronomical lens needs to be given an antireflective coating to improve performance by reducing light loss and increasing transmission. A mirror, on the other hand, requires a reflective layer, typically aluminium, that has been electroplated onto the mirror's ground surface in a vacuum. Taking advantage of the significantly cheaper, but inferior option, I decided to use silver as the reflective coating, which is deposited using a chemical process. Many of the basic chemicals, like ammonia and sodium hydroxide, were cheaply and readily available from supermarkets. However, in those days, chemicals were sold by chemists. The silver nitrate ($AgNO_3$), was ordered and paid for. Sometime later, it arrived as part of the chemist's weekly order. These two different reflective metal coatings used on astronomical mirrors affect the reflectivity of the objective surface in subtle but observable ways, just as different chemical elements affect the reflective colour of an animal's tapetum lucidum.

The light-gathering ability of the human eye is similarly determined by the diameter of the pupil, which is normally around 9 mm in a healthy person.[3] In older or unhealthy individuals the dilator muscle may not fully open the iris, allowing less light to enter the pupil. To achieve night vision the pupil must become fully dilated and the retina dark-adapted, a largely chemical process involving the light-sensitive rods and colour-sensing cones,

---

[3] The human eye collects 1/792 the amount of light collected by a ten-inch telescope mirror.

that may take several hours to fully complete.[4] After becoming fully dark-adapted, the rods are the main sensory cells that make night vision possible but only in greyscale. The cones are mostly used in bright light and are used to differentiate colour. Typically, the rods and cones are unevenly dispersed across the retina, with the rods being mostly located around its outer edge. When attempting to view objects that were at the limit of a telescope's specifications, a technique known as 'indirect observation' was used to improve visual observations. The method involves looking through the eyepiece and centring your vision on the outer edge of the field of view, whilst diverting your conscious attention to objects on the opposing side of the lens. This results in the observer using their peripheral vision, where the highest concentration of rod cells are located within the retina, which allows previously invisible objects to be viewed. Additionally, the rod cells that are located at the periphery of vision are typically linked together to further multiply their light sensitivity but at the cost of poorer image resolution. This was particularly true when indirectly observing faint nebulae where the resolution was usually marginal. This linking of rod cells also improves sensitivity to movement in the dark. I used this indirect observation technique to 'view' Fatfoot on several occasions when at close range and in addition to hearing her shifting her weight, I also detected her swaying from side to side. At that time, I did not know how swaying was being used by Fatfoot to determine striking distance and what the implications were for her pupils.

---

[4] This explains the strong aversion that these hominoids have to light, since restoring night vision may take a long time.

From the time that we started living in our present house, sometime during early 1985, we would occasionally see a 'red glow' in the thick bush, at about the height of a person. Being both smokers, we would always partake outside on the verandah, day and night, regardless of the conditions so that our new house would not be affected by the smoke. Consequently, we tended to see these glows on a habitual basis.

Being familiar with the local wildlife, we typically discussed these apparitions amongst ourselves and with any guests present, speculating on what animal could be responsible for the spectre. Our local wallabies, like kangaroos, tree-dwelling possums and sugar gliders, are all marsupial. Each is capable of producing red eyeshine but with different characteristics. Swamp wallabies can be found grazing in our local habitat, with other species of wallaby and kangaroo being mostly constrained by their environmental niches to other parts of the Blue Mountains. The height of a mature, male swamp wallaby, is about three feet and when standing at full alert, is only about four feet, which is too low for what we were consistently observing. Similarly, the reflective eyes of possums and sugar gliders would appear at any height but not consistently around this narrow height band of six feet.

An alternative explanation was that we were seeing the red glow from another smoker. The height was right. The glow varied in intensity and slightly in colour, as it would when a cigarette is being drawn upon. The observed distance was typically around thirty or more metres or roughly halfway to a neighbour's house. The direction was consistent with the whereabouts of neighbourhood children of an experimental age, who might be seeking the anonymity provided by the bush. However, not all observed aspects of the behaviour were consistent with someone smoking.

We continued to see these glowing eyes for many years and generally in the same location, to the northeast of the house. The height remained fairly constant but the glow intermittently changed from red to yellow and back again. We questioned many aspects of what we were seeing, as inconsistencies frequently arose. Our typical concern was the length of time that the glow persisted for. To us, these 'cigarettes' seemed to be longer lasting, so better valued than our own. Also, if the red glow was the result of cigarette combustion, there was no obvious start and stop, nor was there a period of rest between drags or any obvious displacement from hand movement. The illumination was mostly constant with some minor variations in intensity and colour. On other occasions, the red glow would completely disappear without the residual smouldering of a cigarette being seen, then reappear minutes later at a slightly different location. Sometimes the outside temperature was very low, even for us standing on the verandah, yet, the 'heavily nicotine-dependent teenagers' would persist in chain smoking in the exposed elements. Even when fulfilling our supervisory and duty of care responsibilities as teachers by calling out for the children to stop smoking and return home, we had no apparent disciplinary authority or impact on their smoking behaviour. However, walking towards the red glow always had the effect of making it disappear.

From our experience over the following days and weeks after becoming Yowie-aware, we realised and then confirmed that the red glows that we had been seeing for more than a decade, were the reflective eyes of a tall, 'humanoid', bipedal animal. After speaking to many Aboriginal Elders, they commonly referred to these 'marsupial hominoids' using the anglicised description of 'Old Red Eyes', as well as having specific tribal names for these bipeds, such as 'Dooligahl', the name used by New South Wales,

South Coast tribes. There was also an occasional and divergent reference to 'Old Yellow Eyes', something that our neighbour Gordon came across when questioning members of the local historical society. They mentioned that there were local stories of large red or yellow 'glowing eyes' often being seen at nighttime in the bush. Early last century, and before, when animals were still being used for transportation,[5] horses would frequently shy and refuse to travel further down our local roads. This 'Old Yellow Eyes' reference would appear to be a simple variant of the more widely used name, based upon the commonly observed colour shift of the reflective eyes.

THE THIRD PHOTOGRAPH (see Chapter 8) that showed the red eyes was taken after blinding and then stalking Fatfoot in the bush across from our driveway. After determining the eye and body heights, the next priority was taking physical measurements of reliable objects seen in the photograph, for comparative purposes. Visible in the photograph are several *Banksia* branches at various field depths. The two labelled branches 'a' and 'b' were situated at approximately the same focal range as the eyes. Measuring their diameters simply required a vernier calliper or a digital micrometer, appropriately orientated to take a set of readings at a right angle to the optical path and then averaged.

These branches were chosen because they appeared sharpest in the photograph by having less granular film noise, were closest in physical proximity to the red eyes

---

[5] I often wonder what my great grandfather, John Thomas Frost, would have known of this, when he took bullock trains and horse teams that he owned and drove over the Blue Mountains from Penrith to Bathurst along the convict-built Cox's Road, during the early- to mid-nineteenth century.

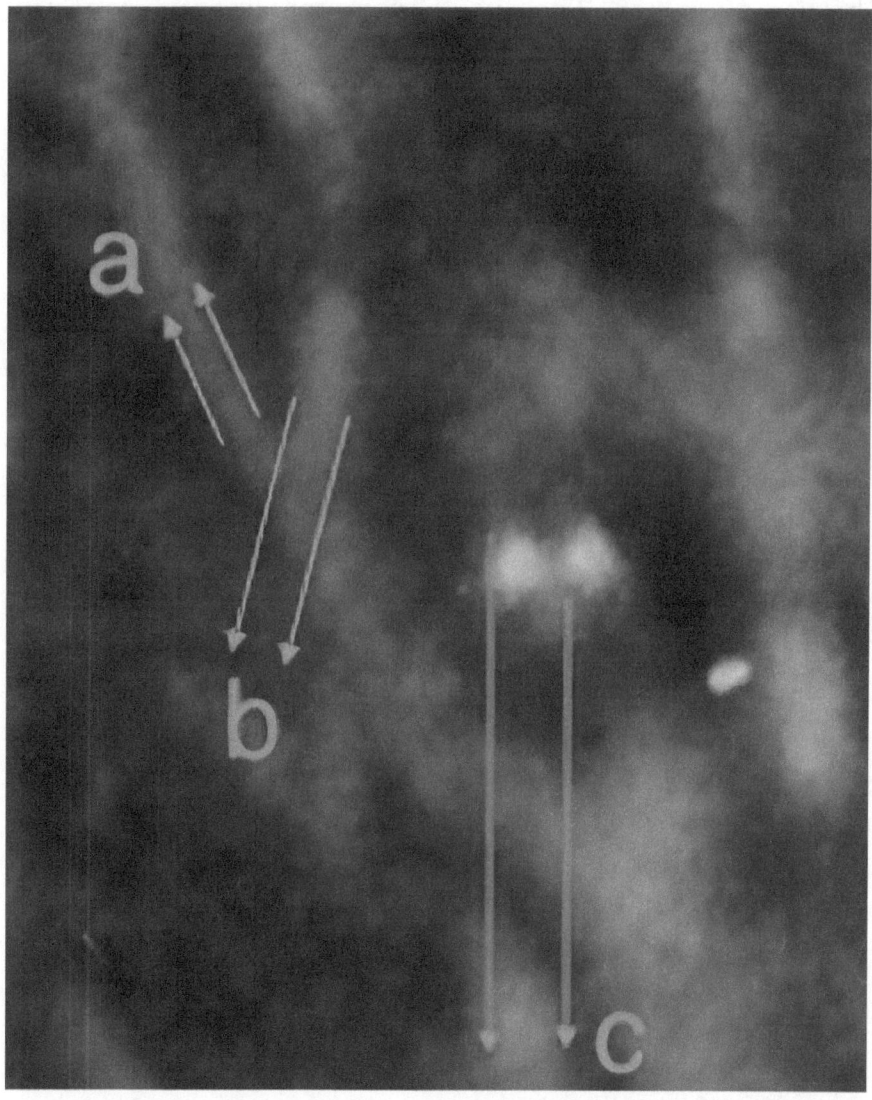

Fig. 13.2 Branch diameters for 'a' = 32 mm and 'b' = 50.4 mm, used to comparatively determine the inter-pupillary distance (IPD) and the eye dilation of Fatfoot. (Photo: Neil Frost)

and had firm, younger branches that had no internal insect damage. The average diameter of 'a' was 32 mm and 'b' was 50.4 mm. The eyes have a representation of the inter-pupillary distance (IPD) shown by 'c'. The inter-pupillary distance is the measurement between the centres of the two eyes. For accuracy, the centres shown are measured from outside edge to outside edge on the same side because determining the actual centres is much more difficult. This is the preferred method of centre measurement used in construction and engineering.

To determine the inter-pupillary distance, 'a' was visually divided into 'c' by taking measurements using vernier callipers directly off the enlarged photographic print and the quotient obtained was multiplied by 32 mm. Similarly, 'b' was visually divided into 'c' and the quotient was multiplied by 50.4 mm. However, there was some difficulty in determining the real visual diameter of the branches shown in the photograph, because of granular noise caused by using a high ISO film and light bleeding into nearby film grains. Consequently, when taking measurements, the halo or ghosting was ignored in favour of the more solid, central image core. This produced a surprisingly narrow range of IPD values, > 66.7 mm to < 68.11 mm, with a mean of 67.4 mm. This mean IPD value closely matches the precisely known canine separation value of 60 mm for Fatfoot, which adds further credibility to the methods used. This potential correlation is possible because the canines or 'eye teeth' are said to descend or erupt from below the eye sockets—although with my personal circumstances, this association does not apply very well because of my narrow mandible profile. Consequently, the association between canine separation and IPD does not always closely correlate, with dietary factors playing an important role. I don't chew trees.

13: Eyes

As Ralph was a local professional photographer who was very interested in our work and knew a great deal about the optics of Olympus cameras and lenses, I asked him if he could accurately determine the angular separation of the eyes in degrees from the negative and using the manufacturer's specifications for the Olympus OM Zuiko Auto-S 50 mm f/1.4 lens. Using the specifications, Ralph calculated the range of angular separations at 0.25 degrees and 0.3 degrees. At the time, I didn't trust my calculations because Ralph's results seemed to correlate closely with mine, so just to be sure I faxed Ralph's results to Mark, a maths colleague. Using Ralph's figures and the known range of 13500 mm, Mark's trigonometric calculations showed that the tighter angle of 0.25 degrees gave an eye separation of 58.9 mm and the broader angle of 0.3 degrees gave a separation of 70.7 mm. This gave an average interpupillary distance (IPD), using Ralph's upper and lower calculations, of 64.8 mm. As I had already done

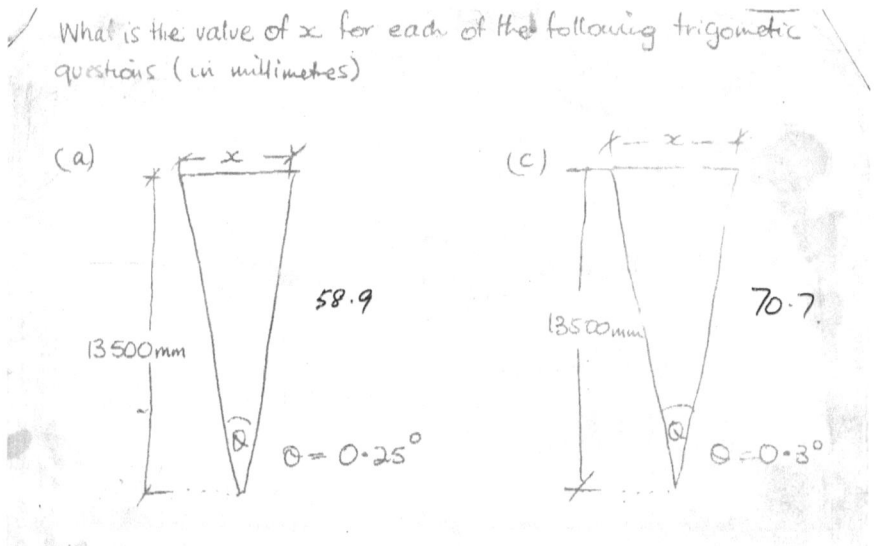

Fig. 13.3 The trigonometric questions that were faxed to Mark for confirmation and his written answers. (Image: Neil Frost)

my calculations, based upon the comparative diameters of *Banksia oblongifolia* branches, these results were very complimentary. Using the two average IPD comparative method results, the mean interpupillary distance (IPD) was calculated to be 66.1 mm. The average IPD of humans is 63 mm. The average measured IPD of our local possums was 40 mm and the maximum pupil dilation was 13 mm, which were determined from measurements made from dozens of 'capture and release' possums over the decades.

In humans, IPD varies according to age, gender and race. It has a typical range between 50 mm and 75 mm. The mean and median IPD for the adult human population is about 63 mm. Fatfoot's IPD was, therefore, slightly wider than average for humans but would be considered normal. However, the animal photographed has a tapetum lucidum, which is not a human trait nor a characteristic of

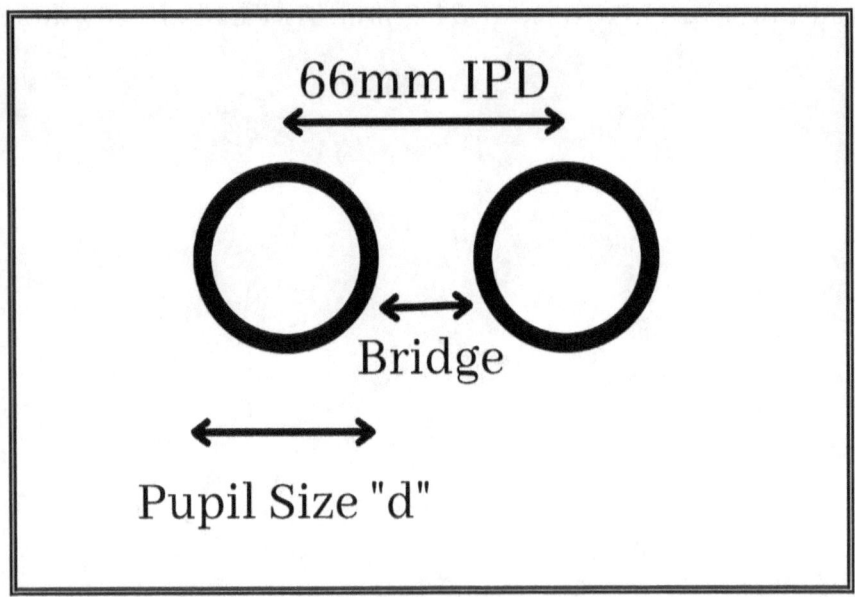

Fig. 13.4 Pupil size and IPD. (Image: Neil Frost)

the other great apes. Clearly, this 'hominoid' is not related to man, ancestral hominins, or the other great apes.

Having forward-facing eyes means that Fatfoot has stereoscopic vision, like humans and other predators, where the vision from one eye overlaps the vision from the other. This allows the brain to construct a three-dimensional map of the viewed area and achieve a sense of visual depth. The greater the IPD, the more pronounced this effect becomes. This ability is crucial for predators, as they must be able to accurately determine the striking distance to prey. However, there are other visual methods that a predator can utilise to complement this range-finding ability, especially in a low-light environment.

From the red eye photograph, it is possible to estimate the pupil size or 'd'. By simple comparison, the pupils appear to be about the diameter of branch 'a', or 32 mm. However, the camera flash that was reflected off the tapetum lucidum has generated a lot of noise interference as the light exits the eye via the pupil, which greatly complicates the measurement. Therefore, it would be reasonable to assume that the maximum pupil size is smaller than 32 mm. If the pupil diameter was 32 mm, then the bridge gap between the eyes would have a maximum available space of only 34 mm. Taking a conservative pupil size of 20 mm or slightly larger is more likely, which gives a maximum bridge space of 46 mm. So, the range of possible pupil dilations is above 20 mm and somewhere below 32 mm. However, like a telescope, the problem with a bigger pupil is that the eye must become appropriately larger to accommodate the increased aperture—more significantly because the focal length would need to be longer too. This enlarged eye would need to be accommodated within a larger eye socket, which causes problems with the structural strength of the orbit. For Neanderthals, the solution was to have a reinforced brow ridge. Consequently, a pupil size closer to the minimum 20 mm diameter seems to be more feasible, although the reader can draw their own conclusion from the evidence.

With a telescope, the aperture determines image resolution, with larger objectives allowing finer detail to be seen. Consequently, a 20 mm pupil when compared to a fully-dilated human eye, should be able to resolve two points less than half the distance apart; a 30 mm pupil, about a third of the angular separation. In animal eyes, the larger the aperture or pupil the better the resolution and the sharper the image will appear.

Similarly, the larger the aperture of a camera lens the greater its resolution becomes. The Olympus Zuiko Auto-S 50 mm f/1.4 lens used to take the film image of the red eyes has a focal length of 50 mm and a full aperture or a maximum diaphragm stop of f/1.4, which gives a calculated aperture diameter of 35.7 mm.[6] This gives the well corrected, seven-element[7] Zuiko lens a theoretical resolution of 3.15 seconds of arc[8] or about one six-hundredth of the 0.5-degree angular diameter of the full moon. Since the minimum and maximum IPD measurements were between 0.25 and 0.3 degrees angular separation, the red eyes were very easily resolved by the Olympus lens at a distance of 13500 mm. In comparison, a typical smartphone camera lens is only a few millimetres in diameter. Consequently, these tiny apertures have very poor resolution, but should theoretically be able to resolve the red eyes, as long as

---

[6] f-ratio = FL/d substituting 1.4 = 50/d, then d = 50/1.4, resulting in d = 35.7 mm

[7] The seven elements of this sophisticated lens are arranged into six groups in order to eliminate most aberrations.

[8] A simple formula for calculating resolution is 11.25 seconds of arc/d (cms).

13: Eyes

there aren't other limiting factors. However, these small lenses demand very high standards and otherwise, tend to amplify their inadequacies resulting in a number of problems such as background blurring, which is exacerbated by lens imperfections that produce softer images and different kinds of image artefacts. Also, depending upon the various processing and storage algorithms used, the image can suffer from spatial and colour resolution loss. Additionally, the autofocus on smartphones can generate difficulties in obtaining a stable image.

Another issue related to the red eye photograph is flash illumination. Currently, the integrated small LED flash unit on a smartphone is unable to produce the very high power levels required to light up large and deep spaces like that seen in the photograph (Fig 8.2), where background trees fifty metres away were lit-up. With an LED flash on a smartphone, its lumen capability would not be achievable to the same degree[9] unless high capacity battery storage was attached, which seems unlikely to happen. I could not imagine a successful outcome from facing off with Fatfoot, armed only with an LED flash from my smartphone.

In astronomy, the size of the telescope aperture also determines the amount of light that is available for viewing or astrophotography. The light gathering power of a telescope is simply determined by the surface area of the lens or mirror that is fixed in size, as is its f-ratio, which are both decided prior to optical manufacture. No mechanism is deployed to restrict the entrance of light into a

---

[9] Our son Drew found a wild dog at the front of our house on returning home from night shift duty at around 2 A.M. We both attempted to photograph it at close range by successfully enticing the dog to come nearer to us with meat. Our photographs were extremely poor and unusable because of problems associated with the cameras and the LED flash units.

telescope, other than with solar astronomy because there is no advantage or benefit gained by controlling it. However, the circular aperture size of the human eye can be adjusted by using the sphincter and dilator muscles of the iris, which alter the diameter of the pupil. This is similar to a camera lens that uses adjustable blades or leaves to produce a round diaphragm that controls the light entering the camera, using predetermined f-stops. Generally speaking, the amount of light entering a camera lens decreases by half with each downward, full f-stop (for example, f/1.4 to f/2) and doubles with each movement to a higher, full f-stop (for example, f/2 to f/1.4).[10]

Controlling the light entering the eye or camera has many advantages for an animal, photographer or astronomer. A basic control of light is the focus, which regulates the incoming angular rays from a terrestrial object to produce a sharp image coming from a particular range or distance. For an astronomer, the incoming rays are virtually parallel with a fixed range and focus set at infinity. With animals, focus is controlled by the muscles inside the eye that distort the lens shape and with cameras by shifting the focal length of the lens cell by rotating the helical focusing ring. Consequently, the focus or lack of it can be used to determine range or distance.

Another control is regulating the light intensity. Adjusting this allows the eye to precisely function during the day or night and similarly, with a camera it allows photographs to be optimally exposed under varying lighting conditions. A fully-dilated, human pupil[11] has a surface area of 64 square mm. In contrast, a dilated 20 mm pupil

---

[10] The words 'downward' and 'higher' refer to relative aperture size, not numerical size.

[11] The maximum dilation of a human pupil is about 9 mm. At my age, this has decreased to about 4 or 5 mm, an area of about 12 to 20 mm^2.

has an area of 314 square mm. By dividing the larger area by the smaller, gives a relative difference of 4.9 times, or simply, the 20 mm Fatfoot pupil collects 4.9 times more light than an optimally adapted human eye. A 30 mm pupil, the equivalent size of a small finderscope used on a telescope, has an area of 707 square mm, or 11 times the light-gathering power of a fully-dilated, human eye. Taking an average pupil estimate of 25 mm, gives a comparative advantage of 7.7 times the light-gathering power of a human eye. Whatever the final pupil size is, it is certainly a major contributing factor to Fatfoot's night vision capability.

For terrestrial use, controlling the light aperture has other consequences. When an optical system does not have a diaphragm or it is fully dilated, there is no influence on the behaviour of the incoming light. However, when the incoming light is restricted by an iris, diaphragm or any other intrusion, apart from limiting its intensity, the rays passing closest to the edge of the adjustable aperture are diffracted or bent because of interference generated by the light passing by the restriction. This is similar to ocean waves bending around a headland and into an otherwise, sheltered harbour. With optics, this diffraction has the effect of increasing the depth of focus with decreasing aperture and since the size is reduced, the resolution also diminishes.

In astronomy, there are many telescope designs. The basic categories are refracting and reflecting. With refracting types, the light passes directly from the objective lens to the observing eyepiece, but usually for convenience and increased viewing comfort a star diagonal and other

---

This gives Fatfoot a basic comparative night vision advantage of 16 to 26 times my sight.

devices may be inserted into the optical path. Some of the advantages of refracting telescopes are their portability, low maintenance, and sharp, bright images. On the other hand, reflecting telescopes bounce light off the primary mirror and back up the main tube assembly where the cone of light is intercepted by a smaller and suspended secondary mirror, reflecting the converging light into the eyepiece. The principal advantages of reflecting telescopes are their lack of chromatic aberration, simplicity and consequently their lower cost. Like most things, there are also hybrids, such as Schmidt-Cassegrain telescopes that are both reflecting and refracting. With this telescope variant, the incoming light first refracts through a Schmidt corrector objective plate at the front of the tube assembly before continuing on to the short f-ratio main mirror which bounces the light back to the convex reflective underside of the corrector lens, where a much narrower, high f-ratio light cone travels through a central hole in the primary mirror to the eyepiece. This folded optical design ultimately achieves a long focal length telescope with a wider field of view and a highly compact size for its objective diameter.

As with the enlarged red eye photograph (Fig 13.3), the image of the star Sirius A shown in Fig 13.5 contains additional information. The image was captured using a reflecting telescope because diffraction rays can be seen emanating away from the bright light source. These rays would have been generated through the light's interaction with the suspended secondary mirror of the telescope. Typically, the arms used to accurately support the secondary mirror are thin metal vanes, that are orientated parallel to the incoming rays along the optical axis and aim to minimise, but not eliminate, any diffraction effects. Usually, three or four vanes are used. The photograph shows four diffraction flares which confirms the number of vanes used

Fig. 13.5 The over-exposed image of the star Sirius A showing four diffraction spikes from the vanes used in the reflecting Hubble Space Telescope to support its secondary mirror. The focussed central star image is called the 'Airy Disc' which is surrounded by concentric diffraction rings known as the 'Airy Pattern'. Diffraction occurs when light passes close to an obstacle or through an aperture, causing aberrations. This effect can be used to detect and measure minute variations in an optical system, which is frequently used with Ronchi interference testing. The small circular dot to the bottom-left of Sirius A is its companion star, Sirius B.

(Image credit: NASA, ESA, H. Bond [STScI], and M. Barstow [University of Leicester]; CC BY 4.0 [http://creativecommons.org/licenses/by/4.0])

to support the Hubble Space Telescope's secondary mirror. Refracting telescopes do not produce diffraction flares because they don't typically contain any internal light obstructions, other than light baffles. Similarly, a Schmidt-Cassegrain telescope does not produce these flares because the secondary mirror is not supported by vanes but is invisibly included on the inner surface of the corrector plate.

However, a camera lens, which is nearly always refractive, will have many diffraction flares visible on photographs of bright objects. This occurs when the lens has been stopped down below full aperture by the diaphragm because the edge of each blade or leaf causes a diffraction effect. As with Hubble photographs, counting the number of flares seen will confirm the number of leaves used in the manufacture of the circular diaphragm.

AT THE TIME that the red eye photograph was taken, there was no alternative to using film. Even today, it remains highly suitable for intensive nighttime photography, although digital cameras are superior across a broad range of features and abilities, except for phone cameras. At full aperture, the 50 mm f/1.4 Zuiko lens was the second fastest available through Olympus and the T20 external flash unit was more than adequate, although, with Ralph's assistance, we later added an additional T40 unit which was capable of generating serious nighttime illumination. The resulting light output from the dual flash was staggeringly deep in range and extremely bright in intensity. This allowed us to photographically cover a much larger territory during pursuits and take full advantage of the motor drive, however, on recognition of the two reflectors, Fatfoot would correspondingly maintain a very safe distance. We often thought that the flash bounce off nearby objects may have forced Fatfoot to retreat further into the bush than before. As I typically held the camera overhead in an attempt to clear the vegetation, it also became apparent that Fatfoot was increasingly moving about as a quadruped during these interactions, as a means of minimising her vertical exposure to the flash.

Using an intense light source to see and identify objects is not the best way to obtain empirical evidence of Wildmen. It involves using an active system that allows operations to be easily monitored and anticipated by others and produces an outcome that has been altered because of its use. Passive systems are much better because they don't interact with their surroundings and their operations are difficult to detect, resulting in more realistic empirical results. It is for this reason that we started experimenting with an early Sony Thermal Imaging Camera in 1997. Although its use was unsuccessful on this preliminary occasion, it showed us that this was the path to take, as it was also a passive detection device for anything in the bush undergoing metabolism. This led to future success as infrared technology improved and more importantly, became more affordable.

Understandably, most animals use passive methods of visual detection with the exception, for example, of some deep ocean predators which use active biochemical means like bioluminescence and fluorescence to generate light for a variety of purposes. Some snakes use their pit organs to see infrared emitted from their prey.

A common method of photo-amplification used by animals uses a reflective layer behind the retina to redirect light back through the optical sensors for a second time. It is believed that this process may involve phase alignment of the light that results in the doubling of the wave crest height, with any misalignment resulting in some cancellation of the combined signals.

The closest equivalent of a tapetum lucidum in astronomy is a photomultiplier or other optoelectronic device.

A tapetum lucidum is nothing more than a reflective layer that is positioned behind the light-sensitive retina. When light from the pupil is focused onto the retina by the lens, it energises the rod, cone and other photoreceptor cells before being reflected back along the same optical path. The intensified electrical signal is sent to the brain, via the optic nerve.[12] This action has the effect of nearly doubling the sensitivity of the eye, so a 20 mm pupil now has about 9.8 times the available light, compared to a human. Similarly, a 25 mm pupil collects 15.4 times more light and a 30 mm pupil gathers about 22 times more. During this process, the outgoing light is coloured through interactions with various reflective agents, such as amino acids, that are present in this layer, with red of various intensities being the most common colour produced in marsupials. Other animals, like some dogs, have green reflective eyes, while horses and cows have blue. Also, as the reflected light is increasingly viewed away from its optical axis, the colour may shift in intensity and frequency. With Dooligahl and Quinkan, the shift is towards yellow. With Junjudee, this effect has not been confirmed.

Rhodopsin is an important sensory protein found in rods that converts light into an electrical signal and becomes bleached as a result of this reaction. The signal amplification made by the capture of a single photon can

---

[12] It is appropriate at this point, to dispel the misconception that the eyes of these hominoids 'glow internally'. Although the eyes seem to glow conspicuously bright in a dim environment, it must be remembered that the pupil is a large light collector that focuses the photons onto a reflective retina that returns the accumulated light back towards the observer under specific conditions. What the observer views is a fully illuminated disc that is the same size as the pupil and coloured by the reemitted light from chemical compounds that are present in the tapetum lucidum. Logically, a self-illuminating eye would only blind itself.

be extremely high, although the amount of amplification varies in response to the eye's degree of dark adaption. In low light levels where the rods have had time to replace their bleached rhodopsin, the photo-amplification is at its greatest. As illumination increases, the sensitivity drops as greater amounts of rhodopsin lose their reactivity before being replaced. This allows the eye to function efficiently across a wide range of lighting conditions. With Fatfoot receiving somewhere between ten and twenty-two times more photons at the retina compared to a human, her night vision after final photo-amplification would be difficult to determine, but it must be excellent. Also, as Fatfoot is predominantly a nocturnal predator, having colour perception would seem to be a squandered ability, whereas greyscale vision with an abundance of rod cells, might be a better option. Furthermore, a real possibility is having the frequency sensitivity of some or most rod cells shifted towards infrared, which would be similar to the highly-detectable images of warm bodies produced by commercial infrared cameras.

EYES, LIKE TELESCOPES, vary in construction and performance according to their intended use. Through our daily interactions with other people, we are most familiar with human eyes. Our eyes have a dark, round central region that we call the pupil. Its light-controlling ability is suited to our evolutionarily determined range of diurnal needs. Surrounding and regulating the pupil is the iris, which varies in colour and patternation, that may serve as a means of attracting a mate. The remaining visible part of the eye is the sclera, the tough supporting wall that protects and maintains the shape of the organ. The sclera is white and it is believed that its high visibility is a social adaptation that is used by others to track where you are looking and what you are interested in.

However, many other animals have differently constructed eyes with dissimilar optical traits that have specifically evolved to suit their environment, diel activity pattern and other needs. Many of the characteristics of an eye are also tailored to suit an animal's dietary preferences. For example, it is often argued that hunters not only need to be camouflaged but also need to have dark eyes that help to conceal their presence and what their predatory gaze is centred upon. For these reasons, the iris should be dark in order to avoid visual attention. Similarly, a white sclera is not desirable.

Although a circular pupil is generally suitable for many animals, its light-controlling ability is not as effective across the very broad range of illumination encountered by cathemeral animals and is particularly ineffective when trying to stop light from entering the eye. Such animals are better served by having a slit pupil which is capable of being fully dilated for low light situations and also, effectively closed off when an intense light is threatening to bleach its rhodopsin and remove any night vision capability. Additionally, the orientation of the slit is important, with vertical pupils being associated with predators[13] and horizontal pupils found with prey. The orientation of the variable width pupil has consequences that affect the eye's depth of focus along a particular vertical or horizontal line of sight. Unlike round pupils, a slit pupil provides a hybrid approach to an animal's vision.

With round pupils and camera lenses, the depth of focus is uniformly increased in all directions by the iris restricting the pupil or stopping the lens diaphragm downward,

---

[13] Koalas are a notable exception. Despite having vertical slit pupils, these marsupials are herbivores. Unless their evolutionary history is hiding a dark secret, it could be argued that their vertical depth of focus is used to more clearly see the available leaves above and below them.

for example, from f/4 to f/16. The increased depth of focus is caused by light being variously diffracted by differing amounts as the optical restriction of the iris tightens which broadens the range of focus. This also reduces the light intensity. On the other hand, when a slit pupil is restricted along its vertical axis it increases the depth of focus along this plane only, while also allowing light to enter. This produces a narrow and vertical depth of focus that is suitable for a predator's need to clearly see its prey through a long central corridor of stalking distances. When vision from both forward-facing and overlapping eyes are combined and processed by the visual cortex, a stereographic image is conceived that allows distance and perspective to be determined. These effects are moderated by the size of the IPD, with a wider IPD causing a more exaggerated stereoscopic effect. Outside this narrow band of sharp vision, the remaining side view is variously blurred according to its relative distance from the central focus, which provides additional or alternative clues for calculating range. For prey with horizontally slit pupils and eyes widely located on opposite sides of the head, these sharp flat images with deep focal range provide a panoramic scene to assist with predator detection across an extremely wide field of view. Mostly, this vision is not stereoscopic except perhaps, for a small frontal view where the two images might overlap.

From the very beginning, we knew that Fatfoot had exceptional eyesight. Although we did not know her exact eye dilation at that time, on rare occasions when we managed to get close enough and had a perfect alignment, the red eyes seemed to be quite large and were well separated. Exceedingly few witnesses have observed the eyes under favourable lighting conditions and at a close distance, being probably around one in a thousand or more cases. However, Susan, who was a canteen assistant at our school, and her father had a very close and persistent encounter with Fatfoot during an early morning pursuit and she described the eyes as being "large and different", being "set a long way back into the head". As the lighting conditions of their sighting were not optimally aligned, as expected, they didn't see any red eye reflection.

Justin, a former senior Industrial Technology student from Wentworth Falls, was able to clearly and closely view the eyes of one of three 'marsupial hominoids' from inside the comparative safety of his vehicle. He told me that during his close encounter on a remote bush road around mid-evening, "They had real big eyes. They looked like big, black, shiny eyes. I could see two big, round discs with a shine around the edge of each eye." In comparable circumstances, there were several bush locations within the mountain settlements, generally at the end of remote roads, where newly licensed teenagers were known to park their vehicles during the evening. Many frequent and similar reports were received by the Octopus where the occupants of these vehicles were harassed by a large animal that violently shook the vehicle, threw objects, walked noisily around, peered through the windows and undertook other frightening activities. On a unique occasion, one of the car's occupants saw a large figure looking through the side window and described the eyes as "very large and black that looked like bottomless pits". No other features of the eye were seen. Some important aspects of these rare opportunistic sightings are the overall 'blackness' of the 'large' eyes that also encompass the area where the iris and sclera are found. Most probably because of the close proximity, the red eyes weren't seen because the volatile alignment conditions couldn't be achieved or maintained. It is for these reasons, that the red eyes are only seen at a distance.

Ian Price possessed expert knowledge and experience with marsupials, amphibians, reptiles and particularly

13: Eyes

snakes. He also had a broader biological understanding, developed from many years of interest and study, as well as having worked as a research assistant at the CSIRO and Macquarie University. It was sometime during our frequent and lengthy afternoon discussions that Ian proposed that Fatfoot probably had slit pupils, in addition to many other optical possibilities. He also thought that, like many snakes that hunt in the dark, it was probable that she had some type of infrared vision because Australian animals were indigenous and so exquisitely adapted to the specific and extreme environmental conditions of this continent, that it was difficult to imagine a situation where infrared would not be part of the nighttime solution. It wasn't until the photograph of the red eyes was taken that we started to believe that we had evidence of any variant possibilities.

When it was dark, Fatfoot had round pupils because the iris had fully dilated and the photograph showed this.

Fig. 13.6 Enlargement of Fatfoot's red eyes showing diffraction spikes at the north and south 'poles' of each pupil. See also *Fig. 13.2*. (Image: Neil Frost)

Being round meant that the optical characteristics of the pupil were now different to those experienced by a slit aperture during the day. Later on, when further enlargements of Ralph's photographs were made, this 'roundness' didn't seem to be uniformly developed. At each of the upper and lower 'poles' of Fatfoot's eye images were a number of apparent flares or light protrusions (Fig. 13.6). Probably resulting from the high magnification and the image being saved in jpeg format, these flares became soft in appearance, but unmistakably similar in form to those found in photographs of bright objects and optical defects that are detectable in ground astronomical mirrors. It seemed that these interference spikes were generated at the points where the slit pupil is tightly coming together at the top and bottom of the aperture.

Having a fully dilated pupil has consequences. As can be seen in the larger red eye photograph (see Fig 8.2), although the focus was set at infinity, the bushes in the foreground are out of focus because the f/1.4 lens was at full aperture and had no depth of focus. The same could be expected with Fatfoot's night vision where some perception within the field may become blurred from being out of focus. Also, having such a large dilation leaves the animal highly susceptible to sudden changes in illumination, leading to a loss of night vision and temporary blindness. This could account for incidents alongside a road where bright headlights can cause hesitation or a freeze in action, similar to what occurs with many other native animals and kangaroos when being hunted with a spotlight.

Fatfoot's photophobia did have some uses. During pursuits that weren't proceeding comfortably, Ian and I always knew that carrying a powerful torch or photographic flash could be used to turn the attitude around. Carrying a recognisable device that had a reflective component certainly gained us some respect but it also placed limitations on what we could achieve.

13: Eyes

ONE UNUSUAL ASPECT of Fatfoot's behaviour was her habit of rocking from side to side during standoffs.[14] I could detect this motion by the alternating sound of vegetation being repeatedly compressed underfoot by her considerable weight and then being released. This happened numerous times when I came upon Fatfoot, usually standing a short distance into the bush by the side of our driveway and only when I slowly approached by myself, without a torch, camera or another visibly shiny object. Under these circumstances, I would talk reassuringly to Fatfoot, asking "How's hunting tonight?" If Ian would have been present, he would have immediately given chase and that would have set the agenda for the remainder of the night. This behaviour was cautiously encountered many times and variously over a period of several years, with the opportunity providing unique circumstances to achieve other, non-photographic goals.

When observing planets and large asteroids over a period of time, it is possible to detect movement resulting from their relative displacement due to the Earth orbiting the Sun. This is known as the 'Parallax Effect'. These observations were important in determining the nature of our Solar System and other aspects of astronomy, particularly in determining distance. The technique does not require a stereoscopic field of view to complete the calculation because the observer moves in order to generate the required trigonometric displacement, so a single telescope is all that is required. The realisation that Fatfoot was applying motion parallax to determine my distance was achieved after watching a wildlife documentary where an owl was similarly moving its head from side to side. In

---

[14] I first mentioned observing this behaviour to Tony Healy, which he briefly recounted in *The Yowie* (2006) on page 95. At that time, I was still uncertain about the purpose of this activity.

addition to using a slow shifting of the head, motion parallax can be achieved through rapid movement, such as by looking out of the window of a moving vehicle. Objects in the foreground will move faster than in the background, which provides depth and distance clues to a three-dimensional scene.

FROM OUR COMBINED research and the information obtained from Octopus encounter reports, Dooligahl have excellent vision that is optimised for nocturnal hunting. The pupils of these Wildmen are very large, tending towards the bottom end of the 20- to 30-mm-diameter range. During very close encounters, no iris or sclera have been identified, which is typical of predators, with all witnesses commenting on the black and large size of the pupils which they also describe as being bottomless, shiny, different, set-back and deeply set. During aligned nighttime encounters, the eyes reflect red light from their tapetum lucidum, which further indicates their night vision abilities through photo-multiplication. Although the pupils fully dilate to almost round in shape, the presence of top and bottom light interference spikes suggests that the pupils are vertically slit in type. Considering their enormous light-gathering ability and rhodopsin-based night vision, to prevent diurnal blindness, a wide-ranging vertical slit configuration would be the only way to fully manage light entering the eye during the day. Having a vertical slit pupil is advantageous for hunting because the stopped vertical plane allows for a very deep depth of focus along the anticipated escape path of prey, which would be particularly advantageous when Dooligahl are closer to the ground and moving quadrupedally. Also, an excellent depth of central and vertical focus results in distance-related blurring clues associated with the unstopped outside peripheral vision, which could assist with determining striking range. To achieve or supplement range finding, motion parallax is certainly used and was conducted by moving the entire body from side to side.

Considering the optimised nocturnal characteristics of Dooligahl, it would seem that light-sensitive rods would predominate in the retina. If this is the case, then the night vision capabilities of these 'marsupial hominoids' would be as formidable as observed. However, as Ian pointed out, being a stereotypical Australian marsupial should hold a few surprises and as with many snakes, one might be an adaptation capable of seeing in infrared.

From the research undertaken by the O'Connors and others, Quinkan have significant morphological differences, however, their vision seems to be fundamentally similar. No physiological or behavioural differences in vision have been observed.

Junjudee differ from the other two larger Wildmen in nearly all respects. Their sight, however, seems to be similar except that their night vision abilities are less intimidated by bright light and they will hold their ground against the flash of a camera, as discovered when confronting three stick-welding Junjudee. They show their dislike for visible and infrared illumination through their repeated attempts to deactivate motion-sensing garden lights and when trees were ripped in response to an infrared illuminator being used at the Pendlebury property. As shown in the sketch of the Junjudee's face (Fig. 11.7), the daylight observation of the eyes showed few details. The central black core of the eye was consistent with a pupil and probably an iris as well, though no iris or sclera were visibly observed. In this regard, the Junjudee's eye looked identical to the nighttime observations of Dooligahl's eyes made by various other witnesses, being black and featureless. The estimated total core diameter was 25 mm. As Junjudee are clearly cathemeral, their eyes are most probably vertical slit as well, in order to effectively control the wide range of lighting conditions encountered.

14: Legs

"I HAVE MUCH pleasure in telling you all I know of the kangaroo-hunting in Van Diemen's Land. The hounds are kept by Mr Gregson and have been bred by him from foxhounds imported from England and though not so fast as most hounds here now are, they are quite as fast as it is possible to ride to in that country.

"The boomer is the only kangaroo which shows good sport, for the strongest brush kangaroo[1] cannot live above twenty minutes before the hounds, but as the two kinds are always found in perfectly different situations, we never were at a loss to find a boomer, and I must say that they seldom failed to show us good sport. We generally found them in a high cover of young wattles but sometimes we found them in the open forest, and then it was really pretty to see the style in which a good kangaroo would go away. I recollect one day in particular, when a very fine boomer jumped up in the very middle of the hounds, in the open forest. He at first took a few high jumps with his head up, looking about him to see on which side the coast was clearest, and then without a moment's hesitation he

---

[1] A 'boomer' is a mature male kangaroo. At that time, 'brush kangaroo' was the common name for a wallaby.

stooped forward and shot away from the hounds, apparently without an effort, and gave us the longest run I ever saw after a kangaroo. He ran fourteen miles by the map from point to point, and if he had had fair play, I have very little doubt but that he would then have beaten us, but he had taken along a tongue of land which ran into the sea, so that, on being pressed, he was forced to try to swim across the arm of the sea, which, at the place where he took the water, cannot have been less than two miles broad. In spite of a fresh breeze and a head sea against him, he got fully halfway over, but he could not make head against the waves any further and was obliged to turn back, when, being quite exhausted, he was killed.

"The distance he ran, taking in the different bends in the line, cannot have been less than eighteen miles, and he certainly swam more than two. I can give no idea of the length of time it took him to run the distance, but it took us something more than two hours and it was evident, from the way in which the hounds were running, that he was a long way before us and it was also plain that he was still fresh, as, quite at the end of the run, he went over the top of a very high hill, which a tired kangaroo never will attempt to do, as dogs gain so much on them in going uphill. His hind-quarters weighed within a pound or two of seventy pounds, which is large for the Van Diemen's Land kangaroo, though I have seen larger.

"We did not measure the length of the hop of this kangaroo, but on another occasion, when the boomer had taken along the beach and left his prints in the sand, the length of each jump was found to be just fifteen feet, and as regular as if they had been stepped by a sergeant. When a boomer is pressed, he is very apt to take the water, and then it requires several good dogs to kill him. If he stands waiting for them and as soon as they swim up to the

14: Legs

attack, he takes hold of them with his fore-feet and holds them underwater.[2] The buck is altogether bold, and will generally make a stout resistance, for if he cannot get to the water, he will place his back against a tree so that he cannot be attacked from behind, and then the best dog will find in him a formidable antagonist."[3]

Account furnished to John Gould by
the Honourable H. Elliot (1842)

KANGAROOS ARE LARGE marsupial herbivores that are primarily bipedal. They are renowned for their speed, endurance and the highly unusual methods that they use to get about. They utilise a wide range of ambulation, from slow-moving pentapedal and quadrupedal gaits, used when grazing on grass, through to bipedal hopping, sometimes referred to as saltatory locomotion, which is used when travelling. They also swim by 'dog paddling'. Kangaroos can use five appendages when grazing, two hind legs, two small forelimbs/arms and their tail, which is used as an intermediate prop when transitioning between stances. Although fundamentally adapted for hopping on two legs, they are also able to walk on four or five appendages—making kangaroos facultative quadrupeds/pentapeds.

Whilst most of these macropods[4] are terrestrial, a few extant species remain arboreal, with tree-kangaroos being found in the rainforests of tropical Northern Australia and

---

[2] Over several decades the Octopus received reports of ducks being seen held underwater by a Dooligahl along the banks of the Nepean River and another where ducks were drowned in a private dam.

[3] Anon. 1842. Gould's Monograph of Kangaroos. *The Tasmanian Journal of Natural Science, Agriculture, Statistics, &c.* 1(5): 381-382.

[4] 'Macropod' is derived from Greek, which literally means 'large footed'.

the island of New Guinea.[5] This suggests that kangaroos evolved in the trees of the Gondwanan Rainforests of Australia (Sahul), before becoming terrestrial by adapting to the harsher conditions found in the evolving, wet and dry sclerophyll forests.

As with other parts of the world, the fossil record of Australia is normally very patchy and possesses many large gaps in its succession. However, despite this, there have been a number of tantalising discoveries from the rich, freshwater, fossiliferous limestone network of the Gregory River in northwestern Queensland, an area that was once part of a much larger, continental rainforest system. A very interesting fossil skull was found at Riversleigh, described as *Balbaroo fangaroo*[6] which lived between 23-10 mya. (See https://www.eurekalert.org/multimedia/665798 for an image of the skull.)

Balbaridae are believed to be an extinct family of early macropodiforms that preceded modern kangaroos (Macropodidae). The highly unusual dentition of the anterior teeth includes the large maxillary canines and the forward-facing mandibular teeth. The upper canines are large and hooked and would have been externally visible. They are an unmistakable trait of a carnivore, although some argue that this early kangaroo was a herbivore. It would seem to me that the purpose of the curved canines was to lock the bite of this predator onto its prey, like vice-grip pliers. Extant kangaroos no longer have canines.

---

[5] During the 1970s, it was very common for Australians, who operated various establishments in PNG, to maintain small private zoos as a business attraction, which typically exhibited tree-kangaroos, cassowaries and exotic rainforest birds.

[6] Black, K.H., et al. 2014. A new species of the basal 'kangaroo' *Balbaroo* and a re-evaluation of stem macropodiform interrelationships. *PLoS One* 9(11): e112705.

14: Legs

Based on the fossil skull, *Balbaroo fangaroo* has been estimated to be about the size of a modern wallaby. Unlike a wallaby, however, its forearms were robust, which suggests that it was more able to easily move about on four legs. As a quadruped, rather than a saltatory biped, it obtained its speed by galloping instead of hopping. Also, unlike modern kangaroos, these ancestral marsupials had an opposable first toe, which is similar to Fatfoot's four-toed grasping foot (Fig. 10.2) and the ability of Dooligahl, according to the Aboriginal Elders, to climb trees.

It has been estimated that modern kangaroos, of which the red kangaroos are the largest, have obtained the maximum weight permitted for bipedal hopping because of the increasing danger of leg and joint failure. A boomer red kangaroo can weigh up to one hundred kilograms or two hundred and twenty pounds and stand about 1.8 metres

Fig. 14.1 Procoptodon goliah reconstruction from the Australian Museum (Photo: Neil Frost)

or six feet. Although they are large by modern standards, as we have seen, they are not the largest extant Australian marsupial.

In the past, there were bigger kangaroos such as the extinct *Procoptodon goliah*, which was the largest and most heavily-built short-faced kangaroo known. At resting height, it was about two metres or six and a half feet and weighed approximately 250 kg or 550 pounds.[7] Its face was flat, with a chin and forward-facing eyes that also allowed excellent peripheral vision. It had unusually long, robust arms[8] and two extra-long fingers with large grasping, curved claws. The legs were big and heavily built with large buttocks. The feet had one clawed toe, instead of the typical four found on most kangaroos.

Due to their large size and weight, *Procoptodon goliah* were not hoppers. Instead, they were bipedal walkers as substantiated by their large buttocks, which are also found on humans. As they were also herbivorous browsers specialising in difficult-to-obtain leafy material, this low-energy diet would have necessitated a more conservative mode of locomotion. Since efficient high-speed hopping was unobtainable, slow and long-striding bipedal walking was the next best option, which can be more efficient compared to the alternative modes of quadrupedal walking and galloping.

It was most probably their slow speed of motion that lead to their easy predation and eventual extinction by Aboriginal hunters, rather than disappearance due to climate-related factors. Based on snapshots from the fossil record, their date of extinction is uncertain and will prob-

---

[7] These physical characteristics closely match those of Fatfoot.

[8] A large number of Octopus witnesses commented on the unusually long, robust arms of Dooligahl.

ably remain so, although Aboriginal oral history tells of relatively recent stories where large bipedal animals with claws killed people. This suggests that these large kangaroos and possibly other variants may have co-existed with the human population for longer than the previously myopic speculation would acknowledge. Furthermore, as the kangaroo population was affected by various hunting practices, they could have been expected to become increasingly savvier, by retreating to more remote and inaccessible locations and learning other avoidance techniques. This may well be the strategy with the thylacine in Tasmania as well.

The skull of *Procoptodon goliah* also had a large nasal area, which suggests that it had an excellent sense of smell. A daytime sighting of a Dooligahl having successfully used its sense of smell after its hearing was proven to be ineffectual, highlights the broad range of sensitive abilities present in these 'marsupial hominoids'. In June 1989, Gary[9] and two other mates bicycled and hiked through very dense and remote terrain for more than six hours to reach their premium trout fishing location at the confluence of the Kowmung and Coxs Rivers, eighteen kilometres (11 miles) south of Katoomba and thirty kilometres (19 miles) west of the Warragamba Dam, in the heart of the Blue Mountains National Park wilderness. Apart from the difficulty in access, this is the catchment area for the Sydney water supply and has a prohibited access status. As they were collecting firewood by the water, they saw what

---

[9] This very detailed (not fully reported here) encounter was uncovered by John Appleton, a teaching colleague. John came across several encounters like this through careful listening and by applying the 'standard question' to promising interview situations. John, who was also a keen fisherman, returned with Gary to this encounter site several decades later and experienced daytime rock throwing and rock smashing with conspicuous walking around their nighttime campsite.

looked like a very hairy man standing with his back toward them on the other side of the river, about thirty metres upstream. One leg was in the deep water and the other was up on the embankment as the hairy man was scooping water with his hand to drink. The sound from the moving river stones must have disguised their approach and they were able to halve the gap when the 'marsupial hominoid' raised its head and began sniffing the wind.[10] It suddenly turned around and immediately raised itself from the water without any difficulty, which would have required enormous strength to achieve. Gary said that it was six-and-a-half feet tall, very muscular with very wide shoulders but no neck. It had black, deeply set eyes, a tall forehead and a pointed cranium. It had unusually long legs and it took off through very thick vegetation at high speed and up a steep incline.

A few years later, in approximately the same area Pat, a highly experienced and respected bushwalker and naturalist, had an encounter with a Dooligahl while solo walking from Wentworth Falls to Mittagong, a straight line distance of eighty kilometres, through the very harsh terrain of the Blue Labyrinth. Pat, who was Yowie-aware, said that he had picked up a Dooligahl during the early afternoon journey and it had followed close behind him for a lengthy time. As sunset was approaching, Pat searched for a rock shelter and built a large fire across its mouth to keep the Dooligahl away during the night. The following morning the Dooligahl had moved on and Pat continued his journey without further incident. This encounter was highly intimidating but non-violent, possibly because of Pat's cool attitude modelled on his previous experiences.

---

[10] This important observation had been passed over during a follow-up interview. Other reported incidents of large hominoids sniffing the air can be read in *The Yowie,* Healy & Cropper, cases 172, 203 and 212.

14: Legs

IN TRADITIONAL ABORIGINAL hunting culture, bones from many animals were used to produce a variety of tools and components. Glue was made from rendered bones, tendons and other boiled animal parts. Spearheads were produced from a range of materials, including bone, timber and flint. A 'woomera' was used to increase the velocity and force of a spear when being thrown and could be made from timber or the lower leg bone or tibia of a kangaroo. The word comes from the Dharug language which was widely spoken throughout the Sydney Basin, including the Blue Mountains. Variations in the design, materials and manufacture of the throwing stick are found across Australia.

Using the tibia, the longest of two lower leg bones of a kangaroo to make a woomera shows a traditional appreciation for the role that these bones play in the locomotion of the biped and its associated application as a potential weapon enhancement. As a spear-throwing aid, a woomera is a mechanical lever that is an extension of the hunter's arm which increases the radial velocity of the throwing motion and also imparts additional kinetic force over an extended period of time. This results in the spear travelling with increased velocity and striking the target with greater force than would be possible if thrown without it. Woomeras were more commonly carved from timber than from bone.

As components of the lower leg, the fibula and tibia support the kangaroo when landing and rebounding. Having evolved in length, these bones have permitted greater bounding height and increased maximum speed. As an analogy, these long leg bones can be likened to the top gear or overdrive in a vehicle gearbox, where additional speed can be extracted by changing up into a 'taller' gear or in the case of kangaroos, by increasing its stride length. In the hunting description given earlier by John Gould, "the length of each jump was found to be just fifteen feet" which emphasises how speed can be increased by 'reaching out' with the legs.

Many witnesses report observing Dooligahl and Quinkan with disproportionately long legs and further comment that the biped approached or fled the scene at very high speed. Continuing with Gary's description, he said that the animal "took off" up this "incredibly steep hill". The animal had about a fifty-metre lead and was loudly running up towards the ridge. It took them a while to cross the river where it had been standing in the water and they had a lot of difficulty getting up onto the bank. The men chased after it and although they thought that they were very fit football players, it was easily outpacing them up the steep slope. To make things worse, "it was running non-stop through spiky bushes" that were cutting the men, so they had to run around the bushes. After they had run about eight or nine hundred metres uphill, they stopped and turned back to the campsite because it was going to become dark soon and they didn't know how many of these things were about.[11]

We also experienced a similar demonstration of Fatfoot's speed and defiance of prickly obstacles when she ran down Ian's land after being startled by the multiple flash units on our newly improved automated camera. We estimated her speed to be 40 km/hr, which is very fast especially considering that, like Gary's biped, she was running through some very spiky hakea bushes. As Ian said on several similar occasions, "The beastie either has thick

---

[11] There is no doubting the veracity of Gary's account. There are too many collaborating details that few could be expected to accurately know. However, he did mention that the footprint in the sand near the river bank showed five toes. I strongly suspect that the toe count evaluation of the Hairy Man may have been rushed and based mostly on expectation, rather than close scrutiny because Gary did not elaborate on other potential foot features.

protective fur, or it is indifferent to the pain." Confirming what Ian had also mentioned several times, the gait of Fatfoot that I heard sounded irregular, which made me think that she may have been favouring a foot because of some past injury. When looking at the photograph (Fig. 10.2) of Fatfoot's right foot cast, an outside midfoot fleshy protrusion can be seen that suggests that it is a site of an injury-related growth. Other four-toed Dooligahl casts and photographs that I have seen show a foot that is long and narrow, without any substantial side bulge. Even Quinkan foot impressions that have been observed in this region, apart from being three-toed, are similarly narrow and without any noticeable appendages.

During the time of that early encounter in 1993, Fatfoot and I were both relatively fast and pain-free runners. However, in April 2021, a neighbour's son bought a highly-powered 5000-lumen LED spotlight so that he could see the red eyes of Fatfoot at any distance and defend himself on walks down to the swamp if required. Nick and his father lived further down the valley and they both had had encounters, individually and together, over the previous six years. Nick's torch was incredibly bright. Some might even say, excessively so. On this occasion, he said that he had detected Fatfoot's red eyes hiding behind a bush in a six-foot-deep bowl depression near the swamp at a distance of between eighty to a hundred metres from their rear home paddock fence. Sometimes Nick would search the swamp with the spotlight from their elevated verandah, an additional eighty metres further away from this lower position. Nick and Michael were well aware of this deep depression because they had observed Fatfoot and some Junjudee using the ridge line of the hollow to secretly peer over during many surreptitious efforts. In response to this attempt, Fatfoot started moving away towards our house, so Michael rang to warn me of her imminent approach.

On opening our back door, I could immediately hear lively walking approaching from the north and see Nick's LED 'death ray' deeply cutting through the nearby bush of the hollow. Behind the door on the main kitchen table, I had a fully charged FLIR Ocean Scout 640, ready for such opportunities.[12] However, the rapidly progressing circumstances required an immediate reaction, so I abandoned any thought of preparing the FLIR and maintained my position on the verandah, listening intently for any auditory clues as Fatfoot stepped cleanly over the wire fence.[13] She continued to rapidly move parallel to our side boundary, at an estimated separation of two metres and for a count of ten seconds until the sound of her movement was completely absorbed by the distant swamp. The following morning, using the audio-related markers, whose final angular positions I had noted, I determined the length of the path taken to be a maximum of 43.2 metres, terminating at the location where the grassed clearing gave way to the swamp. This gave an average speed over that ten-second journey of 15.5 km/hr. Unfortunately, I wasn't able to sufficiently multitask the situation, in order to accurately determine the number of steps taken during that time and distance. As a crude guess, I would say that she was taking about two steps per second, which would mean that each step was about 2.1 metres in length or a stride of 4.3 metres. However, I think that the accuracy of these estimates is poor.

Despite her age and the considerable pressure that Fatfoot was under trying to evade the bright light, she maintained a steady and brisk pace. This calmer pace was in

---

[12] It is important for people with inherently doubtful dispositions to understand that occasionally there is insufficient time to achieve the intended goal.

[13] This inadequately constructed wire fence was only 800 mm in height at this crossing point.

stark contrast to the similarly illuminated situation that she had confronted thirty years earlier, where she had run nearly two and a half times faster. I know for certain that my running performance has been similarly compromised during this same period. This also begs the question regarding Fatfoot's age. We know that we have been dealing with Fatfoot for more than thirty years because of some very unique characteristics, particularly her foot-thumping behaviour in response to our light signals. Our first encounter with Fatfoot occurred when we were excavating our building site in 1983. As the footprints found were of a substantial size, it has been assumed that she was sexually mature at the time. This means that her minimum age, as of 2023, is forty years plus her age from joey to physical maturity.

FOR A VERY long time, it has been traditionally regarded that marsupials are inferior to placentals across a broad spectrum of traits. This was the typical narrative that was circulating when I was growing up. This belief was partly based upon the knowledge that placentals evolved more recently and must, therefore, be an improvement on what has come before, rather than being an alternative evolutionary response to a different set of environmental conditions. Also, fossil evidence shows that placentals outcompeted marsupials around the world, leading to their broad extinction. The notable zoogeographical exception was Sahul, with a small number persisting in North and South America. However, the main drift of the argument was that marsupials may have never become dominant in Australia if placentals had been present from the very beginning. Some controversial evidence suggests that early placentals were present in Sahul but became extinct.

Additionally, the reproductive system of the placentals has been considered superior because it allows for a longer developmental period for offspring to more fully mature inside a protective womb. The reproductive priorities of placentals, therefore, are shifted in favour of the offspring and at the expense of the mother. However, by giving birth to very immature young and raising them externally within a pouch (Fig. 5.9), marsupial mothers are exposed to significantly less reproductive risk, otherwise caused by genetic and birthing complications or changing environmental conditions. The mothers can also influence their fertility by pausing or ending a pregnancy during adverse situations and conversely, they can initiate a second pregnancy to follow closely behind the current joey when conditions are more favourable. It seems instead, that the greater reproductive flexibility of marsupials significantly benefits the welfare of the mother and is in the best potential interest of the offspring as well.

It has also been argued that marsupials are less intelligent than placental mammals. This belief seems to be based on the smaller size and different form of a marsupial brain. As it has been shown, even very small-brained parrots, like kea, are capable of completing complex problem-solving tasks, so it appears that how the brain is wired is more important than its size. Metabolically, the brain is a very expensive organ to maintain, with a smaller but more efficient biological structure being advantageous in a difficult and competitive environment. However, structurally, a marsupial brain does not have a *corpus callosum*, which is a thick bundle of nerve fibres that connect and allow communication between the two cerebral hemispheres. Instead, a marsupial brain has a simple network of neurons that join the two halves. In humans, each cerebral hemisphere controls the movement and feeling on the opposite side of the body, so an important function of the *corpus callosum* is related to coordination. The *corpus callosum* also allows the two hemispheres to work together to process information for complex problem-solving tasks, including language. Similarly, all three 'marsupial hominoids' have demonstrated complex problem-solving abilities, have an acute awareness of the 'Theory of Mind' and have also acquired language, just to name the few characteristics that have been identified so far. Therefore, it could be said that marsupials can accomplish many of the same tasks as placentals but achieve them differently and most likely, more efficiently too.

"Marsupials generally have a low body temperature (T_b) and basal metabolic rate (BMR) compared to placental mammals. Resting T_b is typically 2.5°C lower for marsupials compared with placental mammals and BMR of marsupials is on average only 70% of that of equivalently-sized placental mammals."[14]

By having a lower metabolic rate, it has long been argued that marsupials are potentially less able to actively compete with placentals, particularly involving high-energy activities, such as running. However, from the kangaroo-hunting account noted by John Gould, it is clear that the boomer was easily able to outperform the many placental hunting dogs for the majority of the very long chase. Similarly, young Gary and his two fit football mates were out-matched by the speed and stamina of the Dooligahl that was pulling away from them up the steep slope. It would seem, therefore, that aerobic factors like having a larger heart size[15] and greater lung capacity could be responsible for this enhanced physical performance and endurance. These enhancements could also include the ability to increase speed by outreaching their stride by

---

[14] Cooper, C.E., et al. 2016. Marsupials don't adjust their thermal energetics for life in an alpine environment. *Temperature* 3(3): 484-498.

[15] Dawson, T.J., et al. 2003. Functional capacities of marsupial hearts: Size and mitochondrial parameters indicate higher aerobic capabilities than generally seen in placental mammals. *Journal of Comparative Physiology B* 173(7): 583-90.

using their longer fibulae and tibiae, as well as being able to reclaim previously expended energy through their lengthy and highly elastic leg tendons. There are also many other considerations, including the role of body fat[16] and differences in muscle types and their form. Whatever the combination of enhancement factors that are required to achieve this physical superiority, the superseding evolutionary aim seems to be achieving accumulative energy savings and efficiency.

From our observation of the two larger 'marsupial hominoid' species, in particular, it seemed that they are more receptive to the cooler conditions found at altitude and during the night, despite having a lower metabolism. In fact, unlike smaller marsupials, their greater body mass to surface area ratio would increasingly cause problems with heat retention if it wasn't for their lower metabolism. This would be further compounded by the thermal insulation provided by their hair. For example, after becoming over-exerted and stressed as happened during the intense winter encounter with Robert and me on Friday, 18 June 1993, Fatfoot became very noticeably overheated and began to noisily pant, as kangaroos also do, despite the cold wintery conditions that were concentrated in the lower recesses of the swamp. For both of us, the panting was very similar to the sound made by a very large dog and at a comparably rapid rate. The loud and lively noise coming from the thick and dark scrub a short distance away from us was very off-putting. While the panting lasted, it conspicuously gave away her position before she began to quietly move closer to the swamp.

Although Dooligahl and Quinkan are capable of maintaining very high speeds, they mostly prefer to stay sedentary. When people-watching, Fatfoot would spend large

---

[16] Kangaroo meat contains less than 2% fat.

amounts of time standing motionless behind a tree. Occasionally, she would also find a larvae-infested tree to bite while waiting or an unsuspecting bandicoot to munch on. Although Fatfoot would have been capable of achieving better, her stationary record for inaction was timed at ninety minutes. This noted period of inactivity was probably brought to a premature end because I had been sitting provocatively on the moonlit verandah, drinking beer and smoking cigarettes. I was certain that Fatfoot fully understood my intent and eventually exploded from frustration.[17] Similarly, during many pursuits, Ian and I would often discover that Fatfoot had 'disappeared' ahead of us, only to discover that she had simply gone to ground somewhere along the way. This obviously meant that we had run past her as she lay motionless inside a bush or other hiding place and would have passed her by, within feet or perhaps inches, on too many occasions to count. This reminded me of Aboriginal Elder Merve saying that Dooligahl liked to touch young girls on the leg when the tribe was passing through a swamp.

This frequent inactivity appeared to be part of Dooligahl and Quinkan hunting behaviour. It seemed to us that Fatfoot was usually a solitary hunter[18] and was primarily a nocturnal ambush predator. Having vertically-slit eyes would have been advantageous under these circumstances. By standing or laying motionless, large amounts of energy are conserved by waiting for the prey to pass by. Only when necessary would any rapid action be required. *Thylacoleo carnifex*, sometimes described as the marsupial lion,

---

[17] As Fatfoot afterwards refused to acknowledge my attempts at reconciliation and continued to storm off, this incident demonstrates a strong emotional aspect of her behaviour.

[18] As neighbour Phil discovered, when Fatfoot had joeys they would either stand off at a distance or join the hunt.

was a well-known example of a carnivorous, megafaunal, ambush predator that became extinct, possibly due to the consequences of climate change but more likely hunted by the Aborigines due to the high danger that they posed to family tribe members, or the mob generally.

There were several reasons for believing that Fatfoot was an ambush predator. For example, several ambush sites were found throughout the valley over the decades, where something appeared to have been concealed alongside a waterhole or game trail. Sometimes, the still of the night in the valley would be shattered by the brief high-pitched scream of a distressed animal being dispatched. More significantly, an early morning thermal image video showed Fatfoot waiting for a swamp wallaby to approach her hidden position (Fig. 14.3; also http://tinyurl.com/fatfootwallaby).

Junjudee, on the other hand, when typically working as a 'mischief', seem to use the pushing method of hunting. Fig. 11.11 shows a still from a game camera video of a young joey being pushed by a solitary Junjudee from behind the camera, using the newly-installed fence as a race.[19] Junjudee can recognise game cameras and similar devices, which they frequently interfere with, disable, steal and destroy.

Thermal imaging cameras are a very useful research tool because they can be passively used to detect endothermic animals. Since using a Sony thermal camera in 1997, the cost of these IR cameras has dropped dramatically. The Seek CompactPRO was an affordable device because the video processing was achieved using application software running on Apple iPhone hardware. The interface connecting the devices is a 'Lightning' cable which physically

---

[19] Also known as a chute, run or alley. A corridor that is used to control the movement of livestock or game.

Fig. 14.2 Seek CompactPRO thermal camera attached to a RC pan-and-tilt mount, with two channel sound recording. The video processing and monitoring is achieved using an Apple iPhone, remotely connected by a 'Lightning' cable. (Photo: Neil Frost)

joins the two hardware items together. Consequently, trying to passively use the camera to search the bush for Dooligahl at night is not possible because of the emitted light from the attached iPhone screen. It is like optimistically walking around at night searching for 'marsupial hominoids' with an LED light source above your head.

The Seek IR camera has a small 5 mm diameter lens that is larger than those on most phone cameras and is made out of low IR absorbing glass, usually containing Germanium or other similar elements and compounds. The sensor resolution is 320 x 240 pixels with a slow refresh rate of about nine frames per second. The higher frame rate specification of fifteen frames per second is unavailable in Australia because they are legally forbidden by U.S. export restrictions. The wide field of view is 32°. There are several palettes to choose from, with 'High-low Temperature' being selected as the most appropriate choice for this video (see Fig. 14.3).

In order to make the Seek an adaptive passive device, it was necessary to make some modifications (Fig. 14.3). The IR camera and iPhone were connected using a high-quality, two-metre-long, 'Lightning' extension cable. The cable was usually connected without difficulty over this distance and allowed the camera to be mounted on the external side of the window, with the iPhone freely located on a table inside the house, where the light from the screen could be more easily concealed and the controls conveniently accessed. Infrared devices cannot see through standard glass as IR is absorbed or blocked by it. The camera was fitted onto a radio-controlled pan-and-tilt mount using a short length of 32 mm square section tubing, to which shotgun and mini parabolic microphones were also attached. The pan-and-tilt mount with attachments was then secured to an adjustable ball mount by using the ¼ inch standard camera thread and screwed to the outside wall, next to the window.

14: Legs

From the early evening before 23 May 2019, the dogs had been barking intermittently along the boundaries of the swamp, at spot locations that very slowly migrated up the valley over a period of many hours. This signature behaviour indicated that Fatfoot was moving through the valley and that her dominance had returned after the resident 'mischief' of Junjudee may have moved on. By following the sound of the dogs' barking from my bed for some time after midnight, it seemed that something was taking place a short distance to our north. After getting out of bed, the iPhone was turned on and connected to the infrared camera via the cable. By using the remote control to widely scan around the bush, a distant infrared blob was detected, which unmistakably indicated the presence of a significantly-sized, endothermic animal.

Looking at the phone screen, it was impossible to determine any form from the blob so monitoring was continued for some time. After a while, the white-hot[20] object began to rapidly move towards the camera but even though no recognisable form was visible, it was moving like a swamp wallaby hopping through the scrub. Although the dogs had begun to bark again, I hadn't noticed anything else moving within the same field of view. The swamp wallaby made another move by hopping across the field of view to the left for about twenty metres. The wallaby abruptly halted on landing, as if it had detected something and was forced to reign in its forward rebound momentum before slowly rising to its full alert height to commence looking into the scrub. As I had not been aware of Fatfoot's immediate presence, I discontinued recording shortly afterwards and

---

[20] As with most thermal camera software, there are a number of options that determine how temperature variations are displayed on a screen. A common choice is showing hot objects as white. There are also black-hot and a number of colour palettes to choose from.

returned to bed. Had I known otherwise, I would have remained vigilant in the hope that an ambush could have been observed and recorded.

On reviewing the recording the next morning, a tall and wide figure could be seen as it stepped into a gap in the scrub before stopping for six seconds and then moving on again before disappearing into the thick bush.[21] I had not noticed this grey body at the time, most probably because the surface temperature of the Dooligahl was close to ambient and the range of temperature was only eight degrees Celsius. However, it was around this moment that the dogs had also been briefly barking. Soon after this, the swamp wallaby had hopped towards the position where the large figure had been standing and halted at full alert a short distance away.

Not many details are visible in the infrared video and even fewer can be seen in the reprinted still image, but it is always possible to extract some valuable information. The terrain in the area where the video was taken, gently slopes to the left of the photograph towards the swamp. The ground is slightly uneven. There are no clearly defined environmental landmarks that could be used to make comparative measurements of other FoV objects. However, our mature swamp wallabies are generally about three to three-and-a-half feet tall, though when they are concerned about their security, they can rise to a maximum height of about four feet, so that they can see further afield. If they are part of a mob and feel threatened, like nearly all macropods, they will foot thump to indicate their position to other wallabies or warn them of potential danger.

---

[21] This thermal video can be seen in the 2020 documentary *TRACK—Search for Australia's Bigfoot,* by writer and director Attila Kaldy.

14: Legs

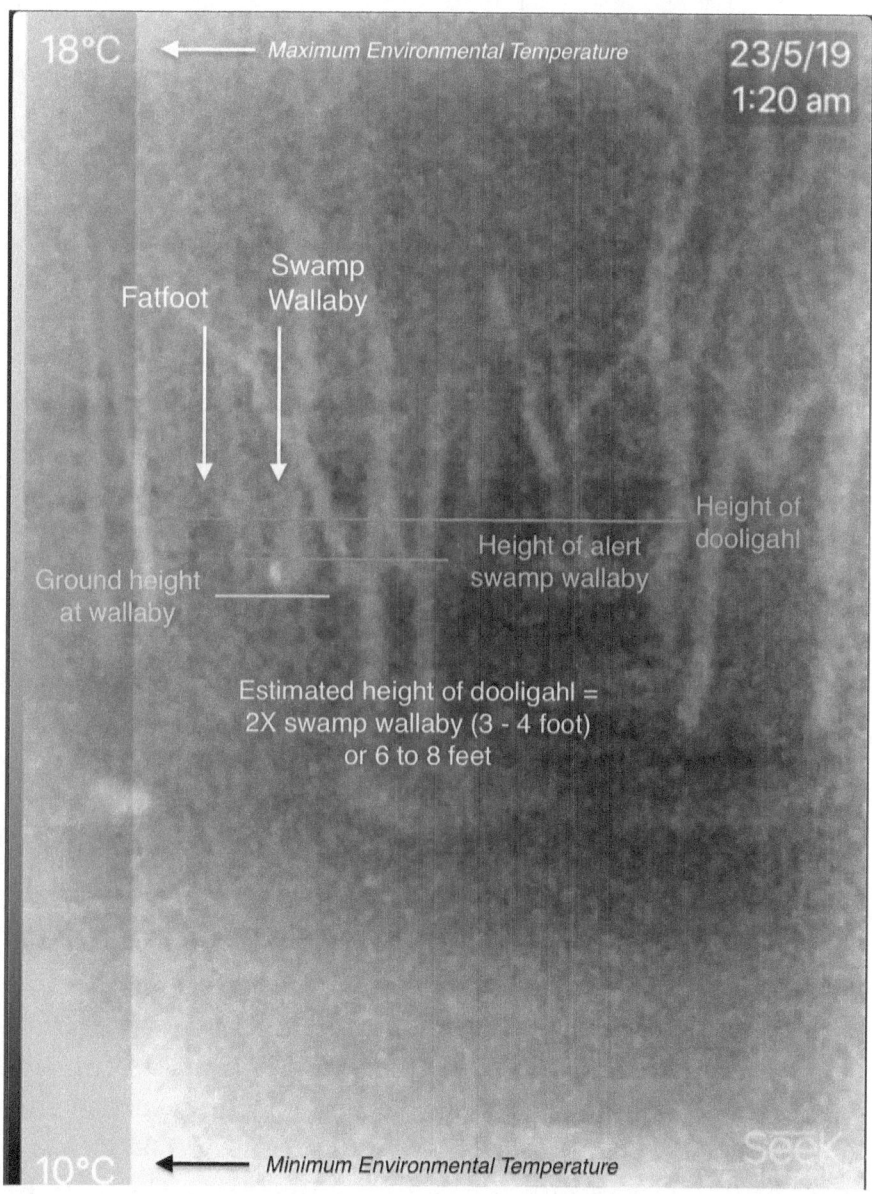

Fig. 14.3 Still image from a Seek thermal video camera 'showing' a blobsquatch image of Fatfoot and a targeted swamp wallaby. A grey-scale temperature scale is shown on the far left side of the image with a range of 8° C. (Image: Neil Frost)
Thermal video can be viewed at http://tinyurl.com/fatfootwallaby.

Determining the apparent height of the alert wallaby was easily obtained from the video and was transferred to the still image. It was assumed that the actual height of the wallaby was between three and four feet. As the height of Fatfoot is known to be seven feet, the height of the wallaby must be three and a half feet, which lies between the lower and upper height estimates. The ground height was determined from where the feet of the wallaby had landed in the video and because of the close proximity to Fatfoot, it was assumed that this could be used as a baseline for both marsupials, despite the presence of a small downward slope.

As the camera software was set to show the 'High-low Temperature' display, the iPhone calculated the temperature of every pixel and determined the range of thermal values. The camera was also set to produce a greyscale to represent the thermal range, with the maximum heat shown as 'white hot' through to the minimum temperature of 'black cold'. The variation in temperature in the photographed environment was eight degrees Celsius. As a result, the greyscale used is shown on the far lefthand side of the thermal image, beginning with the lowest temperature of 10° C (black), through to a maximum of 18° C (white), measured in pixel increments of one hundred steps.

By using a pixel sampler[22], an accurate numerical value from 0 to 100 can be obtained for every pixel in the still image which can be used to determine its spot environmental temperature. The hottest element in the photograph is 18° C and has a corresponding greyscale value of 100—similarly, the lowest temperature of 10° C has a score of 0. After sampling the pixels that form the wallaby image, a value of 71 was obtained, which means that the wallaby's surface temperature is 71% of the range, 10° C to

---

[22] ColorSlurp for the Mac was used to read the greyscale values of the sampled pixels.

18° C or 5.68° C, plus the base temperature of 10° C, giving an external body temperature of 15.68° C. Similarly, by sampling the pixels that form Fatfoot's image, a value of 34 was obtained. This means that Fatfoot's external temperature is 34% of the environmental range or 2.72° C, giving her an external body temperature of 12.72° C.

Having spent a great deal of time scanning our valley and elsewhere with thermal cameras, it was very surprising to see how close to the environmental minimum temperature the body of Fatfoot was.[23] Little wonder that this lack of thermal contrast has caused difficulty in separating her image from the bottom-end noise. Consequently, it seemed prudent to review a number of other thermal videos that had been previously dismissed because there were no conspicuously 'hot' objects moving in the field of view. One of these videos showed a roundish grey (low temperature) object slowly gliding from left to right and briefly pausing within the swamp. To facilitate this view, a narrow corridor had been cut through the intermediate vegetation to allow visual access to the heart of the swamp. Such a visual corridor is, of course, a two-way street. The video had been manually recorded on 25 August 2019 at 3:50 A.M. after dog alarm activity in the valley was heard. The minimum environmental temperature was 10° C and the maximum 16° C.[24] Using the pixel sampler, a value of 38% was obtained for the 'head' which gave a temperature of 2.28° C over the six-degree range, giving an external body temperature of 12.28° C. The amount of variance between the two body temperature measurements is 0.44° C.

---

[23] Some predators, like snakes, owls and house cats have weak to nearly-invisible infrared emission. Baker 2021, *op. cit.*

[24] The minimum temperature of the swamp is usually much lower than this. The lowest temperature recorded on the Sony thermal camera was -4° C. Winter in Australia is from June to August.

Typically, other local marsupials like wallabies and possums tend to glow brightly, but placentals like foxes and rats are very radiant, which clearly demonstrates their higher metabolisms. The image shown in Fig. 14.4a was taken using a different thermal camera[25] that, unfortunately, did not provide a temperature greyscale. However, as can be more clearly seen, the fox and rat[26] were the thermally brightest environmental objects in the field of view, with a relatively hot pixel score of 99/100. Consequently, their actual temperatures could not be calculated from the image, but their metabolisms are obviously higher (around 30% warmer) than the two marsupials from Fig. 14.3.

As Fatfoot and the swamp wallaby are both marsupials and appear in the same thermal image, it is interesting to consider why they have different external body temperatures. The swamp wallaby had been grazing for a considerable period before being videoed. After its grazing was disturbed, the wallaby hopped for a short distance which would have slightly raised its body temperature. Furthermore, the thermal video did detect two areas of the wallaby that were slightly warmer than the rest of the body. These areas were the eyes and gut region of the abdomen. As with the 'red eye' phenomenon, the tapetum lucidum briefly reflected infrared light that was shown in several frames as the wallaby turned its head towards the camera,

---

[25] The sensor resolution of this image is 640 x 512 pixels (although the output display is only 640 x 480 pixels) with a 15 mm diameter lens, 35 mm focal length, f/2.3 and an 18° FoV.

[26] The ecological consequences of having neighbours who keep chickens can be seen in this partial food chain image, where a rat has been caught by the fox. Other food chain participants are not shown. However, using a thermal camera over time provides many insights into the local menagerie and their behaviours.

Fig. 14.4a Fox with a rat in its mouth. (Image: Neil Frost)

Fig. 14.4b Same image, enlargement. (Image: Neil Frost)

giving a warmer internal temperature reading.[27] The lower region of the wallaby's abdomen also shone more brightly, showing heat possibly being given off as a result of the gut fermentation process or from a joey in the marsupium.

Fatfoot, on the other hand, had been making even slower progress through the valley during the night. It would seem that, apart from also having a slow metabolism, Fatfoot's lower temperature may be due to her ambush hunting practices involving minimal detectable activity. It might also be related to her insulating fur. Therefore, it is not surprising that Fatfoot would pant after exertion or stress because of her high retention of heat.

It also seems that this particular area to our north has some special quality or attraction for both predator and prey. For swamp wallabies, the land east of the swamp tends to be more open and grassed, which would provide better grazing opportunities and the ability to more easily detect predators. The grasses found here tend to be soft, introduced varieties, instead of native species that are much harsher and undoubtedly, less palatable and more difficult to fore-gut ferment. Nutritionally, these alternative grasses may offer benefits that are not normally available or extractable, compared to native plants. When wallabies are in the valley, they are more often seen grazing in this area, compared to other locations. Similarly, there is a higher probability in this area of seeing joeys moving about independently.

For Fatfoot, an obvious attraction to this area would have been the higher population density of prey. As with previously-observed local situations and accounts from

---

[27] With their highly dilated pupils and excellent night vision ability, owls produce extremely prominent thermal eyeshine when looking towards the camera.

Octopus witnesses elsewhere, the typical method of solitary hunting is to stand for long periods behind a tree on the edge of the cleared area, waiting for an opportunity to pass by. When more than one animal is involved, particularly with a 'mischief' of Junjudee, prey is pushed into a zone where other predators are waiting. For Fatfoot, the majority of her hunting seems to be ambush-based. On their neighbouring land, which is visible on the far side of the thermal image (Fig. 14.3), Anne found several wallaby heads lying in the bush at various locations at the back of her property in 2019. It seems that the common killing method involves breaking the neck and back or the removal of the head. For Michael and Sandy, who live amongst a similarly grassed area further up the valley, the killing method used on one occasion was dismemberment.

PLACENTAL AND MARSUPIAL mammals are significantly different. Since separating from a common ancestor, each has independently evolved, with Australian marsupials adapting to the uniquely difficult circumstances found here. For early European immigrants, who were very familiar with animal husbandry in their homeland, their unfamiliar Australian experience was like being "transported to another planet" (John Gould, 1863).

The dominant Australian prerequisite for survival in a harsh environment is efficiency in everything. Reproduction is the main difference between the two mammalian groups, with placentals taking their offspring to late development and marsupials giving birth immaturely. By allowing their offspring to more fully develop in a womb, placentals are taking greater risks with their limited resources and the life of the mother, as well as becoming committed to investing heavily in the future success of a pregnancy that might not ultimately succeed for a variety of reasons. Whatever the roll of the dice, marsupials can effectively manipulate a pregnancy to suit the changing environmental circumstances by continuing with, adding to, pausing or terminating a birth, without any significant risk to the mother.

Another highly significant difference lies in their metabolisms. Having a metabolism that is about thirty per cent lower than a placental fox is further evidence that these 'hominoids' are marsupial. A lower body temperature requires less food to be hunted, consumed and assimilated, which increases efficiency and therefore, survivability.

Hunting by stealth, remaining stationary for long periods and ambushing prey when possible, further conserve energy and reduce the need for food. When needed, these Wildmen can turn on their high-speed abilities to run down prey. Kangaroos and other hopping macropods have maximised their movement efficiency through saltatory locomotion, which utilises increased stride length by having longer legs, the ability to reclaim expended kinetic energy using their long leg tendons and various cardiovascular adaptations. Compared to placentals, the greater endurance of marsupials can be attributed to their higher aerobic potential, by having larger hearts[28] and greater lung capacity,[29] in addition to many other physiological advantages. However, saltatory locomotion is limited to lighter animals to reduce or eliminate injuries related to bone fracture and joint damage. Consequently, large carnivorous bipedal macropods are required to walk or run, rather than hop and on rare occasions, move about on four limbs as facultative quadrupeds.

---

[28] Dawson, *op. cit.*

[29] Dr. John, a trauma neurosurgeon estimated the lung capacity of Dooligahl to be 5 litres or twice that of humans.

Epilogue

THE IMPORTANCE OF having a geologically-active planet can be better appreciated when Earth is compared to Mars. Even though life probably got started on an equally hostile Mars, there would be more significant impediments to its nascent and long-term development. With a lower gravity, Mars has difficulty retaining an atmosphere and any gaseous output from volcanism couldn't build a warm, dense and protective layer capable of supporting important systems, like a carbon cycle. Consequently, there would be little weather and no climate of significance that would be capable of storing and redistributing heat and water vapour. Mars currently has no surface water and no water cycle but from photographic evidence, water undoubtedly flowed. With no large moon orbiting Mars, there would be no internal tidal friction capable of generating areological heat, which would leave the depleting fuel of volcanism mainly restricted to residual heat from planetary formation and energy produced by radioactive decay during its early history. There is little recent evidence of tectonic movement, other than a mantle plume that is pushing up a section of the Martian crust. With no convection currents in the mantle, tectonic plate movement could not be maintained and the planet's magnetic field would have collapsed soon after the iron-core dynamo shut down, leaving the planet exposed to excessive amounts of lethal radiation from space. Mars may have developed life during its early history, but was unable to maintain and advance it.

Earth, on the other hand, was more fortunate. From the earliest iteration of life, organisms chemically modified the planet's hostile environment to their initial advantage but in doing so, produced toxic waste products like oxygen, that would instead, become the essential biospheric components for future life form variants and may have even brought about their own demise. It seems highly likely that life on this planet may have had many starts, or been reset at various times and at particular locations due to catastrophes or the consequences of life's terraforming activities.

As a volcanically sustainable Earth, the expelled greenhouse gases warmed the environment and the surface was rearranged by the mantle's conveyor belt of magma as the supercontinent of Pangaea rifted and separated to form Eurasia in the north and Gondwana towards the south. Although separated, the lifeforms on these two lesser supercontinents remained in partial contact as land bridges appeared and went. As with the other large landmass, Gondwana also rifted into smaller continents, with most moving northward, with Australia, South America and Antarctica maintaining their relative positions at the South Pole.

As the Earth's climate was warmed by greenhouse gases and tropical oceanic currents flowing directly from the equator to the pole, the remains of the Gondwanan landmass maintained their temperate rainforests, which were the instigators of biodiversity and major evolutionary change across the planet. For life at high southern latitudes, there were additional challenges brought about by lengthy periods of dark and relative cold—for example, the southern dinosaurs were adaptively different from others across the planet. Several types of mammals were evolving as Australia and South America were preparing to isolate themselves from Antarctica. In the distant future, the Wollemi pine, an ancient relic tree, would be discovered in a remote Blue Mountains valley and the Gondwanan Rainforests of Australia, although fragmented, would continue to survive and support an evolving ecosystem with the Daintree Rainforest in Far North Queensland—becoming the oldest rainforest in the world.

Before the Chicxulub and Nadir asteroid impacts 65 mya, it seemed that intelligent bipedal reptiles were destined to rule the Earth. *Troodon formosus*, a human-size theropod that was related to modern birds and lived in cold and dark northern environments, was probably a good regal contender because of its enhanced physical attributes and implied behavioural tendencies. Having large, forward-facing, binocular eyes that indicate stereoscopic night vision ability, an excellent sense of smell and good hearing would have required a lot of sensory processing to coordinate. Having long, lethally-clawed legs with grasping three-fingered hands meant that it was a swift and dextrous carnivorous hunter that, because of its size, most probably hunted cooperatively in packs. All of these factors would have significantly contributed toward the need for a large brain, particularly demanding hunting skills involving cooperation, coordination and planning. As seen with other avians, like kea and similar parrots, size isn't everything and perhaps how its brain was hard-wired was important too.

As the decimated planet began to revamp and replenish its depleted biota, new opportunities arose in the newly-emptied niches after the global reset. Other natural processes continued, with tectonic movements constantly altering the lithosphere and other complex systems, like climate, being affected as a consequence. Ocean currents in the vicinity of Gondwana, at that time, continued to flow from the tropical Equator to the South Pole, which warmed these southern continents to a temperate climate that supported abundantly variable animal and plant life. With the absence of large ice sheets, global sea levels were at their maximum. However, ongoing tectonic activity was rifting South America and Australia from Antarctica, opening up and disrupting the prevailing warming oceanic current patterns and establishing instead, the Antarctic Circumpolar Current. This cooled the pole and allowed the formation of continental ice, burying the Antarctic rainforests, raising the planetary albedo and further lowering the temperature. The colder air was less humid, which contributed towards the aridification of the southern continents, including Australia, which caused long-term changes in vegetation, habitat occupancy, herbivory and the predatory response. As ice accumulated at the pole, sea levels fell and with the redistribution of the planet's mass, the spinning top wobbled differently.

With Australia now effectively cut off from the rest of the world and beginning its northerly tectonic journey, its predominantly monotreme and marsupial fauna were literally at a fork in the road with placentals that were occupying the remainder of the Earth's surface. Although the Gondwanan Rainforests of Australia were in slow retreat, they continued to provide a rich environment for evolutionary change for a long time to come. Monotremes appear to have gone into decline, while marsupials vigorously radiated throughout the vacant niches of the new continent. This resulted in marsupial species that were convergently adapted to their specific niches while also producing solutions that were subtly different from the rest of the globe. Just as primates were the placental response to the rainforest opportunities provided by the African Rift Valley, 'marsupial hominoids' were similarly influenced by the Gondwanan Rainforests of Australia. Life in the trees initially developed various arboreal skills, which enhanced dexterity and intelligence. With increasing aridification, the importance of rainforests, as with the Rift Valley, shifted towards the grassy plains and the skill set adapted in response. As Australia continued to move further north, away from the other southern continents, it increasingly found itself geographically and biologically isolated, completely surrounded by the Southern, Indian and Pacific Oceans.

With impacting asteroids causing a shift in opportunities and the evolutionary odds, the circumstances for mammals to produce an intelligent being were duplicated when the mammalian classes were divided between two independent worlds. In Australia, marsupial sugar gliders evolved in an environment that was remarkably similar to the placental flying squirrels of America and consequently, they possess very similar convergent traits and behaviours. There are many comparative examples where marsupials are evolutionarily similar to placentals. Kangaroos are large herbivores that are evolutionarily similar to the antelope. Thylacine are large nocturnal predators that are similar to wolves from the northern hemisphere. Quolls are small, highly-aggressive nocturnal predators that are similar to weasels that are widely found around the world. Numbats are diurnal insectivores that have a specialised tongue to feed on ants and termites which is similar to anteaters found in Central and South America. From a seemingly endless list of comparative specimens to correlate with, where are the analogous marsupial examples of the great apes, including man?

As became apparent from our study, it seems to me that life on Earth has shown itself to be irrepressible, particularly in its march towards increasing complexity through intelligence. It didn't arise from a singular linear event but most probably from many instances that restarted the process multiple times until a range of workable solutions were found. Intelligent reptiles were unlucky but by splitting the early mammalian groups and geographically isolating them, the separate experiments converged on similar solutions for an intelligent biped, effectively on their own 'worlds'. Life on Mars also had potential but certain planetary flaws and limitations, such as being unable to retain an atmosphere and having no influential satellite, ultimately condemned any macro-Martian life to certain extinction. However, other possibilities for simple life in our Solar System remain, with Jupiter's Europa and Ganymede and Saturn's Titan and Enceladus satellites. Similarly, spread across the galaxy and the universe are endless simple and complex life possibilities. As stated by Ian Malcolm (Jeff Goldblum) in the 1993 Steven Spielberg *Jurassic Park* movie, "Life finds a way."

AT A FUNDAMENTAL level, the apparent physical and behavioural similarities between North American and Australian Wildmen[1] are commonly used to suggest a close biological association between them and also with the great apes, when in fact, it is simply an inevitable demonstration of convergence. Commonly cited examples of their

---

[1] In early October 2006, Ray Crowe and his daughter Pam visited us at our home. We had a long discussion, mainly concentrating on the similarities between the North American and Australian Wildmen. This was a very rewarding experience because Ray's extensive knowledge allowed us to ask detailed questions regarding specific physical and behavioural traits that we were, otherwise, very ignorant of. Ray confirmed the truth of a discussion that I had with an Indigenous American where Sasquatch were known to bite maple trees for their sweet sap. Ray also said that there should be more than one predator and added that bears, moose and deer were also responsible for the damage, with each having their characteristic bite pattern.

Epilogue

physical likeness include bipedalism, height and mass, hair, grasping hands, sagittal ridge, evidence of speciation and the presence of a tapetum lucidum, in addition to other characteristics. Consequently, these similarities are often used to suggest that these two 'humanoid' groups are biologically related. However, the presence of a tapetum lucidium, or reflective 'glowing' eyes, within both distant groups should immediately exclude them from being a close human relative because the great apes don't have a tapetum. Commonly cited examples of their behavioural likeness include high intelligence, contact avoidance, predominant nocturnalism, communication, rock throwing, tree damage, gift giving, bed making and their perceived solitary interactions. Like gliders and squirrels, their behaviours are convergent.

From the beginning, there were certain niggling inconsistencies between the two segregated Wildmen types from different continents and within the model that was supposed to unite them. I could not accurately say when our doubts had sufficiently risen to rebel against the imported dogma, but it was probably sometime after the visit of Steve Rushton that the process commenced—it took many more years to finalise. The beauty of the new model was that certain previously unreconcilable obstacles, were now less concerning or meaningless. This particularly applied to the highly variable foot morphology involving long thin feet and three, four and five toes because there were plenty of analogous examples to be found within the indigenous animal population, both extant and extinct. More importantly, every consecutive addition to the framework seemed to comfortably fit, as if it was meant to and without any need to make excuses or exceptions.

Having corrected our theoretical path, we moved beyond a flawed and misaligned understanding and began to rework our earlier experience and knowledge in this new light. For example, after confirming the validity of black claws on all three Australian 'hominoids', these independent facts were permitted to override the default North American evidence where flat pink nails are found on primates. Rather than placing doubt on the legitimacy of the research, as would have been the case with advocates of the old paradigm, it further supported the break-away marsupial theory. However, this new position did not make matters easier as we found ourselves increasingly marginalised within the already niche field of Australian cryptozoology.

Observing and testing behaviours was perhaps the most difficult research task. In order to achieve this greater depth of knowledge, the research required some risky and challenging fieldwork with communal support from the Octopus. It seemed that our communal situation and the sustained interactions of the Wildmen were unique and would be very difficult to replicate to the same extent.

Over the decades, our depth of knowledge and confidence that we were on the right track continued to build as specific and highly detailed traits of the Wildmen slotted seamlessly together. There was rarely an occasion when the new information failed to conform, as the completed framework was beginning to take a shape that was usually found in a mature and homogenous model. Finding multiple pieces of physical evidence that supported the presence of vertically-slit pupils complemented Fatfoot's known ambush and side-to-side rocking behaviours that were used in an attempt to establish striking range through motion parallax. The lower metabolism of Fatfoot, compared to a placental fox, was measured using a Compact Pro thermal camera, an iPhone and a ranging greyscale palette which further confirmed her marsupial nature and greater capacity for physical endurance—but it also highlighted the difficulty in achieving infrared observations, compared to placentals, under certain low ambient temperature conditions.

From an Australian perspective, our Hairy Men are marsupials, just like the majority of our indigenous animal population. The North American Hairy Men are undoubtedly primate in origin and possibly include some other more highly evolved indigenous variants. Any similar physical characteristics between the two groups are convergent traits that have independently evolved when adapting to their comparable environmental niches. The two groups do not share an immediate common ancestry.

FOR INDIGENOUS AUSTRALIANS, knowledge of the Hairy Man is a part of their cultural and religious belief. The presence of the Hairy Man, across all sections of Aboriginal society, is taken as given. Many Aborigines believe that these entities are spiritual beings whose supernatural nature is confirmed by, say, sightings of the red eyes, while others consider them to be a physical threat, capable of doing significant harm. It is generally believed that these beings have existed on this continent from a time that preceded the arrival of humans. For many non-indigenous Australians, there has been a perceived shift away from ignorance of the phenomenon, towards uninformed scepticism.[2] This arrogance can arguably be related to our increasingly high degree of urbanisation and our consequent loss of contact with bush reality, as well as the

---

[2] Although it is difficult to survey popular attitudes towards our Hairy Men, a rough indicator of this persistent negative trend can be gauged from the media's response to this phenomenon, in comparison with attitudes related to UAPs. The credibility of UAP research seems to be moving in a positive direction because of the highly-regarded reputational support given by experienced U.S. Naval pilots and the consequent policy shift by the U.S. Navy, which now encourages the safe reporting of incidents. In contrast, NASA, as a scientific body, has adopted a very conservative and noncommittal stance towards this metaphysical challenge, that is reminiscent of the Vulcan Science Academy. (See Spock's refusal to join the VSA in *Star Trek: Into Darkness*, 2009)

compulsive and conformist need to debunk everything that does not fit the dominant societal view of acceptable ideas. However, due to the positive and supportive work of the Octopus during its time, the Blue Mountains has gone against this broader trend.

Aboriginal and Torres Strait Islander people were not counted in any census up to 1971, which was the first opportunity to include them after the constitutional change was made in 1967. This lack of inclusion also applied to their traditional and expert knowledge of the Australian bush, which embraced many broad fields of study, such as indigenous weather knowledge, horticulture, astronomy, cooking, medicine and bushfire management strategies, just to name a few. While much of this traditional wisdom has now gained national acceptance and has received international praise, certain other aspects of the knowledge base have remained compartmentalised and under strict quarantine. Just as the local wisdom of Chinese villagers regarding the presence of Yeren had been trumped by a minibus full of civil engineers, the Aboriginal experience in this matter is similarly overlooked and regarded as quaint by many.

From the very beginning of European settlement, Australia was the world's most urbanised country. With Australia's white population being about 750,000 in the 1850s, the discovery of gold resulted in large wealthy towns and cities being established alongside the immigrants' mines. When my grandparents from both sides of the family were born during the 1880s, Australia's white population was about 2.2 million people. Traditionally, my father's family were dairy farmers who lived in various locations including Harris Park, nineteen kilometres from the centre of Sydney. As a consequence of his early life on the dairy farm, Dad always got up at four in the morning and he hated milk and all dairy products. On my mother's side, the family were originally wealthy merchants who lived in the large and prosperous gold-mining town of Beechworth in Victoria. By 1901, 1.3 million people out of a population of 3.8 million were living in the major capital cities, giving an urbanisation percentage of 34%. Despite Australia being a major primary producer of wool, wheat and other rural commodities, by 1911, the time that my father was born, only 43% of Australians were living in remote rural areas, with the remainder living in the major capital cities and medium-sized urban settlements. By 1944, half of Australia's population was living in the major capital cities alone, which had grown to two-thirds by 2016. More broadly, 86% of Australia's population is currently living inside the densely populated coastal urban band, which lies, on average, within fifty kilometres of the coast and covers an area of 1% of the continent.[3] Despite these limiting statistics, most Australians still like to maintain their strong nationalistic identification with Crocodile Dundee and his highly-tuned bush skills and persona.

Even when those living and working in remote rural settings had encounters, the experiences were not always openly spoken about and few were recorded. These incidents became taboo topics, probably because of the fear of ridicule where this self-censorship resulted in no record being made. What makes some of these hidden encounters surprising is when they involve stories from family members who have kept the incident quiet for a very long time. Sandy's father Bernie and younger brother Michael shared an incident sometime in 1967.

Bernie was born in 1920 and had extensive bush experience working on sheep stations from Bega to Armadale

---

[3] Australian Bureau of Statistics, https://www.abs.gov.au/

in NSW. From a young age, Bernie had worked many difficult and demanding jobs in remote areas. He worked in Outback stations as a shearer's cook and then as a jackaroo. He did this work to gain practical experience so that he could become an overseer or a manager of a sheep station. When the opportunity was finally offered to him, he had to refuse the position of station manager because understandably, his wife Loris didn't want to live in such a remote bush area.

Bernie was renowned for his bush-knowledge, like many others who lived and worked through the Great Depression in the Outback, prior to going to World War II. As with his wartime experiences, Bernie spoke infrequently about some of his bush encounters, but on one occasion when taking a shortcut to avoid roadwork near Dorrigo Mountain[4] on the way to Armidale, he commented on 'signs' that he recognised along the side of the road that he said were made by 'Bunyips'. A Bunyip is an Aboriginal term that was in use up until the mid-1970s, that is synonymous with 'Yowie'. Yowie is a more recent term that seems to be the corrupt European form of the Aboriginal word 'Yourie'. Bunyips are said to live near water, such as swamps or lakes, make loud howling sounds and eat women and children. What these 'signs' were that Bernie saw, no one from the family can remember. Most probably, these conspicuous roadside signs would have been treebites and particularly, trees that had been snapped and broken off.

In 1967, Bernie took his young son camping and fishing along the banks of the Hawkesbury River, in rugged bushland between the Dharug and Marramarra National Parks. At that time the area was extremely remote and

---

[4] Dorrigo and Armidale are adjacent to the Gondwana Rainforests of Australia World Heritage Area.

sparsely populated. They slept in their new Chrysler Valiant Safari Wagon. During the night, something came into their campsite and "trashed everything". The heavy cast iron bush oven and tripod that contained the leftovers of their evening meal were taken and later found about a hundred metres down a bush track. They left the campsite very early that morning to return home. As Bernie didn't elaborate on the incident to anyone, no other information can be remembered by family members.

THE OCTOPUS RARELY found Blue Mountains encounter witnesses who claimed that their 'marsupial hominoid' experience had interconnections with supernatural and paranormal phenomena, unidentified flying objects (UFOs), or unidentified aerial (or anomalous) phenomenon (UAPs). This was probably because the Octopus, when it was fully functioning, provided most local witnesses with fast interpersonal support and a rational community-based explanation for what they were experiencing. There was no need to search for any radical interpretations to fill the encounter vacuum, as long as the support was accepted quickly. There were a few notable witness exceptions over the decades, but most eventually realised that they were dealing with a real physical animal or we never heard from them again. As a long-time amateur astronomer with about fifty-six years of observational experience and having seen UAP on two occasions, I have no doubt about what I saw and the conclusions that I made. Considering any connections between UAP and the Hairy Man, I have not seen any evidence of this. Regarding other supernatural connections, I have not seen any evidence of this either and I would like to think that I wouldn't immediately dismiss alternative explanations out of hand. However, if anyone wanted to pursue 'high strangeness' beliefs there were always websites that cater to this.

The humble etymology of 'paranormal' has increasingly become a very value-laden word since its appearance in the 1920s. Its basic meaning is 'beyond' normal or 'outside of typical experience'. The *Cambridge Dictionary* defines it as "impossible to explain by known natural forces or by science". Other definitions, like that from Dictionary.com, are more ideologically loaded in meaning: "of or relating to the claimed occurrence of an event or perception without scientific explanation, as psychokinesis, extrasensory perception, or other purportedly supernatural phenomena." Broader and more tolerant definitions are suggestive of future explanations being found, such as, "phenomena that are beyond scientific understanding at this point in time". Simply, paranormal is a term for something that we don't understand at present. Having picked up many connotations, as usual, the simplest definition is undoubtedly the best. The word 'supernatural', however, tends to suggest non-physical entities, like demons and spirits together with any of their associated powers, like magic, transportation, precognition, and extrasensory perception, to name a few.

Since nine out of ten Australians live on the coastal fringe that comprises a narrowly-skewed one percent sample of the continent and may only do some casual camping intermittently, most have little understanding of the bush and what it contains, despite what they may believe. This general ignorance permits many people to think that there is nothing out there because if there was, they would surely know about it. Consequently, this general arrogance provides some people with the confidence to ridicule others who surely know better. However, when anyone encounters these exceptional 'marsupial hominoids' from a position of ignorance, their experience is understandably interpreted as paranormal because the circumstances are beyond their personal experience and knowledge and not readily assimilated into their worldview.

All of Australia's Wildmen have physical abilities and behaviours that are atypical when compared to most familiar animals. Their natural talents are superior to placental mammals—when compared to humans, they possess excellent senses, particularly vision, and have very high endurance, speed and strength that are extremely intimidating when encountered. Behaviourally, these Wildmen demonstrate high intelligence which is anomalous in the animal kingdom and can lead some professional hunters and researchers to describe them as "cunningly clever". When this cunning is applied to their superior skills, the resulting demonstration of ability dwarfs the lifetime experiences and expectations of the typical observer which could now be interpreted as 'supernatural'.

Similar to macropods, the two largest Wildmen move about on two legs and occasionally on four, while Junjudee are not as restricted in their locomotion, mostly walking or running on two legs, occasionally on four, crawling, tumbling, climbing and sitting in trees. These traits, and many others, could be described as being 'beyond normal' or 'paranormal'.

It has to be admitted that the first time someone sees red eyes in the bush, the experience is viscerally persuasive. It initiates a primal response of some kind that is felt in unexpected parts of the body. Such autonomic reactions have been honed through millennia of evolutionary testing and refinement to best prepare an animal for a dangerous situation—for example, by raising the heart rate in preparation for 'fight or flight'. For someone who is not very aware of bush animals, any red eyes that are seen at night most likely belong to a possum, glider, fox or kangaroo. For slightly more experienced viewers, the eye height, IPD and size should assist in narrowing identification, as would other clues. Other animals may have differently coloured reflections depending upon the chemical composition of their retina and other factors. The red eye colour of Wildmen may change to yellow, caused by a refractive shift as this tight configuration becomes misaligned and eventually lost. The red colour seen in most Australian animals' eyes makes them seem demonic with potentially malicious intent, which could mislead some towards a supernatural interpretation.

When the eyes of any animal that has a tapetum are seen, they may appear bright and very intense depending upon where you are viewing the eyes from and the lighting conditions. With proper alignment, the pupil of the animal will appear as a fully-illuminated disk. When this occurs, the animal is looking directly at you and you are made consciously aware of it; subconsciously you may experience a piloerection or other autonomic response. Any eye movement away from the central axis should first cause a refractive colour change, followed by a complete loss of the reflected light disk. When this occurs, the animal is looking elsewhere and no longer directly at you. The intensity of the light is determined by the light-gathering power or diameter of the animal's pupil and the reflective characteristics of the tapetum at the back of the retina. Any ambient light source such as the moon, a campfire, vehicle headlights, verandah lights or even starlight will be suitable under the right conditions. The ambient light entering the eye through the pupil is focused onto the retina and then bounced back off the reflective tapetum the way it entered. An alternative explanation that is sometimes put forward for the reflective glow is internal self-illumination, which does not deserve any further discussion or warrant investigation.

While we were learning from Fatfoot during encounters and pursuits, she was simultaneously learning from us. One of the first things that Sandy, Ian and I learnt, was Fatfoot's strong aversion to bright light. Using a spotlight to locate Fatfoot was unlike the behaviour of hunted kangaroos and wallabies that I experienced as a boy on the farm, as they would stare into the light beam and become mesmerised while waiting to be shot. The difference was that Fatfoot was a highly intelligent marsupial, capable of reading our intentions and taking proactive measures to avoid our manoeuvring. Consequently, she was almost impossible to observe. Our early and principal success was mainly due to the passive and largely ineffectual use of the Russian night vision binoculars or when I would stand in the bush at close range and use my peripheral vision by applying the indirect observation technique used in astronomy while listening to her swaying from side-to-side. Fatfoot would have known that, compared to herself, we were slow, clumsy, effectively blind and had the strange habit of staring off-target into the bush. When attempting active methods of observation, Fatfoot never hung around long enough to have a bright light aimed directly at or anywhere near her. If she had, her fully dilated and vertically slit eyes would have caused her retinal rhodopsin to become instantly bleached, which would have required many minutes to recover from, leaving her blind during that time. To avoid this, she learnt to recognise the shiny torch reflector long before the light had been switched on, turning away and running rapidly below our radar and into the dark, thumping each foot noisily into the ground as she fled. Fatfoot would never tempt fate by standing facing us, as long as there was a possibility of a bright light being used, so we never saw her frontal or side profiles as she was preparing to flee. This was why it was necessary to ambush Fatfoot and blind her with the bright camera flash so that a photograph could be taken, though such betrayals, like hiding a torch or flash from view under a coat, tended to raise her ire. Despite audible clues and other barely perceptible indicators of her presence, Fatfoot was effectively invisible and for some, she even seemed ghostly. However, Fatfoot was very corporeal.

Some Yowie commentators suggest that these animals are capable of reading your mind and in extreme cases, of being able to communicate telepathically. It would seem that being highly intelligent makes them more than capable of reading and anticipating human behaviour and responding appropriately, which could be regarded as 'reading your mind'.[5] More significantly, Fatfoot was certainly capable of applying the 'Theory of Mind' by knowing that throwing a rock would result in the targeted person running away. Instances of telepathy, together with some other unconfirmable behaviours were only mentioned to the Octopus by one witness known as Frank, whom I spoke to and met with on many occasions, mainly because of his eclectic experiences, many of which were legitimate and the desire to delve a bit deeper. These claims were impossible to verify, despite multiple interviews involving several questioners. Nonetheless, all three bipeds are capable of highly sophisticated interactions and effective communication without the need for alternative conjecture. Fatfoot or one of her joeys, told me to back off by roaring in my face. Similarly, rock throwing and branch breaking are very effective ways of communicating anger or issuing a warning.

Alternative forms of interspecies communication, on both sides, were used and capable of yielding a response. For example, shining the spotlight into the night sky like the Bat-Signal, often summoned Fatfoot from outside of the valley. Leaving food offerings frequently resulted in their acknowledgement—banana skins conspicuously

---

[5] Most pets can read the behaviour of their owners. Our dog becomes visibly concerned when the towels are gathered in preparation for his bath.

returned to the house or reciprocal offerings given in gratitude, even leaving a wallaby carcass at the front door. Also, Junjudee left a message that I interpreted to mean that they were watching me during the day. A large stick, or 'stick plant', had been inserted through the interconnected branches of a potted citrus tree. The stick plant had been intentionally navigated between the branches and pushed fifty millimetres into the potting soil. As previously demonstrated, this was no natural event where a branch had fallen from an overhead tree. The potted citrus tree was next to the outside table and chair where I regularly sat during the afternoon when the weather was favourable. Furthermore, the large area on which the table and chair were situated was a concrete slab that provided no opportunities for inserting a stick into the ground apart from the few potted plants. At face value, some people might argue that this was a supernatural event, but it was just another form of symbolic interspecies communication.

On one occasion, Fatfoot spoke to Cheryl and me saying 'mook, mook, mook' in the bush between our two houses but, unfortunately, we had no idea what she was trying to tell us on that occasion though the increasing volume suggested anger. From a digital recording, Junjudee can be heard communicating with each other using a continuously high-frequency stream that they obviously understand and reply to. Quinkan have been heard 'talking'. Sue described their 'utterances' as "monosyllabic, consonant/vowel combinations, that were not human". Such vocal presentations might seem demonic or supernatural to a few, particularly when the eyes are also observed.

Over time, it became increasingly apparent that Fatfoot sometimes moved on four limbs as a facultative quadruped. From the very beginning, Ian had repeatedly suggested that Fatfoot's gait was "irregular" or "not quite right".

Also, we would catch vague images that hinted at her transition from vertical to horizontal using the night-vision binoculars, but it was not seeing her seven-foot stature occasionally towering over the shorter vegetation that suspiciously alerted us to ongoing shifts in her gait and stance. There were also occasional references to quadrupedal gait by some other witnesses.

Very few observers had been in a position to see if, about fifty percent or whatever the sex ratio is, of the 'marsupial hominoid' population, had a pouch. However, Dane, a local business owner saw on his way to work what was most probably Fatfoot, a huge seven-foot-tall female Dooligahl, at the top of his driveway, back illuminated by the streetlight at about four in the morning. In silhouette, Dane first saw the beast rise after being on all-fours and then transitioning to two legs. As she turned sideways, Dane also saw a "stomach", similar to a pregnant woman's but lower down and not as large, that reminded him of a "beer belly". As Fatfoot is marsupial, I specifically asked Dane if he saw any breasts, which he hadn't. Fatfoot then ran rapidly on two legs, down the neighbour's vacant land and into the thick bush at the bottom of the valley.

Other witnesses have also referred to the facultative quadrupedal nature of their local Dooligahl. One excellent pair of reports came from thirteen-year-old Peter who saw a Dooligahl on two separate daytime occasions and wrote: "From the deck I saw it moving away through the trees on all fours", and "suddenly I saw this big, black, furry thing. It was about thirty metres away, at the creek at the bottom of the gully. It ran on all fours extremely fast and disappeared into the very tall ferns at the bottom of the slope."

Similar to the other 'marsupial hominoids', Junjudee are not obligate bipeds and occasionally move about differently for various purposes. For a brief time, Michael observed a Junjudee following behind his family as they were walking a short distance ahead of him on their land. The Junjudee had provocatively run in from an outside track on two legs and then it began tumbling on top of the thick but loosely packed swamp vegetation, like an 'inflatable beach ball', probably as a means of avoiding entanglement in the undergrowth but certainly as a goading attempt to gain their collective attention. During a daytime encounter where I observed the face of a Junjudee, it had stalked my position for a total of about half an hour, having first approached from the north and then circling back to where I was working on the fence, after first inspecting the activities of the neighbouring building workers. It was most probably moving bipedally for the majority of the journey but could have been travelling on all fours when passing through shorter vegetation or areas that had a higher perceived risk of visual exposure. However, during the final part of its approach, the Junjudee had been advancing on my position using a commando crawl. Once again, the natural hunting abilities of all these 'marsupial hominoids' allow them to stalk their objective in an 'invisible' and largely undetectable manner when required, even during the day. Such behaviours could account for some commentators wildly suggesting that these hominoids 'teleport' into the area or are 'interdimensional'. These unfounded and conjectural opinions are not necessary or helpful.

It is probably their high intelligence and ability to run on all fours, plus a few other characteristics, like being able to climb trees, that have led some commentators to imagine that these 'marsupial hominoids' possess supernatural powers, such as the ability to suddenly disappear or walk through objects. The problem with multiple witnesses, each having one or more isolated and non-repeatable encounters, is that any collective knowledge is repetitively relearnt and not necessarily built upon through compounded follow-up experiences, experimentation and documentation. It was only after many pursuits that Ian and I realised that Fatfoot was applying a variety of diversionary tactics and various deceptions from a well-rehearsed playlist in her effort to remain hidden and keep us guessing, whilst still encouraging and attempting to maintain our willingness to play the game. With every early pursuit, the length became progressively shorter, causing us to overshoot the mark on subsequent attempts and leaving us with a feeling of abandonment by wondering where she had disappeared to. Pursuits were becoming shorter because, like most animals, there is no desire to expend energy unnecessarily, particularly when your pursuers are incompetent and food is difficult or expensive to obtain. As she was conserving her effort, the chases were also becoming quieter.

Other strategies used by Fatfoot to conceal herself, simply involved deviating from the original course direction and even doubling back during pursuits, two very effective ploys that Sandy and Cheryl made us aware of after tracking the sound of our separate paths through the bush from the driveway steps. By moving as a quadruped, her visibility would be minimised by having a lower profile. Fatfoot's very fast speed together with her very low profile made her daytime movement through the bush seem ghost-like as she moved away. The vegetation would move about substantially in response to the passage of a very large but invisible being.[6] Regardless, I was able to

---

[6] This reminded me of the 'Monster from the Id' that was portrayed in the excellently produced 1956 science fiction movie *Forbidden Planet*, where a single-toed and glowing-eyed psychic manifestation of Dr. Morbius's subconscious mind attempts to protect his naïve and inappropriately-dressed daughter from the leave-deprived, all-male crew of the United

find her hiding place, on one daytime occasion, inside some thick bushes as the lengthy branches remained in motion for long enough after I turned the corner of the track and could see her point of entry, still in the process of stabilising their motion. Needless to say, these locational clues would have been impossible to see in the dark, however, it seemed reasonable to assume that this was one strategy used during our night pursuits.

Living in the trees was one behaviour that Aboriginal Elder Merve warned us to be aware of from the very beginning, although it took me far too long to fully realise that he wasn't speaking allegorically.[7] For Junjudee in particular and Dooligahl to a lesser degree, climbing and residing in a tree at night would be another good way for these 'marsupial hominoids' to avoid detection by deceiving the pursuer with the unfathomable mystery of another dead-end pursuit and the vanishing apparition of a ghost-like figure. Unless the pursuer was aware of the deception and had the ability to immediately identify the ploy and was able to closely inspect the treetops, it is unlikely that the ruse would be readily discovered at night. Aboriginal stories commonly attest to how impossibly difficult these Wildmen are to track.

There were numerous daytime reports of tree climbing received by the Octopus over the decades, mostly describing Junjudee activity and a fewer number associated with

---

Planets Cruiser C-57D. My favourite scene from the movie was Altaira, played by Anne Francis, asking a crew member whilst bathing, "What's a swimsuit?" It also featured the debut appearance of 'Robby the Robot'.

[7] Having spent time on several Melanesian islands interviewing the local people about their beliefs and practices during the 1970s, I was very familiar with the common belief in 'tree spirits' and their cultural role. In Aboriginal society, it is commonly held that these 'marsupial hominoids' are the 'Protectors of the Environment'. Somehow, I had initially misinterpreted Merve's literal statement, "They live in the trees!"

Dooligahl. No identifiable reports involving large Quinkan climbing trees were received, though this does not mean anything because the O'Connor Quinkan was required to climb the ten-metre rock fissure at the back of their property on most nights. Nearly all daylight incidents involved a biped being seen climbing down from a tree and the animal running away, with most incidents being very similarly reported, although the descriptions of the observed biped tended to be inconsistent. It seemed that this behaviour was a panic reaction to being discovered. Climbing up a tree during the day may have been regarded as a risky and potentially inescapable activity, also referred to as being 'treed', where any detection could attract unwanted and unavoidable scrutiny. This could account for there being no daylight reports of bipeds climbing up trees. On the other hand, because of the poor visual conditions, night climbs and descents would have been highly inconspicuous, so perhaps it is not surprising that only a few nighttime reports of bipeds moving through the treetops have been reported, despite them being potentially a safe and common occurrence. There was a nighttime incident involving a 'mischief' of Junjudee which had been observed and heard as they noisily climbed onto a farm roof at Mudgee. Also, there was a good report from Kosta who was told by the park ranger of overhead activity at the Ku-ring-gai Chase National Park. Another rare overhead incident involving Junjudee came from a nurse from Wentworth Falls and a similar case was reported by Gayna from Katoomba.

Many neighbours found that their local trees had been pushed over to achieve a solution to an obvious impasse or difficult circumstance. More often than not, the problem was a fence or deep depression that took a long time to navigate around or required too much effort. If asked, the common explanation for the fallen trees was said to be wind damage but simply, it wasn't. The loss rate of trees along the fence lines seemed to be disproportionately high. Many trees fell when there had been no evidence of wind or structural damage assisting with the fall. In many situations, the tree had managed to defy the basic laws of physics by falling against gravity, with some achieving a very advantageously positioned drop. The strength that would have been required to accomplish these feats was formidable and would have been of a greater magnitude when compared to Fatfoot impressively shaking several medium-sized trees across from our driveway. As many fallen trees had flattened fences, these were sometimes repaired quickly by the landowners. Others that remained unrepaired for longer, showed signs of traffic wear along their trunks with landing marks appearing on the other side of the fence. These choke points should have been ideal intercept waypoints during a pursuit but Ian and I were more frequently diverted off course and required to deliberately climb over, or run into, barbed fences at other locations.

A repetitive ruse carried out by Fatfoot was laying prostrate under a suitably dense bush as a means of avoiding detection as we continued to run past during our pursuit. Elder Merve similarly mentioned the habit amongst these animals of lying in wait along established swamp tracks and in particular, touching the thighs of young women as they passed by. Fatfoot had laid under a bush numerous times after stealing Bess's food and bowl. There seem to have been simple variations of this—for example, squatting behind a bush, as happened when Fatfoot or one of her joeys rose up and roared in my face. The blink comparator photographs showing Fatfoot laying prostrate under some bushes (see Figs. 8.6 and 8.7) probably confirmed this as being her default method of avoiding detection when moving, by simply 'going to ground'. With Dooligahl and

Quinkan being such large bipeds and capable of accessing ground level, reminds me of the Quinkan that growled in my ear when sleeping at the Blue Gum Forest campsite. Such vanishing behaviours could give the unwary an impression of supernatural abilities.

For a few commentators, the variation in the number of toes seems to demand a paranormal explanation, just like cinema's single-toed, glowing-eyed 'Monster from the Id'. It would seem that having fewer than five toes is mainly a concern for North American researchers, since this issue has, for a small subset of Australian researchers, been very effectively resolved. Similarly, the presence of a midtarsal break found in Bigfoot and a small percentage of humans is not apparent amongst Australian footprint samples.

Another anomaly of Bigfoot research that is in need of resolution, is the apparent presence of a tapetum lucidum. For us, this was one of a number of principal indicators suggesting that a course correction was necessary away from the entrenched model. With Australian Wildmen, the presence of a tapetum pointed to a homological problem if hominid ancestry was to be expected. This trait is found in nocturnal species, which further complicates any relationship with humans and other hominids. There would have to be an explanation for this discrepancy but a supernatural interpretation is not one.

For most Australian animals, their strangeness principally derives from their monotreme and marsupial origins and their similar but unique solutions to the same biological challenges found on other parts of the planet. As stated by John Gould, ". . . I arrived in the country, and found myself surrounded by objects as strange as if I had been transported to another planet . . ."[8]

---

[8] John Gould. 1863. *Mammals of Australia, Volume 1*. London.

Epilogue

ANY MENTION OF UFOs or UAPs in Hairy Man reports received by the Octopus over a thirty-year period was non-existent. Only speculative discussion involving this topic was made on websites or occasionally by UFO enthusiasts who were attempting to show a tenuous correlation between sightings and this area of study. From our experience, the main advocates promoting this line of inquiry were mostly documentary producers, their fixers and their principal protagonists. Most of these production houses were North American, German and Australian. A few of these documentary producers had some pre-shaped programmes with a large amount of control oversight, others were more flexible and open to some external influence and a few were simply prepared to follow the facts and the associated story and were a delight to work with. However, after one early experience where we were grossly misled, we tended to decline participation in these projects.

As an amateur astronomer for more than five decades, I maintained a great deal of enthusiasm for this science and attempted to keep up with the latest research. When starting this interest, my parents were unable to purchase anything better than a small, low-quality telescope, but my father introduced me to making my own larger and much more affordable reflecting telescope. Dad was mostly a self-taught mechanical engineer who gained formal experience and qualifications as a Flight Sergeant during World War II. These skills enabled him to fabricate parts at work, that he lovingly and euphemistically referred to as 'foreign orders'. Together we designed and constructed a large equatorial mount for the ten-inch reflector that we mounted in concrete in the backyard. When looking through the telescope it was possible to see some amazing objects and witness some incredible events.

When observing the night sky, it is impossible to believe that there aren't other lifeforms in this vast universe, just as there are other intelligent and sentient 'marsupial hominoids' on this planet that most people still aren't aware of. With our understanding of the universe growing further since the commissioning of the James Webb Space Telescope, the current number of estimated galaxies is more than two trillion. The Fermi Paradox is the stated discrepancy between the lack of conclusive evidence of advanced extraterrestrial civilisations and the apparently high likelihood of their existence. Fermi's calculations predicted that Earth should have been visited by extraterrestrials long ago and many times since. The Drake equation is an attempt to calculate the number of advanced alien civilisations that are currently in existence, which in our Milky Way galaxy, range from a few to many thousands. Our daunting perception of astronomical distance and the supposed light-speed limitation have cast a sceptical outlook on our future ability to travel and to be visited. Considering the primitive state of our scientific knowledge and technology, faster-than-light travel is still regarded as 'magical' and therefore, unobtainable.

One evening after we had moved into our almost-completed house, but before we had become Yowie-aware, I was standing in front of our house scanning the ecliptic for the planet Mars. If the conditions were favourable I was hoping to set up my Celestron C90 for some viewing. The Sun had set a few hours before and it was well past astronomical dusk. The Earth's shadow was now extending high into the night sky, which reduced the possibility of seeing satellites orbiting through sunlight on the eastern half of the sky. As I was studying the northwestern sky I noticed a red dot that seemed to be Mars. After walking around the cleared area I chose a position to set up the telescope tripod and when viewing the red dot a second time, it seemed to have moved. As I continued to watch, it was seen travelling vertically down towards the horizon

Epilogue

at a pace slower than a typical satellite. I was very familiar with artificial satellites, as they became increasingly common since I started observing with a telescope. I had even observed a part of Skylab reenter the atmosphere and burn up over Western Australia in 1979[9] and another large piece of space debris a few years later, which I initially thought was an aeroplane on fire because of its very flat trajectory. My father told me that I had seen Sputnik 1 in 1957 after he had taken me outside to watch it orbit overhead when I was very young, but I don't remember much, apart from hearing the beeping sound coming from the radio.

Artificial satellites can orbit the Earth in any direction but they mostly travel with the direction of the Earth's rotation at an inclination to the equator. This is because, at launch, the rocket is already travelling at the Earth's rotational velocity for that particular latitude, with the Equator imparting the greater orbital boost, compared to the poles which provide none. Launching against the Earth's rotation requires greater counterthrust, more fuel and larger rocket stages for no benefit. The next common orbits are polar, rotating in either direction, with anything else being unlikely. This red dot was moving vertically down in the western sky, which suggested that its orbit was counter to the Earth's rotation and highly unusual. When satellites appear temporarily red, it usually means that they are moving through a section of sunlight that has been spectrally refracted towards the red by the penumbra and atmosphere. Typically this refracted colour does not persist as it fades in intensity as the satellite moves deeper into the total darkness of the umbra. As I continued to watch the red dot for a few more seconds,

---

[9] The Western Australian State Government fined NASA AUD$400 for littering.

it suddenly executed a perfectly sharp ninety-degree turn towards the north. It continued on this heading, parallel to the Earth's surface for about ten seconds before suddenly disappearing. Unlike other orbiting satellites, it didn't fade away as it entered the umbra. It took a few more seconds for me to fully process what I had observed.

After thinking about what I had observed, I realised that I had seen a UFO. There was only one conclusive piece of evidence; everything else was irrelevant because the other observations could be explained away, even if very poorly. According to our basic understanding of physics, an object making such a sharp change in direction would be undergoing massive acceleration. Using the current terminology, the red dot demonstrated at least one of the characteristic 'observables' of 'instantaneous acceleration'. Anything that is capable of achieving instantaneous rates of change like this is not under the influence of natural physics and is not from this planet. One of these observed characteristics is more than enough proof.

When the nights were warm, my father and I regularly sat outside on reclining lounges having a Lager and talking. In January 1982, about five years before seeing the red light in the sky, I moved back to my parent's house after receiving a teaching transfer from the country, back to Sydney. We talked about many things but my favourite topic was asking Dad questions related to his life experiences.

Dad told me many amazing stories that reminded me of talking to my grandfathers and the village Elders throughout Melanesia during the latter half of the 1970s. Dad was born in 1911 at the time when Rutherford discovered the atomic nucleus. The first powered flight by the Wright brothers had occurred eight years earlier and Einstein's Special Theory of Relativity was published six years before. Einstein's General Theory of Relativity was published four years after his birth in 1915 and in 1924 Edwin Hubble identified Andromeda as the first galaxy outside of our own Milky Way. I particularly remember Dad talking about what powered the Sun. He said that most people, at that time, thought that the Sun was a ball of burning coal or some similar chemical reaction. It wasn't until the 1930s that the nuclear process was fully understood. Dad was also fascinated by powered flight and he went to a paddock near Mascot to watch Charles Kingsford-Smith land the *Southern Cross* sometime after completing the first trans-Pacific flight from California to Brisbane in 1928. Dad was fascinated by aeroplanes, which heavily influenced his decision to join the Royal Australian Air Force in 1939.

Sitting on our reclining lounge chairs, Dad and I were looking up at the evening sky towards the zenith or in the direction of the apex of the house roof. It was sometime after nine o'clock and the sky was clear except for some whispy high-altitude cirrus clouds to the North. I think that I was the first to notice that the stars overhead were becoming eclipsed by something moving towards the southeast. Then a small round white light appeared, followed closely by others in straight alignment. Whatever the eclipsing object was, it was moving above the cirrus clouds. By now Dad had also noticed the object and we both expressed our amazement. The object significantly covered a section of the constellation of Orion and continued to maintain its heading. There were nine round white lights in total that were arranged in a 'V' formation. It wasn't a triangle but more like a boomerang in shape, as we could see stars reappearing from behind the array of lights. The two long arms of the body seemed to be squared off at each end. The body of the object was matte black. Dad and I then used our fully outstretched handspans to take measurements which we actively compared between us. From our combined effort, we estimated the angular width of the object to be ten degrees or about twenty moon diameters. The object was silent and was closely following the northern landing approach pattern for Sydney Kingsford Smith International Airport. We were both very familiar with this flight path as arriving planes consistently passed overhead many times per day. However, the speed of the object was slower than a landing aircraft. It was only about three-quarters of what was observed with aeroplanes landing at about 145 knots or 270 km/hr, and it was not descending but maintaining its high altitude. This meant that the craft was travelling at an estimated speed of two hundred kilometres per hour and because of its high altitude, we were able to clearly follow its progress until it suddenly disappeared about three minutes later. Normally, the descending and low-altitude aircraft would begin to disappear behind the terrain and high-rise buildings within about a minute.

While we contained our excitement, we continued to monitor the movement of the craft and had time to contemplate certain aspects of it more thoroughly. As it moved slowly and further away we could better appreciate its size. It was unbelievably large and obviously massive as well. In guessing its width by using the lights as a guide, we thought that it was several kilometres across but its thickness was not easy to determine because of its black finish against a dark background, although it appeared to be thin. The nine circular lights on the underside of the boomerang seemed to be its propulsion and lift system. The lights were constant in their brightness and were part of a single, unified body.

As if this experience could not be bettered, its departure was spectacular. From our experience gained from watching too many landings at Kingsford-Smith Airport, we knew that the flight time from our house to the runway was normally about three minutes. After observing the boomerang in the distance and knowing that the centre of Sydney lay on that bearing, we determined that the object was probably somewhere at a high altitude over the Sydney Harbour area. All of a sudden, the craft instantaneously accelerated leaving ahead of it a trail of white lines, presumably from the white circular disks beneath the craft, which terminated in the far-off convergence of these lines at its vanishing point! As a Trekkie, I seriously tried to over-talk my father as we both attempted to simultaneously communicate the astounding situation that we had just experienced! I thought at that very moment that Gene Roddenberry hadn't made up this warp scenario and must have witnessed something very similar because there was no difference between the studio effect used in *Star Trek* and the reality that we had just witnessed.

The next morning I considered telephoning the radar tower at Sydney Airport to inquire if they had spotted anything but Dad told me not to bother because they would just think that I was a 'nutter'. There was no mention of the incident in any of the morning newspapers, as Dad had already walked down to the shops at four in the morning, a relic of behaviour from the time that he was a dairy farmer, to buy every early edition. Similarly, there was no news coverage on television or any mention of the incident over the following days or weeks. Surely, other people were looking up at the time but it seems that, like me, no one reported the incident—until now.

I rarely mentioned this incident to anyone, particularly after having been associated with 'marsupial hominoid' research since 1993. It is not a credible look. However, I occasionally saw television shows many years later that reported something very similar. The best report that I have seen was the 'Phoenix Lights' incident of 1997. I don't follow UFO stories very closely but there were a few variations—where we observed nine round lights under the boomerang whereas, five or seven were reported with the U.S. incidents. Any of the alternative explanations for the 'Phoenix Lights' incidents and others are total rubbish. We didn't see a balloon, a formation of helicopters or other aircraft, parachute flares, an atmospheric aberration or, an unusual astronomic event. It was just an extremely large extraterrestrial spacecraft!

This situation can be best summed up using a quote from Mary, "I told one or two people about it but you know the kind of reception you get. They think: 'She'll be talking about UFOs next!' So I didn't elaborate to anyone but I was very puzzled and very interested because I assumed it must be a primate—a pre-Aboriginal primate—in Australia. Then, about a year later, I was really quite amazed to see the ABC's *Catalyst* Programme [about Neil and Sandy Frost and their Yowie experiences] and I thought, 'I've seen one of those!'"[10]

I think that like our cryptological research involving the communal support of the Octopus, it is extremely important that witnesses speak up in the face of seemingly overwhelming scepticism. It is very encouraging that, led by U.S. Navy Pilots and others, UAP research is gaining the support that it deserves.

TRACING THE EVOLUTIONARY origins of the three Australian Wildmen is not something that I would be able to competently achieve. It requires someone with extensive Australian knowledge of evolutionary biology, some more information (a body or a few fossils might be helpful) and a different skill set from mine. However, basic facts

---

[10] Healy and Cropper, *op. cit.*, pp. 114, 289.

obtained from this research can indicate a path for investigation.

Of the three Australian wildmen, two are very similar and one is an outlier. Dooligahl and Quinkan are both big, robust omnivores, the latter being larger with three rather than four toes, that are physically and behaviourally similar to large primates, such as gorillas or perhaps at the top end of the scale, something like *Gigantopithecus blacki*. They are mainly found in environments ranging from dense rainforest to woodland. Junjudee are small gracile omnivores that are physically and behaviourally similar to medium-sized primates, like bonobos, and are mainly found in forests and grassland.

The foot morphology of both large bipeds is very macropodoid, being elongated and narrow with a resting plantigrade posture, as shown in the footprint photographs and cast.[11] Consequently, this physical evidence suggests an evolutionary line for Dooligahl and Quinkan through kangaroos. The foot morphology of the five-toed Junjudee is very primate-looking, despite having black claws.

A telling physical characteristic of both Dooligahl and Quinkan is their disproportionate limb sizes, particularly their very long legs. Many daytime witnesses comment on the long slow stride of these humanoids that gives an impression of 'gliding along'. Having long legs allows an animal to travel at a greater speed with less effort which is metabolically more efficient and is a significant benefit for any animal. Hopping is an alternative gait that provides additional efficiencies for those animals that can. However, the maximum weight of some extant kangaroos,

---

[11] Although it has been impossible to capture evidence of this, these 'marsupial hominoids' undoubtedly run using digitigrade locomotion, as moving kangaroos do, because it would be faster and more efficient.

like the hundred-kilogram Red Kangaroo (*Macropus rufus*), places them at the limit of safe saltatory locomotion. In the past, the extinct two-hundred-and-fifty-kilogram *Procoptodon goliah* was a much heavier kangaroo that avoided foot stress and leg failure by being a fast walker and runner. In some ways, Dooligahl and Quinkan are similar to *Procoptodon goliah* and may have shared a common ancestor along the way. Other possible kangaroo-related contenders might include the carnivorous *Ekaltadeta ima* and *Balbaroo fangaroo*.

As recorded in the activity log, Fatfoot was a chronic foot-thumper. Initially, Fatfoot used this acoustic signalling to alert us when encroaching on her boundaries and as an intimidatory warning to stay away. From other witness accounts, foot thumping was sparingly deployed as part of a suite of measures that were all intended to threaten. After failing to maintain its purpose with us, foot thumping became a method of attention-seeking, which Fatfoot used to gain our awareness of her arrival or continued presence, with the desire to play games or interact in some other way. The sound produced by foot thumping was similar to slamming a heavy car door, but could be heard as two separate sound fronts, one arriving through the air and the other through the ground. Low-frequency thumps or infrasound, are transmitted through the ground and may be used to communicate over long distances. Fatfoot frequently used foot thumping from the distant swamp to wake me up in the middle of the night as an invitation to play.

Wildmen aside, foot thumping is used by other macropodoid marsupials to communicate a potential threat from a predator. There are more than fifty-two species of macropodids in Australia and New Guinea that can be studied. From combined sources of information, "foot-thumping was reported to occur in 46 macropodoid species and subspecies out of the 48 for which information was

Epilogue

available."[12] Also, "the presence of foot-thumping in every macropodoid genus except *Hypsiprymnodon* suggests that it is a highly conservative trait and that foot-thumping has arisen alongside or following the evolution of bipedal locomotion in macropodoids."[13] It seems that, apart from physical characteristics, like long legs and elongated feet, foot-thumping behaviour further confirms that Dooligahl and Quinkan are macropodoid species. Furthermore, since Dooligahl and Quinkan are also arboreal, "foot-thumping has been observed in three species of tree-kangaroo: the grizzled tree-kangaroo (*Dendrolagus inustus*); . . . Lumholtz tree-kangaroo (*D. lumholtzi*) . . .; and the Huon tree-kangaroo (*D. matschiei*)."[14] Like early humans from the Rift Valley of Africa, these 'marsupial hominoids' seem to have evolved in the rainforest trees and transitioned to the forest floor.

Looking for any ancestral associations with Junjudee initially seems problematic. In terms of their size, form and behaviours, Junjudee appear to be primates but in having a tapetum lucidum they are clearly not. Having black claws further excludes this possibility—instead, they appear to be marsupial 'monkey' analogues. Junjudee do not foot-thump, but they do give off a powerful olfactory response during stressful situations. From the remaining list of macropodoid marsupials, "only two species had not been observed foot-thumping, Goodfellow's tree-kangaroo (*D. goodfellowi*) and the Musky Rat-kangaroo (*Hypsiprymnodon moschatus*)."[15]

---

[12] Rose, T. A., et al. 2006. Foot-thumping as an alarm signal in macropodoid marsupials: Prevalence and hypotheses of function. *Mammalian Review* 36(4): 281-298.

[13] Ibid., p. 290.

[14] Ibid., p. 290.

[15] Ibid., p. 290.

The musky rat-kangaroo has existed in Australia's rainforests for more than twenty million years and is the smallest and most primitive extant macropodoid. They are ancestral to kangaroos and wallabies and may represent an early evolutionary stage from a tree-dwelling, possum-like ancestor to a ground-dwelling macropodid. They are omnivorous, mainly eating fruit and insects. Their four limbs are of similar size, with each having five digits, which is unlike other macropods. The hind feet have opposable thumbs that are suitable for climbing. The musky rat-kangaroo does not hop but bounds on all four feet as a quadruped. As with the other large macropodoid species, the rainforests have played a very crucial role in the experimentation and evolution of Junjudee and all life on Earth.

THE PURPOSE OF writing this book has been to preserve the knowledge and experience gained over many decades by participants from our local group, the Octopus, and supporting members of the Aboriginal community. It is also hoped that the book will raise awareness by bringing many aspects of this phenomenon into the open. It is hoped that this information can be referenced and used to provide further support for encounter witnesses who are in need of assistance and perhaps, help to eliminate the corrosive scepticism held within the scientific and general population. The information in this book should not be regarded as necessarily definitive and should be revised and extended as new research is acquired. As traditional indigenous Hairy Man knowledge is similarly being lost across the Australian continent, it is hoped that indigenous anthropologists will collect the stories, images and experiences from their land for future generations, before it is lost forever.

Neil Frost

About the Author

NEIL FROST WAS born in Sydney, Australia, in 1954. His father was a mechanical engineer and his mother was a tailor. From an early age, his father encouraged an interest in science and a broad range of practical skills. From age eight, Neil was a Cub, becoming a Scout at eleven years old when he later experienced a terrifying group encounter with an unknown beast whilst camping in a remote valley in the Blue Mountains. As a Scout and Senior Scout, he learned many bush skills and obtained an understanding of Aboriginal culture from *Guraki*, his Scout Master. In high school, he became interested in astronomy, supported by the librarian and his father. He ground telescope mirrors and learned optics, building several telescopes and an equatorial mount. Later, Neil learnt motor mechanics, spray painting and many aspects of engineering and building from his father.

After leaving high school, he was awarded a Teacher's Scholarship at Macquarie University, where he studied anthropology, economics, geology and education, resulting in a B.A. Dip. Ed. He later completed an advanced certificate in technology at Sydney University. Neil wanted to study astronomy, but it was not taught at Macquarie. During the long university summer holidays, he travelled three times to Papua New Guinea and Melanesia during the 1970s, where he emulated aspects of his university studies, visiting the New Guinea Highlands, Port Moresby, Kula Ring coastal settlements, the Trobriand Islands, New Britain and other regions. These trips were paid for by the sale of cars that he restored during the year.

As a bonded teacher, he started permanent teaching in a very remote rural school and living on an isolated sheep station, where he had a second encounter with a large intruder, which local farmers believed was a transient hermit. In 1982, Neil met his partner Sandy and they designed and built a two-storey timber house in the Blue Mountains using their fortnightly wages. After excavating the site, they found a long trail of large and unusual footprints through the disturbed soil, which they could not identify and soon forgot about.

Over the following decade, they experienced many unusual events, which they attributed to local wildlife. Late

Highland 'Big Man' (holding spear) with his supporters and followers, including missionaries, high school graduates and cargo cult members, canvassing for support on the roadside near Mount Hagen, in 1975. (Photo: Neil Frost)

one evening in early 1993, across from their driveway, a very large and heavy beast stood up in front of Neil and ran off. He described the animal to the police as "an elephant on two legs wearing size twenty boots", which eventually resulted in a stakeout, where the police attempted to shoot the beast. The most terrifying aspect of their nightly encounters was knowing that few believed and supported them, apart from members of the Aboriginal community. Typically, Sandy and Neil were facing severe ridicule when they spoke out. Fortunately, their biker neighbour was prepared to physically help them, beginning a close, investigative partnership. This cooperation subsequently expanded over the following thirty years to become a community-based network that undertook research and provided support and advice to encounter witnesses and other researchers. This network was known as the 'Octopus'.

Boys returning from early morning fishing, at Kiriwina, the Trobriand Islands, in 1979. The best fishing lies outside of the reef. They were not afraid of retribution from the ancestral shark spirits, because they followed the traditions and values of their people. (Photo: Neil Frost)

The author, in the 'Dooligahl Observatory'. (Photo: Neil Frost)

Index

Acacia Hills (NT) 32, 216
Acacia melanoxylon (Blackwood) 526
Africa 63, 134, 147-8, 594, 703
African Rift Valley 16, 139, 670-1, 703
Alisterus scapularis (King Parrot) 538-9
Allocasuarina luehmannii (Buloke) 530
Alum Creek Camping Ground (NSW) 302, 452
Angophora 455
   costata (Sydney Red Gum) 570
Antarctica 15-6, 134-6, 138, 142, 144, 146, 165, 176, 668-70
Antarctic Circumpolar Current 16, 135-6, 138, 140, 144, 146, 670
Antechinus agilis 324
Aquila audax (Wedge-tailed Eagle) 58, 347, 601
Armidale (NSW) 678
Australia 15-16, 34-5, 58, 133-8, 140, 142, 147-156, 163, 165, 168, 176, 350, 440, 450, 452, 494, 527, 532, 639-40, 645, 649, 656, 661, 668-70, 676-7

Baiame Cave 436-7

Balbaroo fangaroo 170, 602, 640-1, 702
Banksia 537, 612
   integrifolia 531
   oblongifolia 258, 616
Bass Strait 154
Bells Line of Road 342
Bering Strait 147
Bermagui (NSW) 181
Blackheath (NSW) 23, 36-37
Blink comparator 299-301, 307, 499, 691
Blue Gum Forest 18, 21, 38, 195, 388, 404, 692
Blue Labyrinth 419, 644
Blue Mountains 11, 12, 19, 23, 35, 38, 75, 83, 175-7, 180, 184-5, 192-3, 200-202, 210, 217, 224, 327, 329, 331, 342-6, 378, 381, 391, 410, 419, 445, 449, 452, 471, 490, 506, 514, 517, 520, 527, 532, 534, 536-7, 539, 541-2, 544, 546, 549, 571, 582, 588, 595, 610, 612, 645, 669, 676, 679
Blue Mountains National Park 643
Brindabella Range (ACT) 6
'Brown Jack' 368, 390, 451, 519

Bungonia National Park 59
Burragorang Valley 19, 419

Cacatua galerita (Sulphur-crested Cockatoo) 136, 195, 537, 545, 557
Calyptorhynchus funereus (Eastern Yellow-tailed Black Cockatoo) 121, 398, 527, 531, 535-7, 541-2, 545-7, 552, 554-5, 571, 576, 596-7
Cargo Cult 126-7, 254, 706
Casuarina 537
Catalyst (ABC) 556, 590, 597-8, 700
Cataract Creek (NSW) 412, 417
Cataract Falls (NSW) 411-3
Cerambycidae 530
Commonwealth Scientific and Industrial Research Organisation (CSIRO) 74, 321, 540, 551, 584-6, 588, 631
Convergence 159-171, 430, 672
Corpus callosum 650
Corymbia eximia (Yellow Bloodwood) 528, 530, 543
Corymbia gummifera (Red Bloodwood) 532
Cotter River (ACT) 6
Cowra (NSW) 204-6, 466
Coxs River (NSW) 643
Cracticus torquatus (Grey Butcherbird) 537
Crown Land Reserve 382, 572

Daintree Rainforest 138, 443, 669
Dendrelaphis punctulatus (Green Tree Snake) 245
Dendrolagus (Tree-kangaroos) 143, 404, 639-40, 703
Dharug National Park 678

Dingo 28-9, 31-2, 58, 347
Dingoes Lair Cave 432, 434, 440
Dinosaur Cove (VIC) 146
Diprotodon optimum 16, 169
Dooligahl (appearance) 57, 65, 128, 204-5, 207-9, 216-7, 222-3, 236-7, 241, 315, 334, 339, 364-7, 369-70, 404, 612-8, 646-7, 658, 686, 701
    (behavior) 57, 61-3, 73, 78-81, 84, 86-8, 93, 95, 108-11, 129, 181, 183, 188-91, 197-9, 205-6, 215, 220-1, 223, 238-40, 243, 247-50, 256-9, 277-80, 303-7, 311, 324-7, 343-5, 363-4, 546-50, 554-70, 629-35, 641, 646, 648-9, 652-4, 664, 689, 691, 703
    (encounters) 57, 59-61, 70, 76-7, 81, 88-90, 92-107, 112-4, 116-9, 123-4, 126-31, 179, 181-3, 185, 201-28, 233-7, 265-7, 273-9, 311, 313-4, 336, 338-9, 352-3, 379-86, 388-90, 644, 686
    (track) 354, 403
'Dooligahl Observatory' 490, 604
Dorrigo Mountain (NSW) 678
Drake Passage 138, 146
Dromornis planei 170
Dunedoo (NSW) 178

Ekaltadeta ima 702
Eucalyptus camaldulensis (River Red Gum) 90, 263
    grandis (Flooded Gum) 534, 539, 541-4
    haemastoma (Scribbly Gum) 118, 246, 250, 286, 289, 291, 296, 298, 300, 374

Index

microcorys (Tallowwood Eucalypt) 130, 305-6, 479
 obliqua (Stringybark) 455, 460-2, 493
 oreades (Blue Mountains Ash) 479
Euroka Clearing (NSW) 202

Far North Queensland 17, 138, 368, 376, 390, 430, 439, 442, 444, 669
Faulconbridge (NSW) 506
Feral dog 58, 160, 347, 619
Feral goat 549
Feral pig 120, 549
Fish River 201
Flores 17, 148-50
Foot thumping 104, 181, 188, 241, 264-7, 287, 293, 302, 310, 357, 359, 425, 464, 466-467, 474-7, 481, 658, 702-3
Fox 248, 375, 414, 662-3, 666, 674, 681

Gigantopithecus 153, 701
Glaucomys (Flying Squirrels) 162-5, 671
Glenbrook National Park 202
Gondwana 132-4, 137-40, 144-6, 176, 668, 670
Gondwana Rainforests 137-8, 678
Goulburn (NSW) 59, 236
Govett's Leap 26
Great Australian Bight 176
Great Dividing Range 10, 410, 450
Great Western Highway 176, 182, 204, 210, 212, 234, 417, 420, 487
Green Parade (NSW) 212-3, 215, 217

Grose River 13, 22-3, 176, 404, 420, 488
Grose Valley 13, 23, 41, 176, 338, 343-4, 388, 404, 420
Guraki 21-2, 31

Hakea 277-8, 414, 537, 646
Hassans Walls Lookout 185
Hawkesbury River 176, 586, 678
Hazelbrook (NSW) 211
Homo erectus 17, 35, 63, 149, 153
Homo floresiensis 148-9, 518
Homo neanderthalensis 153, 617
Homo sapiens 17, 153
Hypsiprymnodon moschatus (Musky Rat-kangaroo) 703-4

India 134, 460
Infrared 324, 457, 487, 625, 627, 631, 635, 656-9, 661-3

Jenolan Caves (NSW) 452
Junjudee (appearance) 368, 377-8, 404, 451, 453, 464, 469-470, 493, 500-504, 506, 517-20, 536, 578, 626, 635, 701, 703
 (behavior) 196-200, 248, 292, 392, 454, 463-88, 491, 493, 497-500, 507-13, 516-7, 519-25, 578, 647, 654, 665, 681, 685-7, 689
 (encounter) 178, 199-200, 392, 466, 469, 471, 490, 493, 500-506, 516-7, 647, 690
 (track) 354

Kanangra-Boyd National Park 419
Katoomba (NSW) 94-5, 452, 643, 690
Kings Tableland 419

Kino 161, 530, 532, 543-4, 547, 549
Kow Swamp 594
Kowmung River (NSW) 251, 643
Ku-ring-gai Chase National Park 586-7, 690

Lachlan River 46
Lake Mungo 594
Lapstone 210
Laschamp Excursion Event 168
Leaellynasaura amicagraphica 146
Laura Quinkan Dance Festival 444
Laurasia 133, 139-40
Lawson (NSW) 98, 100-1, 105, 412-3, 417-8
Linden (NSW) 233, 379-80
Lithgow (NSW) 185-7, 210, 453
Lydekker Line 151

Macropus giganteus (Eastern Grey Kangaroo) 51
    robustus (Eastern Wallaroo) 490-2
    rufus (Red Kangaroo) 510, 641, 702
Madagascar 134
Magdala Creek (NSW) 378-9, 449
Main Western Railway 204, 210, 233
Malurus cyaneus (Superb Fairy-wren) 245, 364
Mareeba (FNQ) 439, 443-4
Marramarra National Park 678
Marsupium 142-3, 160, 166, 207, 220, 222, 487, 500, 664
Medlow Bath (NSW) 388
Megalania Prisca 170
Megalong Valley 121, 388, 447, 490

Mid Western Highway 56
Milbrodale (NSW) 436-8
Mittagong (NSW) 211, 644
Mogo (NSW) 78, 84
Morelia spilota (Carpet Python) 245
Mudgee (NSW) 35, 199, 202, 453, 690
Myths and Monsters Conference (2001) 557, 582

Nattai National Park 419
Nepean River 176, 202, 213, 406, 419, 639
Nerrigundah (NSW) 58, 236, 435
Nestor notabilis (Kea) 136, 259, 428, 545, 650, 669
Newell Highway 46-7, 56
New Guinea 134, 147, 149, 151-2, 163, 304, 640, 702
New Zealand 134, 168
North America 140, 147-8, 163-4, 368
Nothofagus gunni (Australian Beech) 145

Oaks Fire Trail 184
'Old Yellow Eyes' 612
Orang Utan Gully 38, 41
Orang Utan Pass 38, 40-1
Ornithorhynchus anatinus (Platypus) 140, 150, 165-6, 207, 430, 432

Palorhestes azeal 169
Pangaea 132-3, 139, 668
Papio anubis (Olive Baboon) 327
Papua New Guinea 126, 134, 443, 517, 640, 705-6
Parallax 633-5, 674

Index

Parramatta Aboriginal Land Council 31, 71, 178, 239
Penrith (NSW) 213, 612
Perameles nasuta (Bandicoot) 160
Perry's Lookdown 23, 38, 40
Perth (WA) 210
Petaurus 163
    australis (Yellow-Bellied Glider) 532
    breviceps (Sugar Glider) 160, 532
Phaner furcifer (Masoala Fork-marked Lemur) 532
Phoracantha acanthocera (Bullseye Borer) 535
Phoracantha semipunctata (Common Eucalypt Longicorn) 534
POME 494
Procoptodon goliah 16, 171, 430, 641-3, 702
Providence (farm) 44

Quinkan (appearance) 368, 370, 378, 387, 401-2, 404, 431-2, 435, 448-9, 551, 565, 626, 635, 646-7, 701
    (art, contemporary) 443
    (behavior) 181, 223, 251, 386-7, 395, 400-1, 405, 407, 448-9, 454-5, 463, 466, 515, 555, 575-6, 652-3, 685, 690, 692, 703
    (distribution) 450
    (encounter) 127, 387, 395, 404, 411, 444-446, 447-9, 456, 575
    (masks for traditional dancers) 444
    (rock art) 401, 432, 434-42
    (track) 354, 402, 431, 433

Quinkan Rock Art 400

Red eyes 33, 71, 91, 121, 155, 197-9, 201, 223, 225, 235, 269, 271, 278, 314, 316, 329, 347, 360-1, 383, 438, 444, 465, 611-2, 618, 629-31, 647, 675, 681
Red or orange hair 124, 205, 237, 317, 320, 585-6
Rhodopsin 33, 271, 315, 626-8, 634, 683

Sahul 17, 147-8, 150-2, 166, 640, 649
Sarcophilus harrisii (Tasmanian Devil) 567
Sassafras Gully (NSW) 378-9, 449
Sirius A 622-3
Stegodon 148
South America 134, 136, 138, 142, 144, 146, 649, 668-71
South Coast (NSW) 17, 84, 181, 236, 238, 368, 390, 451, 612
Springbrook (QLD) 444
Springwood (NSW) 213, 378, 381, 406, 449, 499
'Stick plant' 198, 477-8, 480-1, 485, 512-3, 521-4, 685
Stolen Generations 79, 240
Strepera sp. (Currawong) 92, 115, 275
Sunda 150-2
Sunda Trench 150-2
Sun Valley (NSW) 213
Sydney 43, 48, 52, 139, 176-7, 179, 200, 209, 224, 233-4, 342, 358, 489, 517, 532, 557, 579, 582, 585-6, 595, 602, 606, 643, 676, 696, 699

Sydney Basin 11, 645
Syncarpia glomulifera (Turpentine) 496, 529, 578
Syzygium austral 307

Tapetum lucidum 155, 205, 271, 315, 470, 601, 608, 616-7, 625-6, 634, 662, 673, 682, 692, 703
Tachyglossidae (Echidnas) 140
Taronga Zoo Sydney 322
Tasman Gateway 138, 146
Tasmania 138, 146-7, 149, 151, 154, 166, 643
Throwing stick 645
Thylacinus cynocephalus (Thylacine) 16, 166, 567
Thylacoleo carnifex 16, 170-1, 231, 430, 602, 653
Timor-Leste 148
Tree biting (treebite) 121, 187, 198, 222, 284, 344, 398, 417, 429, 448-9, 465, 536, 542, 544-5, 548, 550-1, 553-9, 561-63, 567, 570-76, 581, 585-6, 588-91, 596, 598, 678
Tree ripping 392, 461, 521, 545
Trichosurus vulpecula (Brushtail Possum) 330-1
Troodon Formosus 669
Tungamah (VIC) 599
'Turramulli, the Giant Quinkan' 370, 429-30, 436

Valley Heights (NSW) 212

Wallabia bicolor (Swamp Wallaby) 64, 143, 270, 345-6, 423, 499, 503, 510, 522, 610, 654, 657-9, 662

Wallace Line 151
Wallacea 151-2
Wang Wauk State Forest 579
Warragamba Dam Catchment 176, 643
Warrimoo (NSW) 212
Weber Line 151
Weja (NSW) 44
Wentworth Falls (NSW) 517, 630, 644, 690
Western Line 43
West Wyalong 49, 56
Winmalee (NSW) 348, 376
Witchetty 527, 529-30, 596
Wollemi National Park 420, 432, 434, 440
Wollemia nobilis (Wollemi Pine) 139, 595, 669
Woodford (NSW) 182, 184, 204, 210, 234, 420, 517
Wood knocking 198, 453, 459, 463, 471, 477, 481, 487, 512, 521-2
Woomera 645

Yellow Rock (NSW) 213, 352, 406
Yengo National Park 420, 436

Zaglossus hacketti 167
Zygomaturus trilobus 169

A treebite (2010) showing an 82 mm canine separation (the same as the O'Connor Quinkan). This was found in the bush about a kilometre to our northeast, at a height of six-and-a-half feet. See Chapter 12. (Photo: Neil Frost)

Also available from Coachwhip Publications

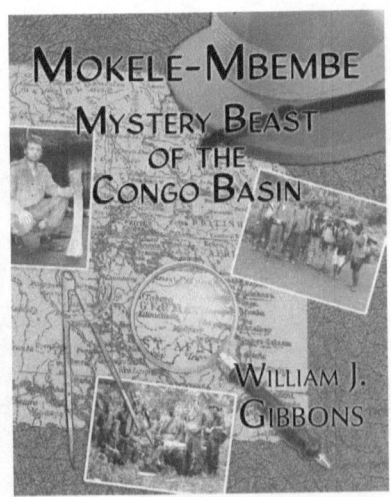
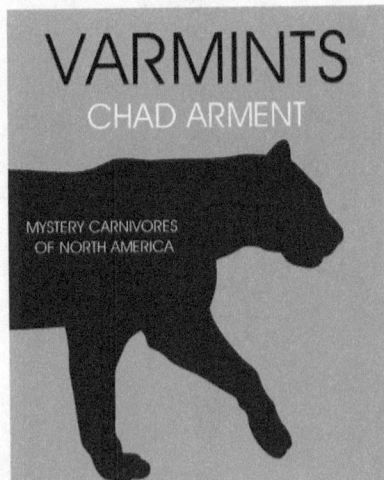

More cryptozoology titles at CoachwhipBooks.com

www.ingramcontent.com/pod-product-compliance
Lightning Source LLC
Chambersburg PA
CBHW031956220426
43664CB00005B/42